P9-CFT-847

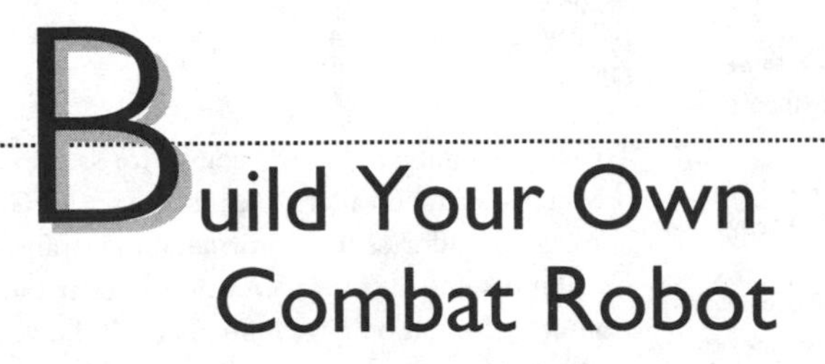

Build Your Own Combat Robot

Pete Miles
Tom Carroll

McGraw-Hill/Osborne

New York Chicago San Francisco Lisbon London Madrid Mexico City
Milan New Delhi San Juan Seoul Singapore Sydney Toronto

Publisher
Brandon A. Nordin

Vice President & Associate Publisher
Scott Rogers

Acquisitions Editor
Marjorie McAneny

Project Editor
Lisa Wolters-Broder

Acquisitions Coordinator
Tana Diminyatz

Technical Editors
Carlo Bertocchini
Grant Imahara

Copy Editor
Lisa Theobald

Proofreader
Linda Medoff

Indexer
Irv Hershman

Computer Designers
Carie Abrew
George Toma Charbak
Lauren McCarthy

Illustrators
Michael Mueller
Lyssa Wald

Cover Design
Jeff Weeks

Series Design
Peter F. Hancik

McGraw-Hill/Osborne
2600 Tenth Street
Berkeley, California 94710
U.S.A.

To arrange bulk purchase discounts for sales promotions, premiums, or fund-raisers, please contact McGraw-Hill/Osborne at the above address. For information on translations or book distributors outside the U.S.A., please see the International Contact Information page immediately following the index of this book.

Build Your Own Combat Robot

Copyright © 2002 by The McGraw-Hill Companies. All rights reserved. Printed in the United States of America. Except as permitted under the Copyright Act of 1976, no part of this publication may be reproduced or distributed in any form or by any means, or stored in a database or retrieval system, without the prior written permission of publisher, with the exception that the program listings may be entered, stored, and executed in a computer system, but they may not be reproduced for publication.

1234567890 QPD QPD 0198765432

ISBN 0-07-219464-2

This book was composed with Corel VENTURA ™ Publisher.

Information has been obtained by McGraw-Hill/Osborne from sources believed to be reliable. However, because of the possibility of human or mechanical error by our sources, McGraw-Hill/Osborne, or others, McGraw-Hill/Osborne does not guarantee the accuracy, adequacy, or completeness of any information and is not responsible for any errors or omissions or the results obtained from the use of such information.

CONTENTS AT A GLANCE

Contents

Acknowledgments

We would like to thank Mike Greene of *Robot Science and Technology* magazine for putting the team together to write this book. Bob Gross, Andrew Lindsey, Ronni Katz, Carlo Bertocchini, and Steve Richards provided a lot of top-quality support and information, as well. Without their help, the quality of this book would not be where it is now. We would also like to thank Carlo Bertocchini and Grant Imahara for taking time out of their busy schedules to serve as technical editors. They provided valuable comments and insights that vastly improved our work. Mark Setrakian, Peter Abrahamson, Christian Carlberg, Peter Menzel, Larry Barello, Dave Owens, Jamie Hyneman, Vincent Blood, Clare Miles, and Ken Gracey were of great help in providing excellent photos. A special thanks goes to Dave Johnson for his help in interviewing Christian Carlberg, Grant Imahara, Jim Smentowski, Stephen Felk, Donald Hudson, and Jamie Hyneman for the "First Person" stories you'll find throughout the book. Additional thanks go out to the people at Vantec, Hawker, IFI Robotics, Parallax, Panasonic, National Power Chair, Acroname, Futaba, and Grainger for their technical support and use of some of their photos. Finally, we would like to thank Margorie McAneny, Lisa Wolters-Broder, Michael Mueller, and the whole team "behind the scenes" at McGraw-Hill/Osborne for their patience and help in putting this book together.

Pete adds: I would like to thank my wife, Kristina Lobb Miles, for all of her tireless help. With her brilliant skills in graphics manipulation, she was able to put together most of the artwork and photos. Without her help, this project would not have happened. She is a wonderful person and deserves a lot of credit. Tom Carroll, too, deserves a lot of credit for putting this together. His infinite knowledge of robotics and ability to write lots of information in a very short time period is greatly appreciated.

Tom adds: I would like to thank my wife, Sue, for her tireless support and encouragement of my many robotics activities for the past 35 years. She has endured my many trips to all over that took me away from home and my family, watched as various robots grew to completion in my shop, patiently waited as I spent many hours in my office typing away at this book, and listened politely as I talked for hours on end about robots. I would also like to thank Pete Miles for his patience, organization, great knowledge, and tremendous effort at spearheading this project. His wife Kris proved to be a most valuable asset at making the graphics and manuscript flow to perfection. These two are a most incredible team, and without them, this book would have been only a pile of papers scattered on the floor.

Introduction

Some kids spend their free time playing sports. I spend mine building robots. You may think that this is not a typical hobby for a teenaged girl, and you're right. I am part of a rapidly-growing community of combat robot builders from all across the U.S., of all ages, and I'm not exactly new to the sport, either. I was at Fort Mason San Francisco in 1994 watching the first robotic combat competition, *Robot Wars*. I saw my dad win match after match with his flimsy, garage-built aluminum contraption, and beyond all reason of my then seven-year-old brain, I was inspired. The next year, when I was eight, I had a flimsy, garage-built aluminum contraption of my own, and I was ready to roll. Since then I've been hooked.

Through my few years of experience in the field of robotic combat, I've come to realize that the actual battles—the end result of all my hard work—are not the only things that I have to look forward to. Just as important to me are the people and friends involved, the familiar sounds and smells of machine maintenance, the ebb and flow of people excitedly preparing for competition, the long but rewarding hours of taking robots apart and putting them back together again, and the feeling you get when you realize you've become a small but integral part of our quirky little robo-community.

I hope this book will help you get started in the unique and exiting sport of robot combat. Robot experts clearly explain everything you need to know to build a bot of your own. For anyone thinking of building a robot, I strongly encourage you to give it a try. You may not wind up with the super-heavyweight champ after your first fight, but I guarantee it will be an experience you'll never forget!

Cassidy Wright,
builder of *Triple Redundancy, Fuzzy Yum Yum,* and *Chiabot*
Orinda, California
January 2002

About the Authors...

Pete Miles has been experimenting with robots since the mid 1970s. He used to scavenge every part he could from dumpsters at radio and TV repair shops, and he still uses parts that he collected back then in his current robot projects. After serving in the U.S. Marine Corps as a tank killer, he obtained bachelor's and master's degrees in mechanical engineering. He currently works as a senior research engineer, developing advanced machining technologies using 55,000 psi abrasive waterjets for Ormond LLC, in Kent, Washington. As he puts it, "There is not a material in the world a waterjet can not cut, including diamonds." Miles is currently an active member of the Seattle Robotics Society, the world's largest robotics club, and was recently appointed to the SRS Board of Directors. He is an avid competitor in autonomous robot sumo, and enjoys building legged robots for various contests to demonstrate that walking robots can be formidable competitors.

Tom Carroll has been involved with robotics for more than 40 years. He built his first robot at age 14, and later worked as a robotics engineer on NASA projects with Rockwell International for nearly 30 years. Carroll co-founded the Robotics Society of Southern California in 1978 and is now active in the Seattle Robotics Society. He designed robots for the International Space Station, to explore the surface of other planets and to assist astronauts in space. He founded Universal Robot Systems to design and build robots for such feature films as *Revenge of the Nerds* and *Buck Rodgers in the 25th Century*. He is presently a novel and technical writer, and spends much of his time developing a truly functional personal robot to assist the "forgotten generation," the elderly, and give them pride in independent living. Carroll moved from Long Beach, California, several years ago and now lives in the Pacific Northwest, on Orcas Island off Washington's coast. Tom enjoys kayaking, hiking, and traveling with his wife.

About the Contributors...

Bob Gross became involved with robotics in 1978 by building a working facsimile of R2D2. For fun, he has built winning autonomous robots for sumo, maze, navigation, wandering, and combat. Later, he produced three autonomous museum robots that would fetch balls, go to various colored columns, or allow teleoperated control. By day, Gross works as a rocket scientist and has a small company that focuses on various aspects of robotics, including machine vision.

Dave Johnson is a technology writer and scuba divemaster. The author of 18 books, Johnson covers popular technology like mobile gadgets, photography, digital music, and robotics. He's also an award-winning wildlife photographer and the author of *The Wild Cookie*, an interactive kids' story on CD-ROM.

Ronni Katz is an adjunct professor of computer science at DeVry College of Technology in North Brunswick, New Jersey. She was an original member of "Team Spike" at the first *Robot Wars* competition and has helped design and build combat robots that have won and placed highly at numerous competitions. Katz is a proud member of the Society Of Robotic Combat and produced the 1998 non-profit instructional video *Introduction to Robotic Combat,* which helped many beginners get their start in the world of sport robotics. Katz writes fiction under the pen name of Ron Karren and has been published in numerous fanzines. Her first military technothriller novel, *Wing Commander*, can be found at bookstores nationwide. You can visit Katz online at QuestPress.com for news of her future publications.

Andrew Lindsey has been competing in robotic combat since 1996. In addition to competing in all three major televised robotics competitions, he was one of four combat judges at the November 2000 *BattleBots* event. Lindsey lives in New Jersey and designs fiberoptic interface electronics for a living. He competes regularly in the North East Robo-Conflict events in the New Jersey/Pennsylvania area.

Steve Richards has been fascinated by the prospect of fully-autonomous robotics since his childhood. He founded and runs the robotics company Acroname, Inc. in an effort to advance robotics through information, parts, and a robotics community. When he isn't milling, coding, wiring, or ranting about robotics, he also enjoys running. He lives in Boulder, Colorado, with his wife, Karen. Richards admits that the only truly successful autonomous creation he has been involved with is his daughter, Annie.

Cassidy Wright has been involved with robotic combat since 1994. She built her first bot when she was just eight years old. She is a teenager now, and the builder of *Triple Redundancy, Fuzzy Yum Yum,* and *Chiabot.*

About the Technical Editors...

Carlo Bertocchini has been building competitive robots since 1993, and he worked as a mechanical engineer until 2001. Now he divides his time between competing in *BattleBots* matches and running his company, RobotBooks.com. He is the designer and builder of *Biohazard*, the world's most successful combat robot. You can learn more about his robots at *www.robotbooks.com/biohazard.htm.* Bertocchini lives in Belmont, California, with his wife, Carol.

Grant Imahara is an animatronics engineer and modelmaker for George Lucas' Industrial Light & Magic in Marin County, California. He specializes in electronics and radio control at the ILM Model Shop and has installed electronics in R2D2 units for *Star Wars: Episodes 1 and 2,* and the famous Energizer Bunny. For fun, Grant competes in *BattleBots* with his robot *Deadblow,* which set a record for the most number of hits in the first season of the show. Grant lives in a loft in Oakland, California, where he also works on his robot in his spare time.

chapter 1

Welcome to Competition Robots

WELCOME to the world of combat robotics. You've watched them on TV. You've seen models of them on toy store shelves. You've seen them featured on the covers of magazines. You might also be among the lucky ones who have actually sat arena-side and watched in person as seemingly sane men and women guided their creations of destruction toward another machine with the express goal of mangling, dismembering, and smashing the opponent.

Television has brought this controlled mayhem into the living rooms of America. You cheer wildly as your favorite robot with its spinning hammers rips the steel skin off its foe. Your robot chases its limping target into a corner, only to have a series of saw blades arise from the floor and send your hero sailing across the arena. The TV cameras pan over to the operators of the losing robot; they are smiling. Even in a moment of havoc, both sides are having fun. Parts and sparks are flying, and smoke wafts upward from the hapless opponent as hazards and weapons reach their targets. The crowd cheers and banners are waving. A winner is announced, and then two new bots start at it.

You can not stop grinning. "This is cool!"

After the program is over, you turn to your friend excitedly and say, "I'm gonna build one of those robots."

"Yeah, right," she says. "You can't even program the VCR. Good luck building a robot."

"Hey, I've got a book on how to build 'em. I'll start small, maybe build one of those little sumo robots. It's a kick to watch those little guys try to shove each other out of a ring. I have some friends who can help me get started. I'm going to do it!"

Robot combat has come a long way from its origins. The founding father of the sport is Marc Thorpe. He came up with the idea for robotic combat while experimenting with attaching a remote-control tank to his vacuum cleaner to make house cleaning more fun. After a few years spent developing the rules for a game where two robots would duel in front of a live audience, a new sport was created: *Robot Wars*. The first official combat robot event was held at Fort Mason Center in San Francisco. It was a huge success. Since *Robot Wars* first came on the scene, thousands of people have participated in building combat robots, and millions have watched and cheered on their favorite bots. Many new combat robot contests—such as *BattleBots, Robotica,* and *BotBash,* to name a few—have been spawned from the original *Robot Wars* competition.

This sport has become so popular, in fact, that many robots have become better known than their human creators. For example, devout followers of robotic combat are familiar with such famous builders as Carlo Bertocchini, Gage Cauchois, and Jamie Hyneman, but these mens' robots—*Biohazard* (pictured in Figure 1-1), *Vlad the Impaler*, and *Blendo*, respectively—are now bona fide household names among the millions of people who watch *BattleBots* on TV.

The various robotic combat events have seen many different types of machines, from two-wheel-drive lightweight robots to six-wheel-drive, gasoline-powered superheavyweights. Even walking robots, more commonly known as *StompBots*, have entered into the mayhem. Probably the most well-known StompBot is the six-legged superheavyweight *Mechadon* built by Mark Setrakian. Setrakian has even built a super heavyweight snake robot. Though his unusual robots have not won any events, they've all been outstanding engineering achievements and great crowd pleasers.

The weapons on these robots range from simple wedges and spikes to jabbing spears, hammers, and axes, to spinning maces and claws, hydraulic crushing pincers, and grinding saw blades of every type, size, and color. The destructive power of these weapons has been used for everything from scratching paint off a rival bot to denting aluminum plates, punching holes through titanium and Kevlar, ripping off another robot's entire armor plating, and completely disintegrating an opponent in a single blow.

One of the most destructive robots the sport has seen to date is *Blendo*. This spinning robot, more commonly known as a *SpinBot-class* robot, totally destroyed

FIGURE 1-1 *Biohazard, a superstar of robotic combat.*

many of its opponents in a matter of seconds. It had such destructive force that it was once banned from continuing to compete in a contest and was automatically declared co-champion for that event.

Today, most combat robots are remote-controlled; but in the early years of *Robot Wars*, there were several fully autonomous combat robots. These robots ran completely on their own, using internal microcontrollers and computers for brains, and sensors to find and attack their opponents. Many people think autonomous combat robots would be too slow to compete because they would require too much time to locate and attack an opponent. This isn't always the case, however. The 1997 *Robot Wars* Autonomous Class champion, *Thumper* (built by Bob Gross), won a match in 10 seconds flat. That's *Thumper* in Figure 1-2.

Today, most autonomous combat robots are found in robot sumo events, where two bots try to find and push each other out of a sumo ring. In this event, bots are not allowed to destroy each other. Sumo builders face a unique challenge, as they design their bots to "see" their opponent and push them out of the ring before getting pushed out themselves. This contest has become increasingly popular in recent years, and new sumo events are popping up all over the world.

In the past, competition divisions consisted of man versus man, or team of men versus team of men (let's face it—it began as a male-dominated sport). Strength, speed, agility, endurance, and strategy were the only factors that determined the winner or loser. Thanks to robot combat, this isn't the case anymore. At robot competitions, ingenuity, creativity, and intelligence now rule the game. No longer are 6-foot 5-inch, 240-pound male "athletes" dominating the game. A 10-year-old girl with excellent engineering skills can now defeat a 250-pound former

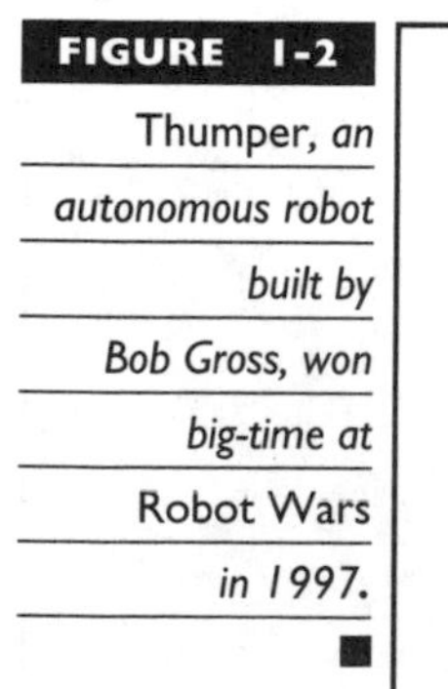

FIGURE 1-2 Thumper, *an autonomous robot built by Bob Gross, won big-time at* Robot Wars *in 1997.*

NFL linebacker, and a wheelchair-bound person can run circles around an Olympic gold medalist. Robot combat has leveled the playing field so that anyone can compete against anyone on equal ground.

What Is a Robot?

Now that you've made up your mind to build a robot, you're probably sitting back wondering just what you've gotten yourself into.

"What *is* a robot?" you ask yourself.

Surprisingly, there are many definitions, depending on whom you ask. *The Robot Institute of America*, an industrial robotics group, gives the following definition: "A robot is a reprogrammable, multifunctional manipulator designed to move material, parts, tools, or specialized devices through variable programmed motions for the performance of a variety of tasks." These people, of course, are thinking only of robots that perform manufacturing tasks.

Now that you're thoroughly confused, *Webster's New World Dictionary* defines robot as "any anthropomorphic mechanical being built to do routine manual work for human beings, or any mechanical device operated automatically, especially by remote control, to perform in a seemingly human way."

Hmmm. Now we seem to be talking about human-formed robots, like in the movies, or it could be the description of a washing machine, or maybe the Space Shuttle's "robot arm."

Where did the term "robot" come from? Back in the 1920s, a Czech playwright by the name of Karel Capek wrote a short play entitled *R.U.R.*, which stands for *Rossum's Universal Robots*. The word *robot* came from the Czech word *robota*, which means indentured servant or slave. In Capek's play, the robots turned on their masters, which became a theme in many movies and stories in later years—robots doing bad things to people. Only in more recent movies have robots become friends of humans and started doing bad things to other robots.

To this day, those in the field of robotics still argue about what exactly constitutes a robot. Many people think that if a machine doesn't have some sort of intelligence (that is, a microcontroller inside), it isn't a robot. Some might look down their noses and claim that only a multiarmed machine driven by a Pentium 4 processor with 512 megs of RAM and fed by 100 sensors is really a robot. Those at NASA might feel the same way about the Space Station's Canada Arm. All this arguing really doesn't matter, because everyone has their own definition of what a robot is—and everybody is right.

Whatever you choose to call a robot is a robot.

Combat Robot Competitions

Before we start talking about types of robot competitions, let's cover a brief history of the events that gave rise to this sport. Organized robot competitions have been around since the late 1980s, and have been rapidly growing ever since. The following is a short history of some of the most popular robot contests around today. There are many other competitions aside from those listed here, and new ones are turning up each year.

- **Late 1980s** The remote control and autonomous robot sumo contest is invented by Hiroshi Nozawa of Fujisoft ABC, Inc., in Japan.
- **1989** Inventor and entrepreneur Dean Kamen founds FIRST. This nonprofit organization, "For Inspiration and Recognition of Science and Technology," pairs up school-age children with local engineers to build robotic projects.
- **1992** Marc Thorpe discovers that his experiments with building a radio-controlled vacuum cleaner to help with the housework can be turned into a new sport called *Robot Wars*.
- **1992** FIRST Robotics hosts its first competition with 28 high-school teams.
- **1994** Marc Thorpe creates Robot Wars. This is the first major competition where robots face off against each other in an arena in front of a live audience. The first event is held at Fort Mason Center in San Francisco.
- **1997** Mentorn Broadcasting produces a six-episode series of *Robot Wars* for BBC television in the U.K.
- **1997** *BotBash,* a similar event to the original *Robot Wars*, holds its first event in Phoenix, Arizona.
- **March 10, 1999** *BattleBots* is founded by Trey Roski and Greg Munson in San Francisco.
- **August 14, 1999** *BattleBots* hosts its first event in Long Beach, California, with 70 robots competing.
- **January 29, 2000** *BattleBots* appears on pay-per-view television, and airs the second *BattleBots* event from November 1999.
- **August 23, 2000** *BattleBots* begins airing as a television series on Comedy Central. The show quickly shoots up in ratings and finishes its first season as one of the most popular shows on cable TV.
- **April 2, 2001** *BattleBots* registers over 650 robots at its Spring 2001 competition.
- **April 4, 2001** *Robotica* begins airing as a television series on the Learning Channel. Early indications show the program is a hit among viewers.
- **August 20, 2001** The new *Robot Wars Extreme Warriors*, a spin-off from *Robot Wars,* premieres as a new television series on TNN.

As you can see, the history of robot combat is relatively short in comparison with baseball or football, but all sports have to start somewhere. With its current growth rate, it won't be long before this becomes one of the most popular sports in the world.

As with any game, there are different rules and goals for each event. Following are brief descriptions of some of these contests. The exact details of the events should be obtained directly from the event organizers.

BattleBots

BattleBots is probably the most popular robotics event in the United States. A large fan base has been accumulating ever since these competitions started airing on cable TV. *BattleBots* is a single elimination fight-to-the-death contest where one robot tries to destroy another in a 3-minute time frame. If one of the robots becomes incapacitated for 30 continuous seconds, or is destroyed, that robot loses the match. If both robots are still fighting at the end of the 3-minute time frame, the winning robot is declared by how many points they scored. There are three official judges who award up to 5 points each for aggressiveness, damage, and strategy, for a total of 45 points. The robot with the most points wins the match.

If your robot is fortunate enough to survive the match, it has only 20 minutes to undergo any repairs before the next match. If the robot faces another fight soon afterward and cannot be repaired in the 20-minute time frame, it must forfeit the next match.

The main BattleBots arena is called the *BattleBox*. Weighing in at 35 tons, this "box" consists of a steel floor measuring 48-feet-by-48-feet, and walls that tower 24ft high. The walls of the BattleBox are made out of Lexan (a highly resilient polycarbonate) ranging in thickness from one inch at the base of the walls to 3/16 inches at the top. There are two 8-foot-by-8-foot entry doors where the robots enter. Within the BattleBox there are a set of hazards and weapons, which are as follows:

- **Kill Saws** These are 20-inch-diameter carbide-tipped SystiMatic saw blades that can cut through virtually any material. They can spring up with many pounds of force, easily tossing 340-pound superheavyweight robots into the air.
- **Pulverizers** These monster aluminum hammers are used to smash any unfortunate robot that gets under them.
- **Hell Raisers** *BattleBots* competitions occasionally employ these 3-foot-by-4foot plates that move up 6 inches, wreaking havoc in a robot's motion.
- **Ram Rods** The ram rods are a set of six carbide-tipped spears that shoot up 6 inches from the BattleBox floor with over 60 pounds of force.

FIGURE 1-3 *Two-wheel-drive, spike-weilding* Toe Crusher, *built by Christian Carlberg.* ■

- **Spike Strip** Around the perimeter of the BattleBox is a strip of 180 metal spikes—each one 1-inch in diameter and 3 inches long—that point toward the center of the BattleBox.
- **The Vortex** This is a 3-foot-diameter disk that will spin the robot around if it rolls on top of the vortex.
- **The Augers** These huge rotating screws mangle any robot unlucky enough to get caught in their grip.

There are four different weight classes for wheeled *BattleBots*, as shown in Table 1-1.

	More Than	Maximum
Lightweight	25 pounds	60 pounds
Middleweight	60 pounds	120 pounds
Heavyweight	120 pounds	220 pounds
Super heavyweight	220 pounds	340 pounds

TABLE 1-1 BattleBot *Weight Classes* ■

Walking robots get an extra 20-percent weight increase bonus, so the weight classes for walking bots are 72 pounds for lightweights, 144 pounds for middleweights, 264 pounds for heavyweights, and 408 pounds for superheavyweights.

All of the details about *BattleBots,* including rules and regulations, can be found online at *www.battlebots.com.*

Robot Wars

Robot Wars is where it all began—two robots fighting to the death. In the early days of *Robot Wars,* there was an arena filled with hazards, including spikes, buzz saws, and a swinging bowling ball. Robots fighting in this competition had to avoid the hazards while attacking opponents. Not only were there remote-control robots fighting, there were also autonomous machines competing.

Since *Robot Wars* moved to the United Kingdom in 1997, the event has changed quite a bit. Before the bots get a chance to go to the big fight, they now have to pass a series of obstacle course tests. These obstacles include crashing through brick walls, climbing over teeter-totters, passing between two closing walls with spikes, avoiding large pendulums, knocking over large metal drums, and steering clear of fiery pits.

FIGURE 1-4 *The vicious-looking* Razer *has been a crowd favorite for several years running at the* U.K. Robot Wars.

To make the events a little more challenging, the contestant bots have to contend with "house" bots whose main purpose is to destroy anything fool enough to come near them. The smallest house robot is *Shunt*. At 231 pounds., this powerhouse can pull a Land Rover *and* wield a deadly axe. *Dead Metal*, weighing in at 247 pounds., is very effective at using its buzz saw and deadly pincers. The 256 pound titanium-armored *Matilda* wields a chain saw on her rear, and the 264 pound *Sergeant Bash* with his deadly flamethrower can cook his victim when it gets caught in his front pincers. Finally, there is *Sir Killalot*, at a massive 617 pounds. His pincer claws can cut through the toughest armor and then lift a 220- pound hapless victim—to be dropped into the fiery pit.

The lucky winners of the obstacle courses get to move on to bigger and better fights. Below is a list of three of the most popular events that bots must pass in more advanced *Robot Wars* competitions, prior to moving on to the final round:

Pinball

In the pinball tournament, bots must navigate around a course and hit certain objects, each of which is worth a different number of points. The bot with the most points wins the tournament. Bots score 5 points for hitting barrels, 10 points for the multiball, and 5 points for each multiball in the pit. Crossing over the ramp is worth 20 points, going through the car door gate is worth 25 points, and moving the sphere out of the pit is worth 25 points. Hitting *Matilda*'s and *Sergeant Bash*'s guarded targets are worth 50 points each, and getting past *Dead Metal* to its target is worth 75 points. All of this must be accomplished in 5 minutes.

Sumo

The Sumo event is held on an elevated ring, and the contestant bot goes up against a house bot. This is a timed event to see how long a bot can stay in the ring before being pushed off by the house bot. Most of the time, the house robot wins this event, but once in a while a challenger will be successful in pushing a house bot to its doom. The bot with the longest time on the sumo ring wins that event.

Soccer

Robot Soccer is an event where two bots try to push a white ball into the other bot's goal. A house bot is positioned in the arena to assist in the game. "Assist" is a relative term because the house bots have a tendency to capture the ball, thus leaving the other two bots to fight. Once the time limit expires, a judge determines which robot is the winner.

Robot Wars has several other events that are less common, one of which is the Grudge Match. In this competition, if your bot has a grudge against another bot—including a house bot—it gets the opportunity to fight that bot one on one.

Another event is the Tag Team match, where two bots team up against two other bots. A popular event is the Tug-Of-War, where a contestant bot is attached to a house bot via a rope. Between the two bots is a pit. As you guessed, the contestant bot must pull the other bot into the pit. Yet another popular event is the Melee. Here, three or more robots fight against each other and the last one standing wins the melee. (*BattleBots* has a similar event to the Melee, which is called the Robot Rumble.)

Table 1-2 lists the weight classes for *Robot Wars*.

The official *Robot Wars* Web site is at *www.robotwars.co.uk*.

BotBash

BotBash is a smaller-scale version of *BattleBots*. The rules of the contest are very similar to *BattleBots*, with the big difference being that *BotBash* is a double elimination tournament. This means your bot can lose one round and still be able to fight on. This is a nice change for bot builders because if a battery connector falls off, or some other unforeseen problem arises in a match that causes you to lose, you can still prove that your bot is the best by winning the remaining rounds. Another big difference is that the *BotBash* bots have lower weight limits. Tables 1-3 and 1-4 list the *BotBash* weight classes for the wheeled and walking robot classes. As with *BattleBots*, there is a 3-minute time limit; and if both bots are still fighting, a winner is declared by points. Here, the three judges award one point each for aggression, strategy, and damage, for a total of nine points.

Each year, the *BotBash* tournament offers different events aside from one-on-one battle. In the past, they've featured a Capture the Flag event where two cones (flags) are placed at opposite sides of the arena and the bots race to capture the opposing bot's flag. The bots can plan either an offensive or defensive role to attack or protect the flag. The bot that touches the other bot's flag first wins the match. Other events at *BotBash* include obstacle courses and sumo events. Occasionally, *BotBash* tournaments feature autonomous events. Because the rules and events for each tournament change each year, builders must keep up-to-date on the rules and regulations. The official *BotBash* Web site is at *www.botbash.com*.

	More Than	Maximum
Featherweight	0 pounds	25 pounds
Lightweight	25 pounds	50 pounds
Middleweight	50 pounds	100 pounds
Heavyweight	100 pounds	175 pounds

TABLE 1-2 Robot Wars *Weight Classes*

FIGURE 1-5 Spike III, *a third-generation robot built by Andrew Lindsey, a long-time combat robot competitor.*

	More Than	Maximum
Class A	0 pounds	12.9 pounds
Class B	13 pounds	30.9 pounds
Class C	31 pounds	58.9 pounds
Class D	59 pounds	115.9 pounds

TABLE 1-3 BotBash *Wheeled Robot Weight Classes*

	More Than	Maximum
Class A	0 pounds	24.9 pounds
Class B	25 pounds	55.9 pounds
Class C	56 pounds	87.9 pounds
Class D	88 pounds	172.9 pounds

TABLE 1-4 BotBash *Walking Robot Weight Classes*

Robotica

Robotica is a new type of robot combat where bots must complete several courses before they can fight each other. This type of contest has different design requirements; brute strength doesn't guarantee that the bot will win the contest. Bots need to be more agile and creative to solve each challenge. In this contest, you must keep up-to-date on the rules because the challenges change dramatically each year.

There is only one weight class for the *Robotica* robots. The maximum weight is 210 pounds., and the robot must fit inside a 4-foot-by-4-foot-by-4-foot cube at the start of the match.

To give you an idea of the different types of events *Robotica* contestants face, the following are details on qualifying obstacle courses from the first two television seasons.

Season One

In the first season of *Robotica*, bots had to survive three different preliminary rounds. The first event was the Speed Demons race, where two bots raced around a figure-8–shaped track in opposite directions. The first bot that finished eight laps won the race. If the 2-minute time limit expired with both bots on the track, the race was ended. Points were given to each bot for each lap finished. The bots were allowed to crash into each other when their paths crossed.

The second event was the Maze event. Here, the bots had to navigate to the center of a maze and overcome several obstacles, which included a teeter-totter ramp, a weighted box, spiked paddles, speed bumps, a guillotine, and a waterfall. The first robot to the center won the event. Points were given to each bot for each obstacle successfully navigated.

The final event was the Gauntlet event. Each bot had to crash through five increasingly difficult obstructions. The obstacles included a pane of glass, a wall made of pint-sized metal cans, small bricks, stacked cement blocks, and a large weighted box. Two bots ran identical parallel courses, and the first bot that moved the weighted box won the event. Points were also awarded for each obstacle the bot went through.

The bot with the most points after the three events won the preliminary round and got to fight the winner of another set of events. The final match, called Fight to the Finish, took place on a 16-foot diameter ring 8 feet off the ground. To win this event, your bot had to push the opponent off the ring to its death on spikes below the ring.

Season Two

During the second season, the preliminary events changed from three events to two events. The first event was the Gauntlet. In this new version of the Gauntlet, the bots had to run through a diamond-shaped track. Both bots started at the same point but went in opposite directions. They had to crash through a number of obstacles on the first two legs of the diamond track, including a wall of wood, weighted cans, a wall of bricks, and then a cement wall. After all this destruction, the bot then had to crash through the debris field created by the other bot. Once the bot completed the diamond track, it then climbed a ramp to destroy a series of glass columns. When all the glass columns were destroyed, the bots had to climb a final ramp to the victory zone. Bots got points for each obstacle successfully navigated. The bot with the most points won the event.

The second event was the Labyrinth. The bots had to navigate through a series of challenges, after each of which was a glass wall to be broken through by the bot. The challenges included a 20-pound box, a suspension bridge, spikes shooting up from the floor, a flip ramp, a sand pit, and a set of steel cargo rollers. When all challenges were successfully navigated and all six glass walls were broken, a seventh glass wall was revealed. The first bot to break the final glass wall received bonus points. To make things more difficult, a set of *Robotica* "rats" with buzz saws are constantly attacking the bots to impede their progress. Points are awarded for each obstacle successfully navigated, and the bot with the most points wins that event.

The bot with the most points after the two preliminary events moves onto the Fight to the Finish event. As with the first season, the bots try to push each other off the ring. The first one falling out of the ring loses the overall match.

As you can see by the different events, *Robotica* is more challenging than a purely destroy-your-opponent type of robot combat. But in order to win *Robotica,* it still comes down to having the strongest and most powerful bot.

The official *Robotica* Web site is at *www.robotica.com.*

FIRST (For Inspiration and Recognition of Science and Technology)

FIRST does not condone competitions where two bots try to destroy each other. However, we are including FIRST in this list because their competitions are very intense and aggressive, and are becoming extremely popular among robot enthusiasts.

The FIRST Robotics Competition is an annual design competition that brings professionals and high-school students together in teams to solve an engineering design problem. One of the goals of competition is show students that science, engineering, and inventions are fun and exciting, so they will be inspired to pursue careers in engineering, technology and science. A big part of the event is having students work directly with corporations, businesses, colleges, and professionals to help support them in building bots for the competition. This is a fast-paced competition that starts shortly after the beginning of a new year. Each team has only six weeks to design and build their bot. After that time, they compete in regional contests and later move on to the final championship.

In 1992, the inaugural year of the FIRST competition, there was only one contest with 23 teams entered. Since then, the contest has grown significantly. In 2001, there were 14 competitions with a total of 535 teams entered. FIRST has grown to include Canadian and Brazilian teams, as well.

Each year the goal of the contest changes, and nobody knows this goal until the first day of the six-week countdown. During this six-week time period, teams must figure out the rules and goals of the contest *and* design and build their bot. During the actual contest, a team is paired up with another team, and those two groups of people must work together to solve the prescribed challenge against two other teams. The particular contests are designed so that teamwork is required in order to score enough points. During most of the preliminary rounds, the contest officials decide team pairings. In the finals, a team is allowed to choose its partners. The FIRST organizers believe this helps promote teamwork and cooperation.

FIRST robotics is an extremely challenging and exciting contest. Many of today's famous combat robot warriors cut their teeth in competition robotics by competing in FIRST, either by participating as a member of a high-school team or serving as a mentor to a FIRST team. A lot of the technologies and skills needed for building combat robots are used in designing FIRST robots.

The official FIRST Web site is *www.usfirst.org*.

FIGURE 1-6 *Team Titan Robotics from the International School in Bellevue, Washington, built* Prometheus *for a FIRST competition.*

Robot Soccer

Probably the most difficult robot sport is Robot Soccer. This is an autonomous game where a team of bots works together to score goals against another team of bots. The rules of the game are similar to those in actual soccer games. Bots use advanced vision systems to track the soccer ball, monitor the location of the opposing team's bots, and know where their own teammates are. All of the bots play their positions just as human players do. There is a lot of cross-communication between all of the bots playing. This contest is usually performed by university students developing algorithms for artificial intelligence. We reference this contest because a lot of the technologies being developed for Robot Soccer players may soon migrate down to combat robots. At some point in the future, there may even be autonomous soccer teams in popular competitions like *BattleBot*.

More information on Robot Soccer can be found at *www.robocup.org*.

Before you start building a bot for a particular contest, you should get a copy of that contest's current rules and regulations. You can usually find this information on the organization's official Web site. Keep in mind that some of these competitions have long and complex regulations for builders to follow, and the rules do change from time to time because the contests are evolving into a mature sport. You need to be very familiar with the robot specifications and safety requirements for the contest you have in mind, as they'll have a significant effect on your bot's design.

First Person

The sport of robotic combat has been called "American Gladiators for people with brains" and the "sport of the future." However, back when I first signed on board with my armored harbinger of destruction, it was just a small bunch of guys getting together in San Francisco's Fort Mason Center for what could only be described as Rockem' Sockem' Robots for grownups.

The crowd was small but enthusiastic. The hazards in the arena were walls that pushed in and out, some spinning blades that popped up whenever the guy running them was alert enough to press the lever, and a large metal ball looming from on high that swung like a giant pendulum of death from a chain on the ceiling. Lexan walls separated the audience from the inevitable flying shrapnel and sparks. The floor of the arena was so dented, dinged, and pitted by the last day that you were sorry your robot wasn't equipped with off-road capabilities.

Someone was nice enough to set up a primitive closed-circuit TV so that we in the backstage "pit" area could see what was happening in the arena and know when we should get on-deck for our matches. While we toiled away on our bots, our spot in the pit was so close to the action that we could almost watch the battles if we stood on our chairs. The sound of saws grinding metal and the smell of overcooked batteries, fried wires, and oil filled the air.

It was heaven.

It was also my robot Spike's first time competing as a lightweight. We came in third, but where we wound up didn't matter. Just being a part of the action was thrilling enough. If you needed a screwdriver or blew a gasket, someone was there with a spare to help you get your bot back into the fray. When our Tekin speed control turned into a smoking slagpile, we got a loaner from the guy we were going to be up against in the next match. In the pit, we were all on the same team, working toward a common goal. However, once our bot was in the arena, all bets were off, and it was mano a mano: let the best-made machine win.

First Person continued

The hardest part for us was just getting there. We had no sponsors and had to pay our own way for everything. It was tough, and it took months to pay off that credit card, but I would do it all again in the blink of an eye! We met some of the most incredible (and nicest) people. The designs we saw and the creativity of the engineers and imagineers behind their bots inspired us. The generosity in the sharing of ideas, tools, and even parts amazed us. We became part of this amazing community of robot builders and battlers and the camaraderie warmed us. It was one of the best weeks of my life.

—Ronni Katz

The Scope of This Book

Building a bot is not that difficult—if you've done your homework on the basic elements involved. It may take you a while to figure out how to do new things, and it might take a long time before you build your dream machine, but consider your first project a learning process—patience and persistence are key when you're building a bot.

Robotics is one of those fields where you need to be able to wear a lot of different hats. That means you must know a little bit about a lot of things, including motors, electronics, wiring, computers, radio transmitters and receivers, batteries, gears, belts, bearings, chains, sprockets, metals, plastics, drilling, cutting, threading, bending, and welding—just to name a few.

You don't have to be an expert in all of these categories—you just need to understand the basics behind each one. Most combat robots are built by a team of people. Each team member is knowledgeable about certain areas of robot building. When you get a group of people together who all know different pieces of the process, it reduces the burden on each individual for having to be an expert on everything. After you have built a couple of bots and competed in a few contests, you'll become something of an expert in all of the different categories because you will have been involved to some degree with every part of building the bot.

Probably the number-one question that gets asked of a bot warrior is, "How do I build a robot?" Well, nobody can give you a quick answer. It usually takes months to years to learn how to build a bot. There is just too much stuff you need to know. Most of the time, people learn just by doing it. We all make mistakes, and we learn from them.

The scope of this book is to help you, the new robot builder, get started in the exciting field of constructing combat robots. After reading this book, you will have an understanding of all the elements that go into building a bot. Usually, the new robot builder is surprised to find out that there are so many different things that go into this process. This is because most people only see the finished product—the beautiful, gleaming *El Diablo* or *Nightmare* or *Deadblow*—they don't see the blood, sweat, and tears that went into building it.

In this book, you're going to learn how to lay out your ideas and come up with a good plan before starting to build your bot. You'll learn the basics behind a lot of technical subjects, some of which are listed here:

- How electric motors work, how to pick the right motor, and how to use it
- The various locomotion methods and the various methods to get your motors to drive your bot's wheels
- Different types of batteries and how to size them for the right job
- What's required to actually drive a motor, and how to choose the right radio control system
- How to minimize radio interference so your bot will do what you want, when you want
- Wiring issues to keep in mind when building a bot
- Materials and how to assemble them into your bot's body
- Armor for your bot
- Weapons for your bot
- Sensors you can build into your bot for use of automatic weapons, or to create a fully autonomous bot
- How microcontrollers can help you control your bot and allow it to run on its own

In this book, you'll learn about two different bots that were actually built for *Robot Wars* and *Robotica,* and you'll even learn how to build a working mini sumo bot. As you read the stories behind the building of each of these bots, you will learn what the builders did to construct them and why they chose their own particular approaches, what worked, and what didn't.

What this book *doesn't* cover is the explicit step-by-step details of building combat robots. The main reason we chose not to do this is that we don't want to prescribe an exact kind of bot for you to build. There are so many different types of bots to choose from, and an infinite variety of designs you could adopt, and the last thing we want is to see hundreds of the same identical bot competing in different contests. We want you to use your imagination! Do something different. Have fun. Be creative. Make a six-legged mama robot that deploys a half-dozen baby robots. That would sure be fun to watch!

For those of you who would like more explicit details anyway, we have included a set of appendixes with references to other outstanding books and sources for information and robot parts. These lists should give you all the information you ever wanted about robot books and resources.

Okay, now let's get started!

chapter

2

Getting Started

As we said in Chapter 1, it's good to let your imagination run wild when you begin making plans to build a bot. However, while you can dream up all kinds of crazy ideas for a robotic creation, keep in mind that you may not have the time—or even the technology—to build most of them. We can't begin to tell you how to design the "perfect bot," any more than we can convince you of what the perfect car or television set is. Everybody has their own idea of what's best. Yes, we authors have our biases and feel comfortable with certain techniques and designs that have been tested over a number of years, but a prospective bot builder can easily arrive at a better idea than anything we've come up with in the past. Read this book and others, talk with respected people and experienced combat warriors, sketch out your ideas, and then just go for it.

Start your design process by deciding on exactly what you want your bot to do. If you're planning to build a machine for *BattleBots,* you're going to have to take an approach quite different from the one used for making small autonomous machines designed to run a maze or blow out a candle in the popular Trinity College Robot Firefighting Contest. A bot designed to act as a servant in your home may be every bit as heavy and complex as a warrior bot, but it doesn't need to be able to survive the blows of a weapon of another machine or travel nearly as fast.

Experience has shown that electronics and computing power are not the limiting factors in bot construction; it's the mechanics, sensors, and related software development that choke a project to a stop. "How do I physically build the thing? What type of sensors can I use? How do I write the code and what language should I use?" are the questions that flood experienced builders' minds.

Of course, if you're building a *BattleBots*-style (radio control) machine, you probably won't need any software, and the "sensors" are your own eyes as you guide it across the floor of the battle arena. Physical and mechanical design are most critical in these large bots. They require more sophisticated machining techniques than most bots because they must endure an environment that is far more hostile than the average home.

First Person

Like I said in Chapter 1, I got started in robot combat for the fun. When I came on board, there was no TV coverage or anything fancy. Tickets were sold locally, and it was promoted through grass-roots efforts. A friend and I happened to learn about it via the Internet and were two of only a handful of people who came to the competition from outside California.

Back in those early days, getting people involved was a challenge because everything was so new and no one was really sure how to promote the idea. Now, of course, there are lots of popular organizations where robot builders can compete, such as BattleBots, Robotica, *or* Robot Wars.

The sport has changed a lot in five years. Because robot combat has gotten more commercial, the standards by which entries are judged have gotten far more stringent. When I first competed, the rule book was maybe five to seven pages of safety tips. Now, the rule book for competing in any of the major contests is 60 pages of dos and don'ts, plus another 50 pages of technical specifications that competing bots must adhere to. It isn't just a game anymore. It has become serious business for the people involved, and the promoters expect those who enter to bring a robot that is both safe and exciting to see in action.

If you're going to build a bot, let it be your love of the sport—not a desire for glory or fame—that brings you into the arena. People thinking of getting into this with visions of becoming "The Rock" of BattleBots *had better check their servos at the door. Chances are your first entry will die a quick, smoldering death, so keep your ego in line. As long as you're there for the joy of the game, you will have as much fun bashing, smashing, and chopping your opponents into miniscule metallic bits as I did!*

—Ronni Katz

The Robot Design Approach

The first step in designing a new bot is deciding which contest the bot will be built for and getting a copy of that contest's current rules and regulations. The rules outline the weight and size limits for each weight class, as mentioned in Chapter 1, and list weapon types that are allowed and not allowed. They also list safety requirements, electrical requirements and restrictions, and radio control restrictions. Read and understand the rules thoroughly. This will set the initial physical constraints in your bot's design.

If you're designing a robot for multiple contests, you should obtain sets of rules for all of them and make a list of all the common rules and non-common rules. When you have this information put together, you'll be able to create a list of the most restrictive rules for each of the contests, which will help you guide your overall bot design. Building a bot to the most restrictive rules will allow your machine to be entered into each contest without significant modifications.

Even if you're just building a bot for fun, we recommend getting a copy of one of the main contest's rules. A good example of rules and regulations can be found on the *BattleBots* Web site (*www.battlebots.com*). Their safety guidelines and restrictions should be followed in all bot building. Most of the rules are there for the safety of builders and spectators alike.

Once you have the physical constraints written down, you can start laying out the conceptual design of your bot. Sketch out what you would like your bot to look like and do. Include the unique features and weapons you would like your bot to have. A lot of this is paper-and-pencil or CAD (computer aided design) work. Next, make a list of performance goals you'd like to achieve, such as how fast you want your bot to go or how much weight you want it to be able to push. How much must the armor withstand in punishment, and how will your bot's weapon attack the enemy? This is all top-level generic design information; you don't need to get into nitty-gritty details like miles per hour or pounds of pushing force yet. That comes later.

The second list includes what you are aiming for—the ultimate goal. Some people call this the brainstorming part of the design process. The ideas come out here. As is the case with any brainstorming session, there is no such thing as a bad idea. Let the ideas flow, and come up with some cool bot concepts. It is usually good to come up with a handful of them.

After this, the conceptual ideas must be trimmed down to meet the physical constraints of the contest. Yes, this means you're going to have to toss out your idea for a laser-guided rocket launcher. (It's a *great* idea, but it's not allowed in any combat robot event.)

In all competition robots, the following subsystems are part of each bot. Each of these subsystems relates to the others and affects the overall design of the bot:

- Robot frame
- Drive motors
- Power transmission
- Batteries
- Wheels
- Electronics
- Radio control system
- Weapons
- Armor

Probably the first consideration in your robot's design is how you're going to make it move. Your choices are many, and could include slithering, swimming, floating in the air, or even climbing up a wall or rope. More than likely, though, you're going to want a mobile bot that travels across a floor, and this will mean legs, "tank" treads and tracks, or wheels.

Wheels are the most effective way of providing propulsion to a bot. They are cheap, and easy to mount, control, and steer, and there are several methods you can use. We'll discuss all this in Chapter 3. There are many sources of bot wheels, from toys for the smaller bots to small trailer tires for larger machines. Some builders have used wheels from industrial casters, lawnmowers, go-karts, and even small bicycles. Your choice depends on the size and steering configuration of your bot's design.

The majority of bots use differential or tank-type steering (also known as "skid steering"). This means that the bot uses different speeds for left and right wheels (or sets of wheels), causing the bot to go straight, or to one side or the other. Having one wheel stopped and the other moving makes the bot pivot on the stopped wheel, and vice versa. Having one wheel move forward and the other in reverse makes the bot spin about its center axis. (We'll discuss this in more detail in Chapter 3.)

Once you choose your locomotion method, the first set of major components you need to identify are the motors. Most motors operate at speeds that are way too fast to control the robot. So, you'll need a gear reduction. Some motors have built-in gearboxes, while others require a speed reduction system. This can be in the form of gears, sprockets, belts, or even gearboxes. Chapter 6 will talk about these various power transmission methods. The advantage of a gear reduction is an increase in the torque to the wheels, which gives your bot more pushing power. Another reason you should select your motors first is that they will dictate your electrical power requirements, which affects the battery and motor speed controller selections.

FIGURE 2-1 *The welded frame structure of* Minion. (courtesy of Christian Carlberg)

Chapter 4 will discuss motor performance requirements, and Chapter 7 will describe various motor speed controllers.

The next step is designing the bot's frame. This is the core structure of the bot that holds the motors, drive shafts, bearings, gearboxes, wheels, batteries, and motor controllers. The core structure should be solid and rigid, as the rest of the bot will be attached to it. Remember when you're designing the frame to leave space for the batteries, motor controllers, and weapon actuators. Another point to keep in mind is your robot's center of gravity. Keep it as low as possible to improve stability.

Okay, so you've determined your power requirements. Next, you need to know the current draw specifications from the robot motors. It is best to estimate this based on worse-case situations. The last thing you want to see happen is your bot stop in the middle of a match because it ran out of energy. Assuming that your bot is running at stall-current conditions all the time is the absolute worse-case scenario, but this estimate is unrealistic since stalling the motor for 5 minutes will destroy the motor. However, assuming your robot is running at 100-percent stall current draw for 20 percent of the match time, and at 50 percent the stall current for the remaining amount of time in the match, should give you a good estimate on the maximum amount of current that you will need. Select your batteries based on the information contained in Chapter 5. Once the batteries are selected and the dimensions of the batteries are determined, a battery housing should be designed for the bot. The battery housing holds the batteries in place and protects them inside the bot.

Knowing what the current requirements are for your bot determines the motor speed controller. You'll find information about motor speed controllers in Chapter 7. When you're installing the motor speed controllers, you should have features in the design to allow for cooling. Motor controllers get very hot when near-maximum currents are running through them. You may even need multiple-speed controllers, depending on how many motors you're using.

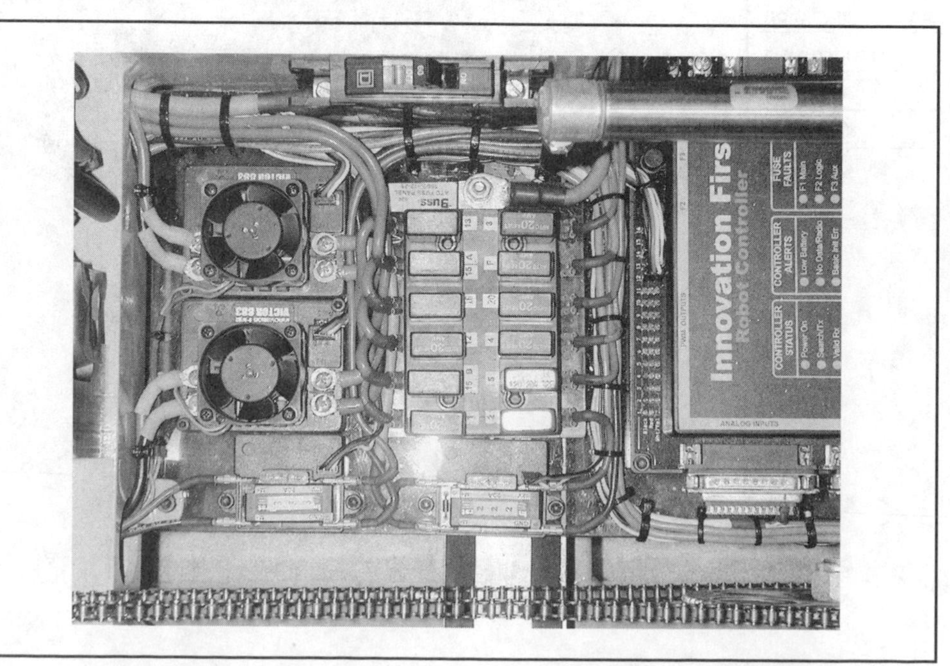

FIGURE 2-2 A robot using two Victor 883 motor controllers and the Innovation First Robot Controller for motion control. (courtesy of Larry Barello)

Now, it is time to add the weapons to the design process. You need to design a support structure to support the weapons and their actuators. The support structure should be mounted to the main frame, and the support structure needs to be very strong. As Newton's Second Law says, "For every action, there is an equal and opposite reaction." In other words, any force your weapon imparts onto an opponent will elicit equal reaction from the opponent onto your bot. Thus, the weapon support structure needs to be able to withstand those forces. Chapters 9 and 10 discuss construction and weapons techniques.

The last part of the mechanical design process is the armor. You should design your armor to be replaced, because it will inevitably get damaged during combat. You don't want to damage your own bot just trying to replace the armor, so it needs to come off fairly easily—when you want it to. Sometimes the armor and the frame are the same thing. In other words, there is no armor other than the frame itself. Chapter 9 discusses the various materials that make good armor.

At any time during the mechanical design process, you can select which radio control system and "robot brains" you want to use. For driving a bot, you need at least two control channels—one for forward and reverse, and the other for turning left and right. This is true for bots that have channel mixing. With no mixing, you would use one channel for the left wheels and one for the right wheels. Additional channels are for controlling the special features.

You might want to automate some bot functions, like shooting a spike when the opponent gets within 1 foot of your bot. Here is where you specify the types of sensors for detecting the opponent and figure out how to mount them inside your bot. You'll probably need to have a microcontroller inside the bot to process and interpret the sensor results in order to control the weapon. Before you implement

FIGURE 2-3 *A robot showing how badly its armor was damaged at a BotBash tournament.* (courtesy of Andrew Lindsey)

any computer-assisted functions, your bot should be built and tested with all manual control. Once the bot works to your satisfaction, then you can add the automatic features.

All of the preceding design steps should be done, as much as possible, on paper or CAD before you start cutting parts to assemble the bot. This will save you from having to remake parts due to design changes. You don't absolutely need to have CAD software to do this, but CAD does give you more professional-looking results. You can use regular old-fashioned graph paper, too. Some people have even used chalk on their garage floors to design bots in full scale. Do whatever you're most comfortable with.

tip *Expert machine designers use CAD (computer-aided design) software; so if you want professional-looking results, you should consider getting a CAD program. CAD is so widely used among roboteers, in fact, that PTC (makers of Pro/E CAD software) has sponsored the last three seasons of* BattleBots. *Each team who showed up at the competition and asked for it got a free one-year license of the software, which normally retails for $21,000. Other CAD packages are available for a lot less money.*

FIGURE 2-4

This robot, Slap Happy, *was built using plywood as templates before metal parts were fabricated.* (courtesy of Dave Owens)

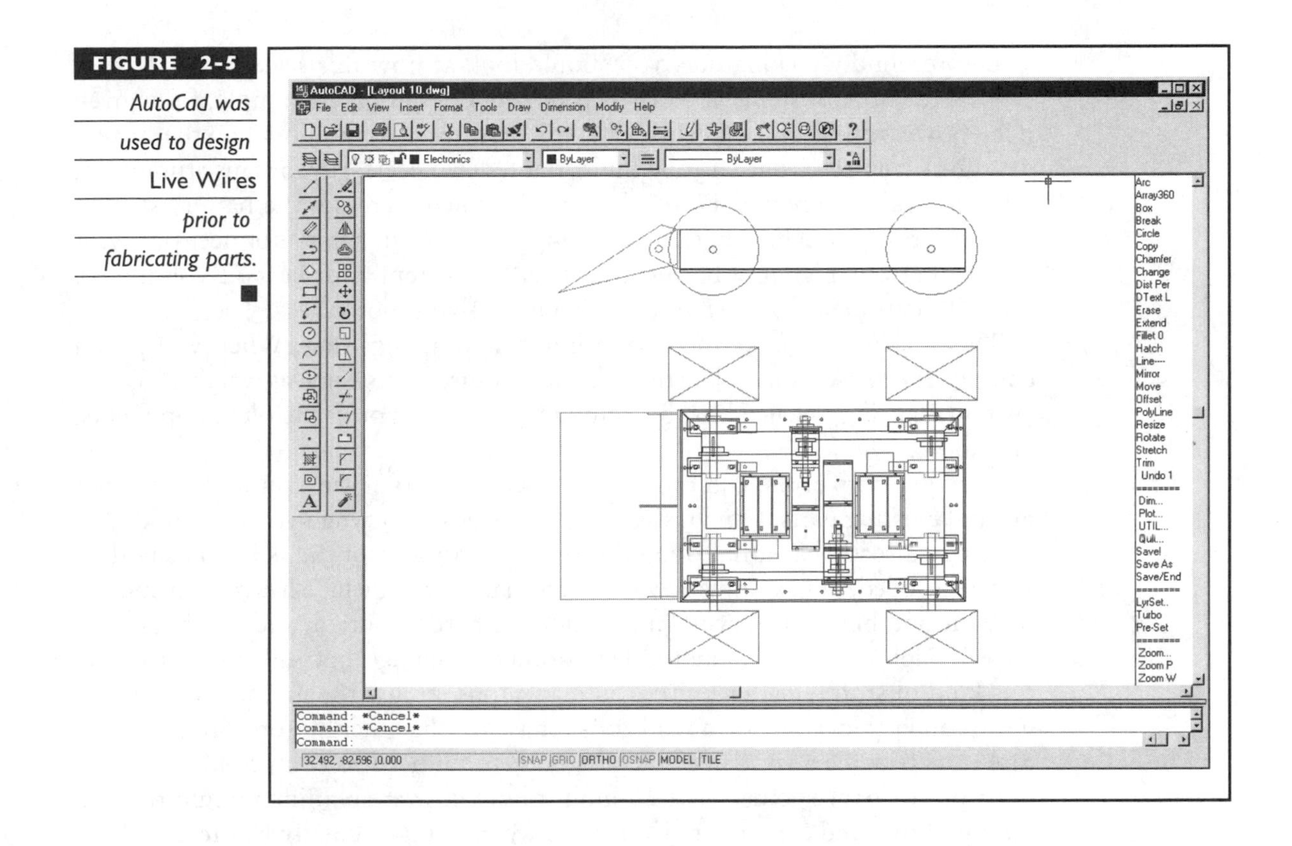

FIGURE 2-5 *AutoCad was used to design Live Wires prior to fabricating parts.*

The Game of Compromise

There has probably never been a bot made that didn't involve some level of compromise on the part of the builder. This is where your time-, money-, performance-, and availability-related trade-offs occur. We builders rarely get the chance to use the best parts available, and therefore must settle for what we can get. This is where you need to let go of your idea for a dream bot and start looking at your project more realistically.

For example, say you want your bot to move at 20 mph and you want to use 8-inch diameter go-kart wheels. To move at this speed, the wheels need to turn at 840 rpm. Now you have to find a motor that can deliver that speed. You search all of the magazines and catalogs you can find, scour the Internet, and you still can't find a motor that will give you the speed you want. This means you'll need to build a gearbox that can change the motor speed to the desired 840 rpm wheel speed. Here you will be faced with lots of options, such as spur gears, sprockets, belts, worm drives, and so on. In your search for motors, say you also found some gear motors—you pick a few motors, and then calculate what gear reductions you need to get the right wheel speed. At this point, you have several motor and gear options to choose from to get your robot to move at 20 mph. So, now you have to choose which combination you want to use.

Before blindly picking one, you should look at how this selection will affect each of the other systems at work in your robot design. For example, for a given horsepower rating on the motors, a 24-volt motor will draw about half the current as a 12-volt motor. That's a good thing, right? Not necessarily, because running at 24-volts will require two 12-volt batteries—which increases the battery storage area and robot weight. That's a bad thing, right? Well, again, not necessarily. A 12-volt battery might not be able to deliver the current to drive a 12-volt motor, but will have plenty of current for driving a 24-volt motor.

This is why you make the system interface drawings first. When you pick a component to use, you update the interface requirements, such as weight, voltage, current, spacing, the need to add new subcomponents or delete old components, and so on.

A bigger part of the compromising process occurs when you build your bot around existing parts. Obviously, life gets a little easier when you can build with stuff you already have, but often this means getting a bot that's less flashy than you envisioned. For example, say you were planning to include heavy-duty motors on your bot, but the ones you had in mind are hard to obtain, and you happen to have a couple of wheelchair motors lying around the garage (bot builders tend to have this kind of stuff lying around). You may choose to use the motors you already have, rather than going on a wild goose chase for the other motors. So, these motors now become a fixed specification, and you'll need to compromise on your bot's performance goals. That 20-mph robot you were planning might only go 10 mph now, and can only push half the weight you originally wanted.

Probably the biggest area of compromise comes with cost considerations. Say you found the ideal motors you want, but they cost $800 each and you need four of them for your four-wheel-drive bot. Like most beginners, you can't really justify spending $3,200 for motors. So you either find different motors, such as $100 cordless drill motors, or change the design from a four-wheel-drive bot to a two-wheel-drive bot.

Again, ideally, you should design the entire bot on paper or CAD before you start constructing it, although this usually isn't as much fun. Most people find designing and building at the same time more enjoyable because it allows you to see the progression of the bot from day one. Other people enjoy the design process more than the actual building. If you enjoy building, team up with a good designer. If designing is your thing, then find yourself a good builder to partner with. When your bot is completed, you should create a new set of drawings showing how the bot was actually built—especially all of the electrical wiring. These drawings will come in handy when you need to repair or improve the bot at later dates. It's easy to remember everything that went into building the bot when we first finish building it. But we soon forget certain details, which can create problems when maintenance is needed. These as-built drawings will save you a lot of headaches down the road.

Design for Maintenance

Part of the whole design process for combat robots is the design for maintenance. In competition, you have about half an hour to make any repairs to the bot. This really isn't a lot of time. So you must design your bot to allow for rapid replacement of parts. This usually means there are more bolted-on components than welded-on components. You need to have quick access to the electronics and batteries so they can be replaced or recharged in a matter of minutes.

Wheels should be designed to be replaced between contests because a lot of weapons and hazards will destroy the wheels. If you break a chain in the transmission, then it should be quick and easy to move the motors to replace the chain and retighten it back in place. The components inside the bot should be laid out in a manner such that you don't have to remove a lot of parts just to get at whatever is broken. The design should also allow for accessibility to the components. You will need to have room to get your hands and tools inside the bot. Think about the length of a screwdriver, or the length of a wrench. When you are designing the bot, imagine yourself having to fix it quickly, and then alter your design for that. This will require a little up-front thinking. The last thing you want is to be disqualified because you didn't have enough time to replace a dead battery. Of course, this is another one of those things that you may have to compromise on. Some of the top bots are difficult to work on. In a design like *BioHazard*'s, for example, the low profile and small internal volume of the bot make things hard to repair. *BioHazard* is held together by 700 screws, so getting inside him requires a lot of work with the electric screwdriver before repairs can even begin.

FIGURE 2-6 *A robot being repaired between matches at a* BotBash *tournament.* (courtesy of Andrew Lindsey)

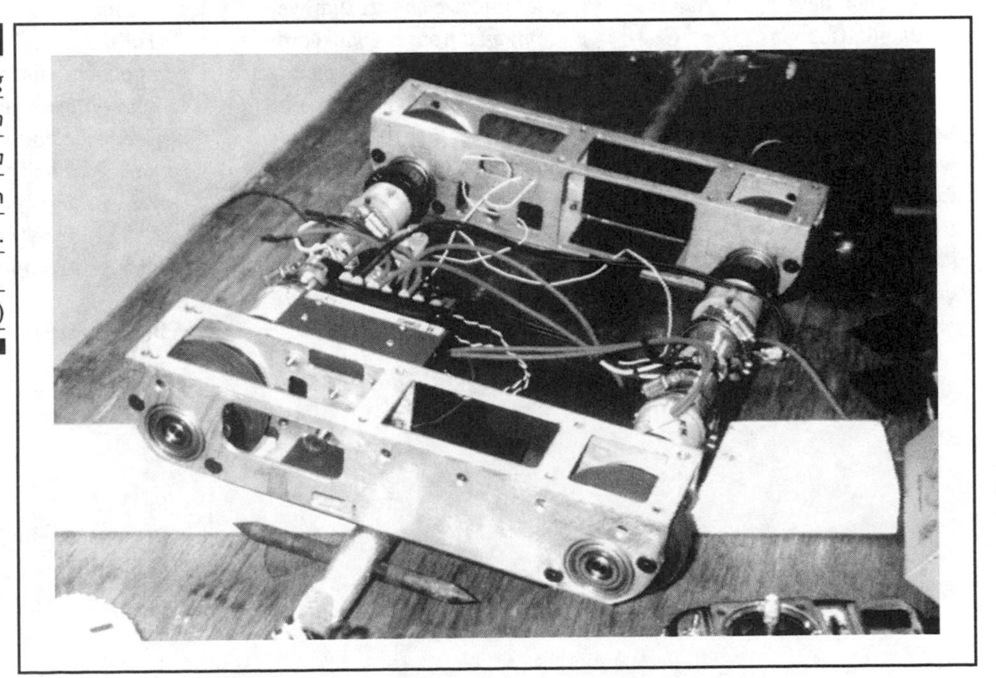

First Person

What got me into combat robotics in the first place was my friend, Andrew. He had read online about Robot Wars*, a gladiator-style competition between remote-control robots of varying weights, and needed help building a bot. I checked out the* Robot Wars *Web site and was very impressed. Andrew showed me his design for a bot called* Spike*. It looked very cool, and the whole idea of dueling bots fascinated me, so I said I wanted in. It wasn't long before I was hooked.*

Andrew and I worked for several months building Spike *version 1.0. This prototype never made it to* Robot Wars*. We thought it would be a good idea to test the design before bringing it to California, so we entered the machine in a* Robot Wars*-style competition held at DragonCon in Atlanta, Georgia.*

We got creamed.

Our ideas were good, but we had some more work to do before we brought this bot warrior to California in August. The drive train had to be reworked to make it more maneuverable and less difficult to control on rugged surfaces. Also, the fact that if Spike *got flipped on his back, the fight was over, was one problem that needed to be addressed in a hurry.*

It was back to the garage and back to work. A sleeker, slimmer bot came out of the six weeks of time and work invested. Spike *version 1.1 could right himself if flipped using the spike that was his namesake, the driveability issues were taken care of by changing the wheels, and Andrew and I made an overall change in body styling. We tested how well* Spike *could handle the road by test-driving him on the poorly maintained street I lived on. If* Spike *could handle those lumps, bumps, and debris and still move well, we were sure he could handle the arena.* Spike *passed the drive test. Next came the weapon test. The "spike" itself did well against cans and other metallic objects that we rummaged from the junkyard, so we were pretty confident it would handle itself well against what would be its first* Robot Wars *competition in the Lightweight Division.*

Because we had no sponsors, Andrew and I tapped our bank accounts to pay for the airfare to San Francisco. Spike *actually did a pretty good job in his first time out. He won a few matches, got dinged up a bit, and even burned out an electronic speed controller. In the end, he lost by getting pinned up against the wall. Even so, we had a great time.*

I came home with a bot in need of repairs and over a dozen rolls of film to be developed. We vowed to return the next year for more, and spent our time wisely revising the design and looking for sponsorship. Frank at Central Metals Fabricators in Red Bank, New Jersey, agreed to help machine Spike *Version 2.0 for free as his way of sponsoring us. Frank thought the whole idea of battling bots was cool, too, and viewed his work on* Spike *as a portfolio-building item. Andrew got a few parts suppliers to give us items for free or at a heavy discount, so we managed to save some money on construction.*

For the next Robot Wars *competition, Andrew managed to talk his employer into paying his airfare to California, but I was not so lucky and had to pay my own way. I was willing to do whatever it took to get myself back to California for the next competition.*

—Ronni Katz

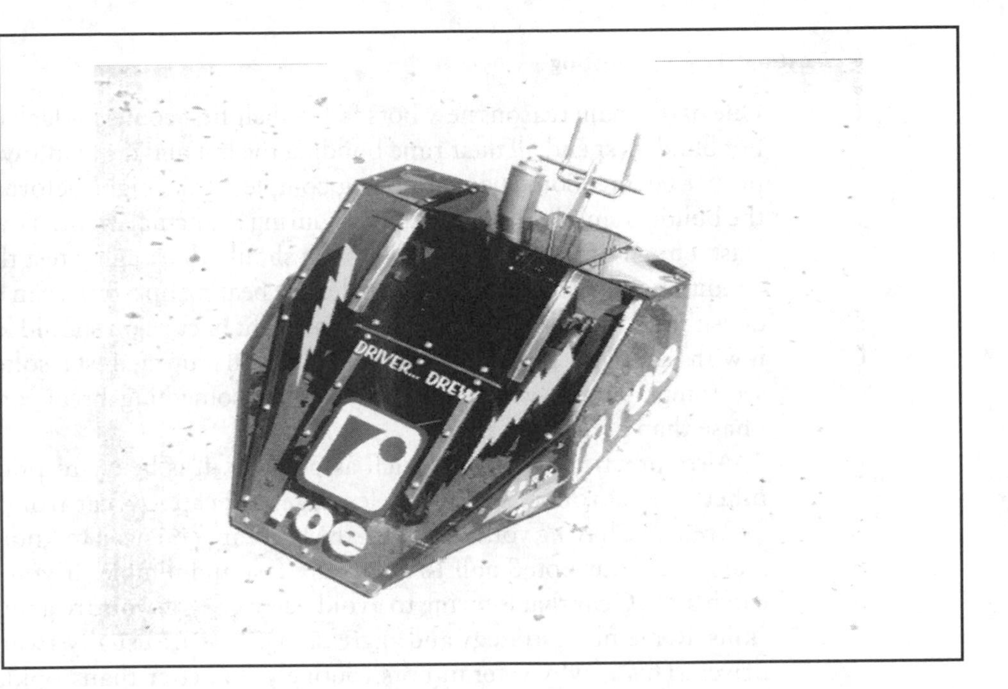

FIGURE 2-7 Spike *before heading off to* Robot Wars. (courtesy of Andrew Lindsey)

Start Building Now

This can't be stated any more clearly: If you want to compete in a contest that is six months away, you've got to start building, *now*. Bots always takes longer to build than you think they will or allow time for, and other, less important time commitments like school or work easily get in the way. There's always something that doesn't work and needs to be redesigned. Things break and have to be fixed. Things don't fit together like you planned and need to be modified. Murphy's Law always comes into play when building bots, especially when you are in a time crunch. There are some people who have successfully built a bot in as little as two weeks, and others who've spent over a year on their projects. Plan to spend at least six months to build your machine. It can be done in less time, but you'll have to work a lot harder to do it.

Testing, Testing, Testing

One of the main reasons new bots fail in their first contest is lack of testing. Often, bot builders spend all their time building the bot and don't allow enough time for proper testing. Some bots are being completed the night before the contest, and the builders simply hope it will work during the actual event. You should allow at least a month for testing your bot. You should thoroughly test the bot in combat conditions, as realistically as possible. But beating up a trash can or a wooden box doesn't test the bot. Garbage cans don't fight back. You should kick your bot, hit it with hammers, flip it upside down, and stall it up against a solid wall. Expect to see things break—you would rather have something break during the testing phase than at a competition.

Also, practice driving as much as possible. It is better to practice against another combat robot. At the very least, get a cheap R/C car from a local toy store and practice having your bot catch the R/C car. You need to know how to rapidly maneuver your bot. Small R/C cars are fast and nimble. If you can consistently catch an R/C car that is trying to avoid being caught, you are gaining good driving skills. Remember, strategy and aggression points are usually awarded to the better driver. This is why veteran bots routinely do better than rookie bots. They are thoroughly tested, and the drivers are excellent, experienced drivers.

Top Ten Reasons Why a Robot Fails

When designing your bot, think about what can go wrong during a contest, and then design your creation so these things can't go wrong. Many bots lose matches not because they're beaten by opponents, but because something broke. Below is a list of the 10 most common failures seen in combat robotics, all of which should be considered in your design process:

1. Wires coming loose, especially battery and radio control connections
2. Improper charging or using insufficient-capacity batteries
3. Speed controllers too small to handle the motor current requirements
4. Motors, transmission, and batteries poorly mounted
5. Belts and chains falling off
6. Motors overheating
7. Radio control interference
8. Shearing and breaking fasteners
9. Using homemade motor speed controllers
10. Wheels becoming damaged by weapon or hazards, or jammed because of the body getting bent into them

Sources of Robot Parts

There are a few hobbyist robot companies that offer parts for smaller machines; but for builders of larger combat robots, it's not that easy to find parts. Some companies—like C&H Sales, Grainger, McMaster Carr, and Servo Systems—offer many items that are ideal for robot construction, and other sources listed in the appendixes at the end of this book offer more choices. However, most of us find we've got to be creative and use local sources to complete our designs.

Before going out to find parts, think about the motions you'll require. What types of things move? Old washing machines have great transmissions. Electric wheelchairs have motors and controls that have design requirements similar to the requirements for large bots. Bicycles and motorcycles have many usable parts, especially chains and sprockets. Power lawnmowers and rototillers have good parts, as does furniture made with movable sections or parts. Car power seats, power windows, electric door locks, and windshield wiper motors are good items. And don't forget garage door openers, car jacks, car "gas springs," cordless power tools (especially drills), office equipment, computer printers, and even drawer slides.

The best sources are old production equipment that may have all types of premachined metal forms, chain and gear drives, bearings, shafts, and motors. Any type of machinery can be used in some way—farm equipment, dairy machines, food processing machines, even items off heavy construction equipment. Any time you see something that's being thrown out or cheap, just think, "Can I use this for a bot?" The famous Blendo has a shell made from industrial-sized cooking woks.

Some people can't afford to buy brand-new parts directly from the manufacturers. So, surplus stores, garage sales, thrift stores, junk yards, and stuff hidden in the basement make great bot parts. Some bots are built from parts that have been used for other purposes, and a lot of those have won competitions. You don't have to have brand-new parts to make a robot, but the parts you do use should be durable and reliable. Sometimes, however, you have to buy new parts. When you are using recycled components, you should find out where to get replacement parts for each component in case it breaks.

Cost Factors in Large Robot Construction

An experimental robot can cost anywhere from nothing to well over $100,000. Mark Tilden, the creator of the BEAM (Biology Electronics Aesthetics Mechanics) robots, can build a walking bot out of an old discarded Walkman radio in one evening without spending a single penny. A simple microcontroller-driven tabletop line, following robot will cost about $200, and a top competitor *BattleBot* can easily exceed $20,000.

Building a combat robot is not a cheap venture, and you should be prepared to spend a lot of money to build something competitive. Most builders spend several thousand dollars building their bots. You might be the lucky individual with a home machine shop (or have a friend with one) and an uncle who owns a junkyard and a surplus store. However, most of us aren't this fortunate and must hunt through countless stores and catalogs to find what we want. Appendixes A-C at the end of this book will lead you to many proven sources of robot parts. No matter how full your junk boxes may be, you'll probably find yourself purchasing a lot of the parts to build the robot—especially the electronics and controls.

Safety

Before you start building your bot, you must also address safety issues. If you've watched *BattleBots*, chances are you've heard the announcers stressing the use of safety glasses and proper supervision. As adults, most of us have already learned the basics in shop safety. But the construction of combat robots extends way beyond what is normally considered a hazard in a home shop, and severe injuries are possible with even the smallest combat robot—both in operation and in the construction process.

Before we delve into safety issues, we should mention gaining knowledge in the use of shop tools. All the safety equipment in the world won't protect you from unsafe shop practices. If you haven't been instructed in the use of shop tools through a shop class at school, or through instruction at your job, you should consult a friend or acquaintance to instruct you, or leave the work to those who know how to do it safely. This cannot be stated strongly enough!

A chuck key left in a drill press when it is turned can be thrown at high speed right through safety glasses. A slight slip with a band saw can turn you into a nine-fingered bot builder in a fraction of a second. Misuse of a bench grinder can cause a grinding wheel to literally explode into shrapnel, riddling your body, face, and eyes with hundreds of rock-shaped bullets. A loose piece of clothing can be sucked into a metal lathe in a second, and you along with it. If this scares you, then we authors have done our job here. You're welcome.

Safety glasses are a must when using any power tool for any purpose. Even the tiniest particle in your eye can ruin your day, and a metal particle traveling at high speed can destroy your eye or eyes. Buy and wear the good, tempered glass kind with side shields. Keep those glasses on even when working with batteries and with high-amperage cables. A sealed electrolyte battery when dropped on a floor can crack and splash acid everywhere. Sparking cables can make you feel as if you placed your face on a welding table.

Okay, enough said on these issues.

Safety in the Use of Shop Tools

There are many power tools available to the robot experimenter. One of the first items you should purchase outside of handheld tools is a bench drill press. In itself, this is not a dangerous tool, but it can still cause injuries. The belts and pulleys at the top, if left exposed, can cause injuries to the hands. The drill chuck generally runs at a low enough speed when drilling to not cause flying bits of metal, but the use of other metal-cutting tools can cause metal to fly everywhere. Again: use safety glasses. Tighten the bit or tool securely and then remove the chuck key. Feed the tool or bit into metal slowly, using a lubricant, and using a lower speed for larger drill bits. Be sure to have the work piece securely clamped to the drill press table to prevent it from rotating.

Many of the same safety tips apply to all power tools when working with metal. Be careful of the placement of your hand when using your other hand to hold a workpiece. Bench grinders, metal and wood band saws, routers, and saws all require you use common sense when operating. Most hand-power tools have an internal blower to cool the motor, and this wind can sometimes blow chips and dust into your eyes. Always have a complete first aid kit on hand and know how to use it.

The larger shop tools such as metal lathes, milling machines, and the various types of welders all require special knowledge that cannot be obtained from any "manual," and it is recommended that you obtain special instruction in their use. Community colleges usually have shop courses, and even a local machinist can give you help in this area.

Safety with Your Robot

Safety is also critical when dealing with your bot. This should come as no surprise, because often these machines are 350 lb. warriors designed to obliterate other machines their own size. You can just imagine what a bot like this can do to the tender skin of a human being. Be extremely careful when you power up your machine for testing. Always remember Murphy's Law: "If something can go wrong, it will." Always assume that any part of your bot will fly off at any time, and plan accordingly. *Never, ever operate a combat robot in the presence of children.* Even a seemingly benign machine such as a wedge can go out of control and quickly smash into someone, breaking legs or doing even worse damage.

No amount of body armor and safety glasses can protect a person from a large spike that is accidentally thrown from a spinning robot. A pneumatic weapon arm can accidentally deploy upward and sever a person's head. Sharpened weapon edges can still cut you severely, even when you're not in the middle of operating your machine. A 1,500-psi gas line can break away and whip about like a mad cobra. The use of a full-face mask is recommended when dealing with high-pressure pneumatic systems.

There are many more ways to be injured while building and operating a combat robot—far too many to list. The authors and publisher of this book cannot take responsibility for injuries sustained during any construction, testing, or use of the bot. Use common sense, then plan, and then work carefully and slowly. Watch out for others. When working on your bot, make sure the batteries are disconnected. And above all, never leave a functional bot unattended. If you follow these simple safety suggestions, you should not be injured. Save the "hurting" for an opponent's bot in a combat contest!

BUILD YOUR OWN COMBAT ROBOT

chapter 3

Robot Locomotion

MOVING is what many might call a robot's primary objective; it's what separates a robot from a plain old computer sitting on the floor. Whether you use wheels, legs, tank treads, or any other means of locomotion, you've got to figure out a way for your machine to traverse across the floor or ground, unless you're trying to build a flying or marine-based machine. The way you make your robot move will be one of the most important considerations in the design of your combat robot.

In this chapter, we'll concentrate on locomotion methods that are easy to construct and most effective for large robots and combat machines. We'll also discuss the drawbacks of some methods for combat robot applications. Several methods of locomotion have been successfully used in combat and other large robots. These are legs, tank-type treads, and various other configurations and styles of wheels. Yes, some really cool machines have used other means to get across the floor, but "cool" and effective are sometimes very different.

Legs are often one of the first types of locomotion we envision when we think of robots. For most people, *robot* means a walking bot like C3P0 in *Star Wars* or Robby from *Forbidden Planet*. However, we must remember that these creatures were just actors wearing robot suits to make them appear as walking machines. Walking is actually a difficult task for any creature to perform, whether its human or humanoid. It takes babies nine months or longer to master the act; and for several years after that, they're tagged with the title of "toddler." A child's brain is constantly learning and improving this complex process each day. Bipedal (two legs) walking is really controlled falling—stop in the middle of taking a step and we'd fall over. Impede the process with a few beers too many, and our built-in accelerometers (our ears' semi-circular canals) feed us wrong information and we stumble.

Robots with Legs

Watch a person walking and you see them swaying from side to side with each step to keep balanced. Try race walking and see how exaggerated you must twist your body to speed up walking. While walking, we always strive to keep our center of gravity over one foot if only for a fraction of a second. If you count the number of joints and motions in a person's leg, you'll realize that these joints are multi-axis joints—not just single-axis joints that we might have in a robot. Many human joints have three degrees of freedom (DOF), in that they can move fore and aft, move side to side, and rotate.

Bipedal robots have been constructed, and a few Japanese companies are demonstrating these in science news shows. Most robotics experimenters, however, soon learn the complexities of two-legged robots, and quickly move to quadrupeds (four legs)—and then just as quickly to hexapods (six legs) for their inherent stability. Sony has sold many of its popular AIBO dogs and cats with four legs, and the same for the much cheaper i-Cybie; but these machines have many motors for each leg and are not being attacked by killer robots, as are combat robots.

Hexapods are a popular robot style for robotics experimenters because, with six legs, the robot can keep three feet on the floor at all times—thus presenting a stable platform that won't tip over. Compare this with a quadruped, which can lift one leg and easily tip over, depending upon the location of its center of gravity. The six-legged "hex-walkers," as they are sometimes called, can be programmed to have their fore and aft legs on one side of the body and the center leg on the opposite side all raise and take a step forward, while the other three "feet" are on the floor. In the next step, the other three legs raise and move forward, and so on. More complex walking motions needed for turning use different leg combinations selected by an on-board microcontroller. Each leg can use as few as two axes of motion or two DOF, and some builders have used two model airplane R/C servos to control all six legs. These types of robots are excellent platforms for experimentation and for carrying basic sensors, but they are difficult to control and might present an added complexity for a combat robot's operator.

Although many of the robot organizations you'll find on the Internet focus a lot of attention on the construction of legged robots, the basic fragile nature of legs makes them an extra challenge for builders of combat robots. Don't get us wrong—walking combat robots have been built, and some have done very well in competition. If you want to build a legged combat robot, go for it. Many popular robot competitions, including *BattleBots* and *BotBash*, even allow an extra weight advantage for walking bots. Figure 3-1 shows a photo of *Mechadon* built by Mark Setrakian. *Mechadon* weighs in at 480 pounds. This robot is the largest and most impressive walking robot ever built for any combat robot event. The robot can roll over, and it can crush its opponents between its legs

If you're a beginning-level robot builder, you'll probably find it easiest to work with one of the more battle-proven methods of locomotion when designing and constructing your combat robot. Since we're assuming that a lot of our readers are still at the beginner level, we'll be focusing on other, less complicated forms of locomotion for competition robots. If you're interested in learning more about walking robots, many Web sites and reference books can provide helpful information. Some of our recommended books and sites are listed in the appendixes in this book.

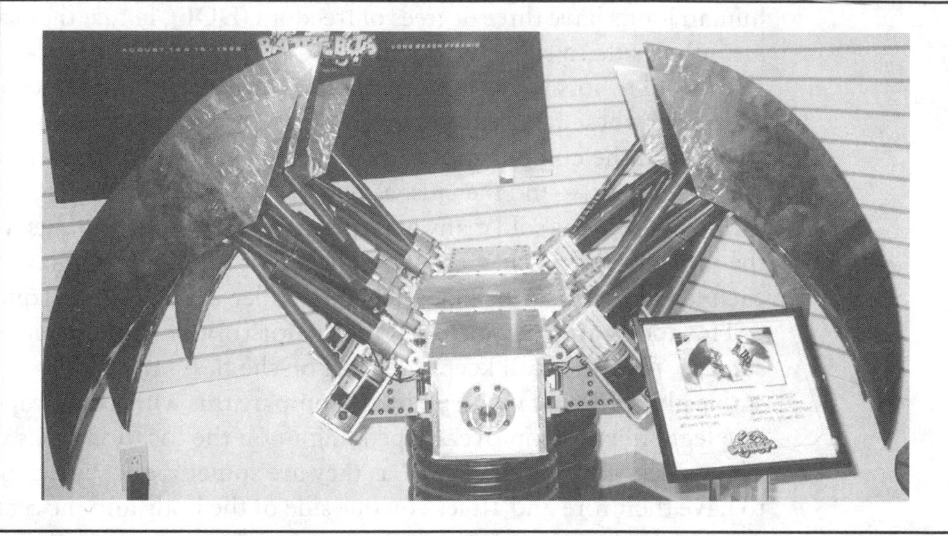

FIGURE 3-1 Mechadon, *the largest walking combat robot ever built.* (courtesy of Peter Abrahamson)

True Story: Christian Carlberg and Minion

"I have been building mechanical devices since I was a kid," says Christian Carlberg, founder and captain of Team Coolrobots. Christian is well-known for robot designs like OverKill, Minion, and Dreadnought. "Erector Sets, Lincoln Logs, LEGOs," he adds, "I used them all."

That early experience with building toys paid off for Christian, who further honed his mechanical skills at Cornell through mechanical competitions ("build an electric motor in a couple of hours with these common house hold items," he says). But LEGOs were—and remain—important. "If you can't build the premise of your robot with LEGOs then it's not simple enough to withstand the BattleBox."

What competition stands out in Christian's mind?

"My favorite fight was the Super Heavyweight rumble for the first season of Comedy Central's BattleBots."

Minion's story actually begins in September of 1999, when BattleBots announced the new Super Heavyweight class. "The idea of building a 325 pound robot really appealed to me, especially considering it was a brand new weight class and there wouldn't be a lot of competition."

For that event, BattleBots placed ten 300-pound robots into a box for five minutes. "I was driving Minion for that fight," Christian recalls. "As the fight progressed it was clear that Minion was the strongest robot in the BattleBox. I was pushing three robots at a time, slamming other robots up against the wall. It was so much fun and totally worth all the hours spent on building the robot."

Indeed, Team Coolrobots exudes bravado about Minion's power. "Minion will not break or be broken. The only way to defeat Minion is to overpower it. This used to be impossible but has been known to happen." Christian admits that there's a secret to that raw locomotive power. "The weapon was always last on my list of priorities. You can still win as long as you are moving, which is why the frame and drive train will always be a higher priority for me."

Tank Treads: The Power of a Caterpillar Bulldozer in a Robot

Tank treads seem to be the ideal way to make sure your robot has the pushing power to allow it to decimate an opponent in combat. Hey, they're called "tracks" because they provide a lot of traction, right?

We'll call the ones robot builders have used "treads" from here on. The military uses treads in tanks to demolish a much larger and more menacing enemy on a rugged battlefield. Earth-moving equipment can bounce across rocky ground pushing many tons of dirt, as the two sets of treads dig in with all their might. These things seem to be the ultimate means of locomotion for a winning combat robot. This could well be the situation if the contests were held in a rocky and hilly locale, but most competitions take place on fairly smooth industrial surfaces. All the same, let's examine the construction and use of tank-type treads or tracks.

Many first-time robot builders are drawn to treads because they look so menacing. Treads come in two basic sizes—massive off-road and toy sizes, and there is no similarity between the two. The toy variety is just a rubber ring with "teeth" molded into the rubber. The larger off-road–size treads consist of a series of interconnected metal plates, supported by a row of independently sprung idler wheels. The construction of interconnected plate treads is complex and should be left to experts with large machine shops. Peter Abrahamson has built a very impressive 305-pound robot named *Ronin*. The aluminum tank treads were custom machined for this robot. Each side of *Ronin* can rotate relative to the other, thus improving the overall traction capability of this robot. Figure 3-2 shows a photo of *Ronin* climbing a log.

FIGURE 3-2 Ronin—*a true tank-driven robot with an independent suspension system.* (courtesy of Peter Abrahamson)

Bot experimenters usually opt for the rubber tracks removed from a child's toy bulldozer, and then start piling batteries, extra motors, sensors, and arms onto the new machine. When the first test run is started, the rubber tips of the tread surface begin to bend as they push onto the floor. The robot chugs along just fine until it has to make a turn. If the operator happens to be monitoring the current drawn by the drive motors, he'll see a sharp increase as the turn begins. This is one of the major drawbacks of tank-style treads: they must skid while making a turn, and energy is wasted in this skid. Only the center points of each "track" are *not* skidding in a turn. For this reason, many robotics engineers opt not to use tank-style treads in their machines.

However, the efficiency of the propulsion system is a less significant factor in combat robots than in other types of bots. Because a combat robot's "moment of truth" is limited to a 3-to 5-minute match, builders can easily recharge or install new batteries between matches, making the issue of wasted energy less of a consideration. With this fact in mind, many builders opt for tank-style treads, so let's examine another feature of treads: they're complex and hard to mount.

The toy rubber ring tank tread seems anything but complex. It's just a toothy rubber ring strung between two pulleys. The experimenter with his toy bulldozer treads might be so preoccupied with the current draw of his drive motors or with maneuvering the machine that he doesn't notice one of the treads working its way off the drive spindle. And if the tread slips off your heavyweight bot in a robot combat match, chances are you'll lose.

Building Tank Treads for a Robot

You've probably realized by now that even the largest toy tracks you can find are too small for a combat robot or any other type of large robot. The smallest of the real metal treads are ones you've seen on a garden tractor, and these are too big for your machine. So, if you're dead set on making your robot move with tank treads, you're probably wondering what to do next. You might start to look at wide-toothed belts, which work much like the timing belt on your car. The only trick to using these is that you need to make sure whatever belt you choose has enough traction to stay competitive on the arena floor. Some successful builders have used snow-blower tracks, which seem to be just the right size for many types of combat robots. Flipping a large industrial belt with softer rubber teeth inside out is another option for builders who want tank treads on their bots. These are ready-made teeth to dig into the floor, flexible and cheap—what a way to go!

In this case, you go to a friend and have him machine two spindles out of aluminum that fit the width of the belt. After mounting one of the spindles on a free-turning shaft and the other to a driven shaft, you try out one of your timing-belt treads. Almost at once you notice the driving spindle spinning on the belt's surface when you apply a load to the bottom of the tread. You remember seeing that the driving spindle on a real tractor has teeth that engage the back of the tracks. You decide to machine two new drive spindles out of rubber. You're back at your friend's shop and he tells you that he'll have to grind the rubber down, rather than machine it like metal. After a few hours of experimentation, he hands you two rubber drive spindles.

Now you have four spindles to mount both belts for a complete robot base, two rubber and two aluminum. After assembly, you find that the new drive spindles work pretty well. The rough ground surface of the spindle does a decent job of gripping the smooth rubber belt's surface. After trying the base out on the floor, you find that the turning is erratic and decide that you need a row of idler wheels to keep the entire length of each belt firmly on the floor. Your friend patiently machines for you 10 idler wheels, which you mount to a series of spring-loaded lever arms. Wow, this robot is beginning to be a bit complicated! After a few tries on your garage floor, you begin to notice that the teeth are wearing down. You smile at your creation and decide to put it away. It was a good learning experience.

Wheels: A Tried and True Method of Locomotion

Many people in the field of experimental robots would not think of any way to make their robot move other than using tank-type treads. Others feel the same way about legs, whether two, four, or six. As mentioned earlier, many other means of locomotion and propulsion for robots are out there, including flying or swimming, but we'll concentrate on wheels from this point on. Wheels are pretty much proven in all types of robot applications, from the smallest desktop Sumo machine to the largest mobile industrial robots. Even designers for NASA's Mars-exploration robots gave up on legs and other means of locomotion in favor of wheels.

Types of Steering

Wheels are generally categorized by steering method and mounting technique. The two types of steering that are used with wheels are Ackerman steering and differential steering.

Ackerman Steering

Ackerman steering, also known as car-type steering, is familiar to all of us. Figure 3-3 illustrates several variations of Ackerman steering. Note that only a single motor source drives the wheels, and a separate motor controls the steering. This method uses two wheels in the front turning together to accomplish the turn. Sometimes a single wheel is used, as in some golf carts, or the rear wheels can turn, as in forklifts. A child pedaling a tricycle is powering the front wheel, but she is also using that same front wheel to control the direction of movement of the vehicle. This turning method has been used in robot applications, but it is not as popular as the differential drive method that we'll discuss in a moment.

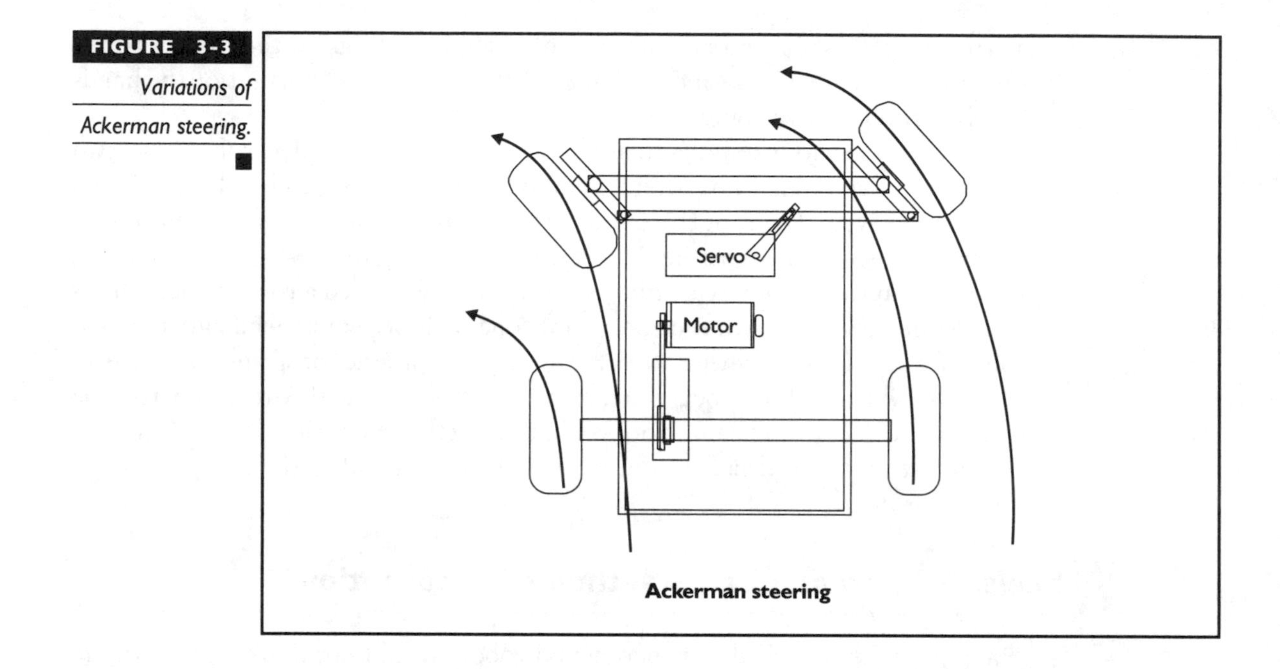

FIGURE 3-3 *Variations of Ackerman steering.*

Ackerman steering is used in radio controlled (R/C) model race cars and in most children's toys. It requires two sets of commands for control. Quite often, a model race car R/C system will have a small steering wheel on the hand-held transmitter to control the steering direction and another joy stick to control the speed, either forward or reverse. This type of steering has the capacity to be more precise than differential steering in following a specific path. It also works best for higher speeds, such as that of real cars of all types and model race cars. Its major disadvantage is its inability to "turn on a dime," or spin about its axis. This type of steering has a turning radius that can be only *so* small; it's limited by the front-rear wheel separation and angle that the front wheels can turn.

Differential Steering

Differential steering, sometimes called "tank-type" steering, is not to be confused with tank treads. The similarity is in the way an operator can separately control the speeds of the left and right wheels to cause a directional change in the motion of the robot. Figure 3-4 illustrates how controlling the speed and direction of both wheels with differential steering can result in all types of directional motion for the robot. Note that each of the two separately driven side wheels has its own motor, and no motor is required to turn any wheels to steer.

With differential steering, spinning on the robot's axis is accomplished by moving one wheel in one direction and the other in the opposite direction. A sharp turn is accomplished by stopping one wheel while moving the other forward or backward, and the result is a turn about the axis of the stopped wheel. Shallower turns are accomplished by moving one wheel at a slower speed than the other wheel,

FIGURE 3-4

Differential steering

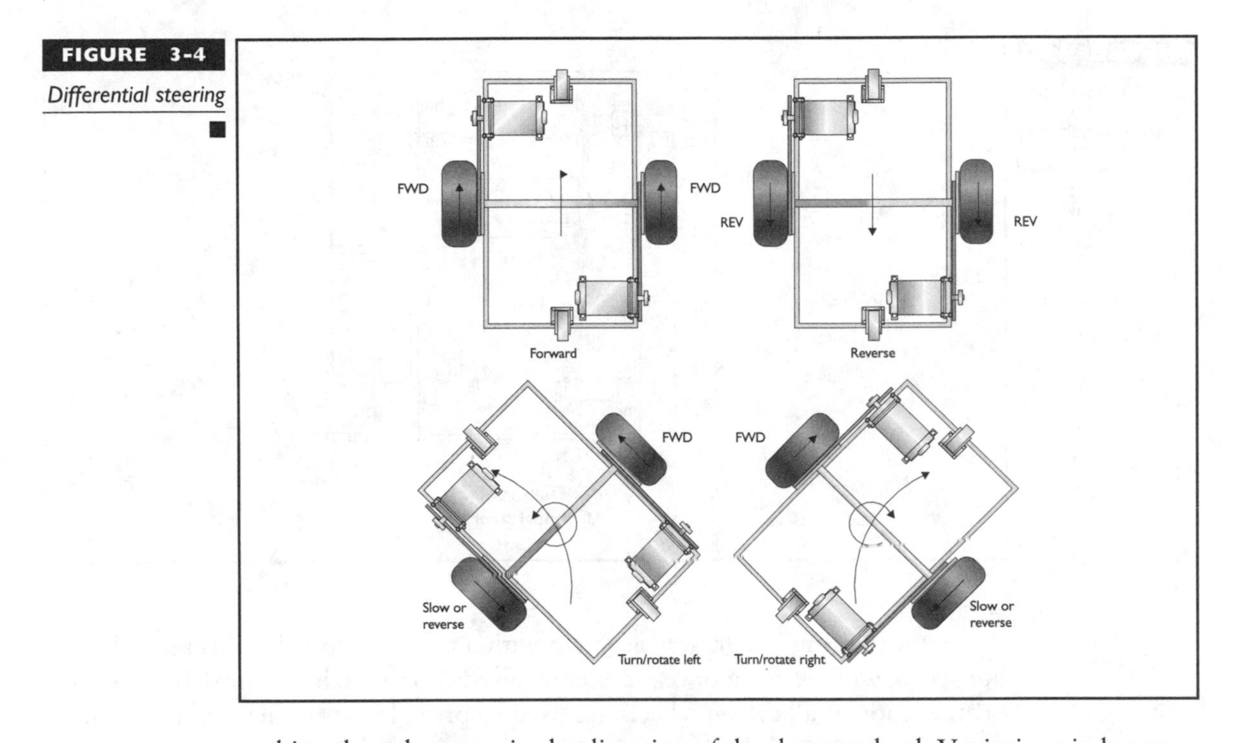

making the robot turn in the direction of the slower wheel. Variations in between can cause an infinite variety of turns. This type of control is most favored by remote-controlled robots on the battle floor and by promotional robots you might see in advertising. The wheels versus treads controversy has produced a design variation that does not use the free-moving caster illustrated in Figure 3-4, but instead uses a series of side-mounted wheels, similar to the idlers pressing downward on the inside of tank treads. See Figure 3-5. Some or all of the wheels on each side may be powered with a separate motor attached to each wheel, or with each set of wheels on either side interconnected by a single chain or belt drive, and a single motor per side. Yes, this method is not energy efficient for the same reason tank treads eat batteries—the front and rear wheels must skid in turns.

Chapter 13 shows you the construction techniques that were used to build the robot *Live Wires*. This four-wheeled combat robot was built on two cordless drill motors, one for each of its sides. For safety purposes, two drive sprockets on each drill motor were used with a separate chain going to each of the two racing go-kart wheels on either side of the motor. If one chain was broken, *Live Wires* still had mobility, and the differential steering capability was left mostly intact.

The multi-wheel platform does have an advantage: it can provide a lot of traction with a low-profile robot fitted with small wheels. To achieve this traction, however, the builder should independently spring each wheel a small amount to prevent high-centering, which can occur when the bottom of the robot gets caught on some obstruction, leaving the wheels lifted off the ground. For example, a four-wheel-drive vehicle can get high-centered after driving the front wheels over a large tree. If the vehicle gets stuck on the tree between the wheels, the wheels can't get the traction needed to get off the tree.

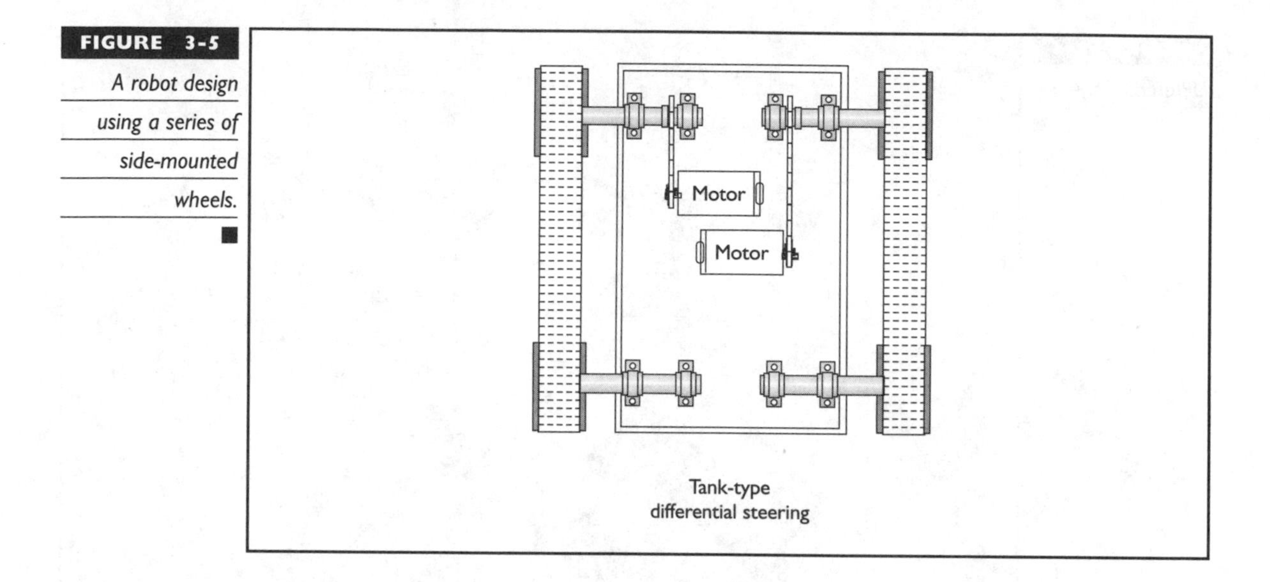

FIGURE 3-5 A robot design using a series of side-mounted wheels.

High-centering is a greater problem with a typical two-side-wheel differential bot setup, where a front or rear caster is raised enough to bring the driving wheels off the floor. If all driven wheels are used to provide extra traction, accidentally raising one or more wheels reduces the available traction that a combat robot may need to defeat its opponent. When using casters in the front and rear of a differentially driven robot, you should have each of them spring-loaded to prevent the robot from rocking back and forth, but not too much so that the robot might be lifted off its drive wheels.

Wheel Configurations

Some of the several methods and configurations of wheel mounting are more applicable to unique terrain conditions such as the "rocker bogie" system used on some of the Mars robot rovers developed at NASA's Jet Propulsion Labs. The predecessors to the famous Sojourner robot that roved about Mars's surface were named various forms of "Rocky," after the wheel-mounting system used. This system employs two pairs of wheels mounted on swivel bars that can help the wheels conform to uneven surfaces.

In smaller robots, many experimenters mount the wheels directly to the output shaft of the gearmotor. This works fine for the light robots that are designed to follow lines on the floor or run mazes, but it doesn't work well for larger machines, especially combat robots that take a lot of abuse in their operation. The output shaft of most gearmotors may have a sintered bronze bushing on the output side, and many times such a shaft does not have any sort of bearing on the internal side of the gearcase. This type of shaft support is not made to take the side-bending moment placed upon it by wheels and heavy loads. *Bending moment* is the name of the force that is trying to snap the shaft in two when one bearing is pressed downward as the other bearing is forced upward. Bending moment forces on a robot's wheel in combat are sometimes so severe that a gearmotor's gearcase can be shattered, even if ball bearings are on both sides of the gearcase.

One unique configuration of wheel mounting can possibly save you if your machine is ever flipped onto its back. Several robots have used identical sets of wheels on both the top and bottom, with mirror-image sets of top and bottom body shells; this allows the robot to continue its mobility while "upside-down." The other, more popular, method is to add wheels of sufficient diameter to protrude equally above the top surface, thus allowing continued mobility while "upside-down." This system works well for the low-profile machines; but for larger machines, it obviously gets a bit more complicated because huge monster truck-style wheels might obstruct a robot's mobility. For these types of bots, a top-flipping actuator can be used to right the robot after a flip.

Selecting Wheels for Your Combat Robot

Wheels are one of the most important considerations in the design of your robot. They are your robot's contact with the rest of the real world—namely, the battle area's floor. They allow your robot to move, maneuver, and attack its opponent, as well as retreat from an unfavorable position. Knowing this, your opponent will do everything he can to remove your robot's maneuvering ability, something you should also do to his robot at every opportune moment. So the words "sturdy," "tough," "puncture-proof," and "reliable" should all come to mind when you select wheels for your combat robot. And sometimes a wheel just looks too cool not to be used on the robot—take a look at Figure 3-6.

FIGURE 3-6 *A 14-inch diameter, flat proof, treaded wheel.* (courtesy of National Power Chair, Inc.)

You must also remember that the floor in a combat robot arena is not exactly like Grandma's living room floor. It includes some of the most destructive and devious hazards the contest producers can conjure up in their sadistic minds. Metal-cutting saw blades, spikes, hammers and even water can all come together to ruin your robot's day. You shouldn't waste time worrying that another machine or the hazards operator will attack your pride and joy in a contest. *It will happen.* Prepare for the worst. Have a wheel configuration and tire construction that will survive far more abuse than you can deliver in your garage tests, as you will be shocked at what a full-blown match can do to your machine.

You might be looking at a set of 20-inch bicycle tires for possible use in your robot, thinking, "If a 150-pound bike rider can jump over curbs and logs for days on end, tires like these should survive a 3-minute robot battle." If you watch a few robot combats, though, you'll see that wheel failure is not caused by downward force or even force from the front of the machine. What kills wheels is force from the side, hitting one side of the wheel, and bending or breaking the shaft or hub. A killer robot will "taco" a bike tire in seconds, or shred its spokes. Leave bike tires for benign robot designs.

Another favorite wheel of the beginning robot builder is the kind found on lawnmowers and other garden tools. Their ability to bounce over rough ground may seem to make them good potential robot wheels, but the same applies here as in bike tires. They cannot take side-bending forces. Most of the newer types use cheap plastic rims instead of metal. You find wheels and tires from so many sources—such as toys, disability equipment, hand-held golf carts, and barbecues—that we will not further elaborate. Consider the original intended use of the equipment and the expected loads the design team might have considered. Many companies have cut quality in areas to compete in the market pricewise. Look at all parts of the wheels you intend to use. Be cautious and use good sense here.

One of the best sources of tires and wheels for combat robots is from industrial applications. The hard rubber tires used in industrial parts carts made to handle thousands of pounds are among the best. Aerospace surplus yards generally have several varieties of these wheels, both mounted and unmounted. These wheels have stout rims and extremely tough tires. Some are non-rotating types and others are mounted in swivel assemblies as large casters. Most of these industrial wheels do not have any sort of tread, as they are used in passive applications that do not require traction.

Figure 3-7 shows a heavy-duty drive wheel.

One of the most popular wheel types used in combat robots are go-kart wheels, which come in a wide variety of rim and wheel types and shapes. They are readily available and easy to mount to a robot. Many top competitive robots use go-kart wheels.

FIGURE 3-7 *An 8-inch diameter heavy-duty drive wheel.* (courtesy of National Power Chair, Inc.)

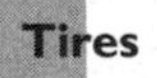

Tires

In addition to wheels, you need to carefully consider the rim and tire of your robot's assembly. The tire or rubber part of the wheel is probably the most critical consideration, because it is the most exposed part and takes the most abuse. It is the part that will encounter the kill saws at some point in a *BattleBot* competition. Tire hazards wreck more robots than all the rest of the hazards combined. Imagine what an opponent's weapon or a kill saw can do to your intended wheels. How secure is the rubber mounted to the rim? Will the rubber stay on the rim if it's partially shredded? How easily can the rubber be shredded? Are the tires pneumatic and can they be "popped?" If one or more wheels have a series of gashes in them, can you still maneuver your robot or allow it to escape your opponent or the hazard to regroup? Can the tire be struck from the side and be knocked off? You must ask yourself these and many other questions before you select the tires used.

You may like a particular wheel/tire combination that you've located and want to make it a bit more resilient to the onslaught it will be facing on the battle floor. You see a pneumatic that is the right size and has good traction, but you realize that it can easily be punctured and flattened, or it can be shredded by some weapon or hazard. In this case, consider filling the tire with a pliable rubber epoxy instead of air. The epoxy will bond to the inner part of the rim, and at the same time hold the inner part of the tire together, resulting in a puncture-proof combination. Another option is to fill the tires with foam, which a lot of experienced robot builders use to keep down the weight of their robots.

Traction on the combat floor is important. Go-kart tires are made for extremely hard use, and their fairly soft surface has pretty good traction (see Figure 3-8). Many of the pneumatic tires you might find in surplus houses or hardware stores have molded treads for traction purposes. The industrial cart tires mentioned earlier with the hard rubber tires are not pneumatic, but they can be modified with grooves, which some builders believe give traction to the wheels. Cutting with a knife or saw is not recommended, though, as any sharp cuts or gouges can easily propagate into a crack that can eventually sever the tire. Grinding the grooves is recommended instead.

Mounting and Supporting the Wheels and Axles

The mounting of the wheel to the axle and other parts of the locomotive system is the next important consideration. Not only must the complete wheel assembly be securely attached to the axle, but the wheel should ideally be able to be rapidly removed if repairs and replacements are necessary between matches. An easily removable wheel can make the difference in winning or losing a competition. You can attach wheels to robot platforms in numerous ways. Attachment methods depend on the wheel configuration desired. A typical arrangement might be the one illustrated in Figure 3-9. Many robot designs involve some sort of metal box chassis with internal motors and associated equipment, and external wheels attached

FIGURE 3-8 *Go-kart wheels give a robot the look of a racing car.*

FIGURE 3-9 *A typical wheel configuration arrangement where an axle is supported by two pillow block bearings. A sprocket is located between the pillow blocks, and the wheel is located to one side of the pillow blocks.*

to axles protrude from the "box." Fortunately for the combat robot designer, the terrain that the robot is to traverse is usually a flat floor with little deviations from level. A few bumps may result from joining floor surfaces, and some of the hazards present an uneven surface area in small spots. However, for the most part, the floor is flat in virtually all of the popular contests.

Such surfaces may not remain the case for future events, though, so a prospective designer may want to take into consideration possible variations in floor flatness. Some present-day contests, such as *Robotica*, have ramps for the competing robots to traverse, so builders must plan for a sudden change of the operating plane. The robot may be high-centered as it starts up a ramp or reaches the top, so flexible wheel mounting (where wheels can adapt to severely differing floor angles) is a must in these scenarios. Quite often, placing the driving wheels at the extreme ends can allow a robot to start *up* a ramp, but this same arrangement might not prevent high-centering as the robot reaches the top and teeters in that position. A series of driven middle wheels would give the robot the final push out of such a situation, but many of the machines rely on inertia built up from speed to "dive" over such obstacles.

Mounting Axles Using Various Types of Bearings

Certain styles of bearings seem to be a bit more popular than other types for robot use, especially in mounting axles for wheels. These are the pillow block and flange mount bearings. Some catalogs refer to pillow block bearings as those with a base mount, while other companies call pillow block bearings any configuration that has holes in a flange or base to bolt onto a surface.

Throughout this text, we'll refer to pillow block bearings as those with a rigid mount or base mount that supports the shaft in a position parallel to the surface on which the bearing is mounted. We'll use the term "flange mount" bearings for those that have two or four holes, and mount the shaft perpendicular to and penetrating the surface upon which the bearing is mounted. Most of the ball bearing varieties of these mounted bearing assemblies cannot change the axis of rotation of the shaft. Certain non–ball bearing types have a bronze bushing or bearings mounted in a spherical "ball" assembly that allows the shaft to rotate from 20 to 30 degrees or more off-center. These types of bearing assemblies are useful when mounting drive components that are not quite aligned with other shafts and components.

The Pillow Block Bearing The pillow block method of mounting wheel shafts is probably the most popular way to attach wheels to a combat robot. The pillow block bearings can be mounted below the bottom surface of the robot with the shafts exposed, or the same bearings can be placed above the bottom plate with the shafts enclosed in the interior of the robot. In the second configuration, the outside bearing can be a flange mounted bearing on the wall of the robot's chassis.

The advantage of using these types of bearings is the ease of mounting. A typical ball bearing race assembly still must have a machined hole in which to insert the bearing. Either the bearing must be tightly pressed in and held in place by friction, or a small slot must be cut into the circumference of the hole in which to insert a retaining ring. The ready-made assemblies of the pillow block or flange mounted bearing are far simpler to install. In most cases, the builder will want the shafts used with these bearings to be securely held within the rotating part of the bearing, so bearing assemblies with set screws are recommended. Grainger, McMaster-Carr, and other suppliers have many varieties of these bearings in stock. These and other suppliers are listed in Appendix B of this book. The McMaster-Carr catalog also has useful data on maximum dynamic load capacities in pounds, as well as maximum rotational speed in RPM.

Either of these types of bearings has applications in other areas of robot design. Large swivel joints that may be used for weapons can make use of pairs of these bearing assemblies in conjunction with a high-strength bolt or multiples of bolts as the "hinge pin." Configurations like these make for high strength hinges and are preferable to standard door hinges for applications of high stress. Such a hinge mechanism is shown in Figure 3-10. In this figure, a flipping mechanism is supported by two pillow blocks. The left-hand side of the figure shows the robot prior to a combat match and the lifting prongs have not been attached. The right-hand side of the photo shows the robot after a round of combat. Note how much damage this robot took, but the shaft and pillows blocks are still intact. This is one of the great advantages of pillow blocks—their durability.

FIGURE 3-10 A weapon hinge mechanism using pillow block bearings. (courtesy of Andrew Lindsey)

Wheel Drive Types

Another important consideration is what method of wheel driving you'll choose for your robot: passive wheel drive or powered axle drive.

Passive Wheel Drive

Many of the wheels you might find in surplus markets and catalogs are of the "passive" type, which means that they are not powered but provide only a rolling support. They are not designed for the attachment of a powered shaft and might have two sets of ball bearings inserted into each side of the rim. A non-rotating axle is inserted through both holes; and a nut, or washer and cotter pin, keeps the wheel on the axle. The wheel on a wheelbarrow is an example of a passive wheel. Many robot builders have used these types of wheels as powered wheels by adding a large sprocket on the inside of the rim. In some cases, the center of the sprocket is bored out with a lathe to accommodate the non-powered axle.

A chain drive is connected from this wheel sprocket to another sprocket on the drive motor or gearmotor's shaft protruding out of the robot's shell. This method provides a simple way to power a wheel, but it exposes the drive chain and power system to damage. Figure 3-11 illustrates this type of arrangement.

Powered Axle Drive

The powered axle drive system requires the robot designer to provide a way to fasten a wheel assembly securely to a powered shaft. Figure 3-12 illustrates a method to power a shaft.

FIGURE 3-11 *A passive wheel arrangement*

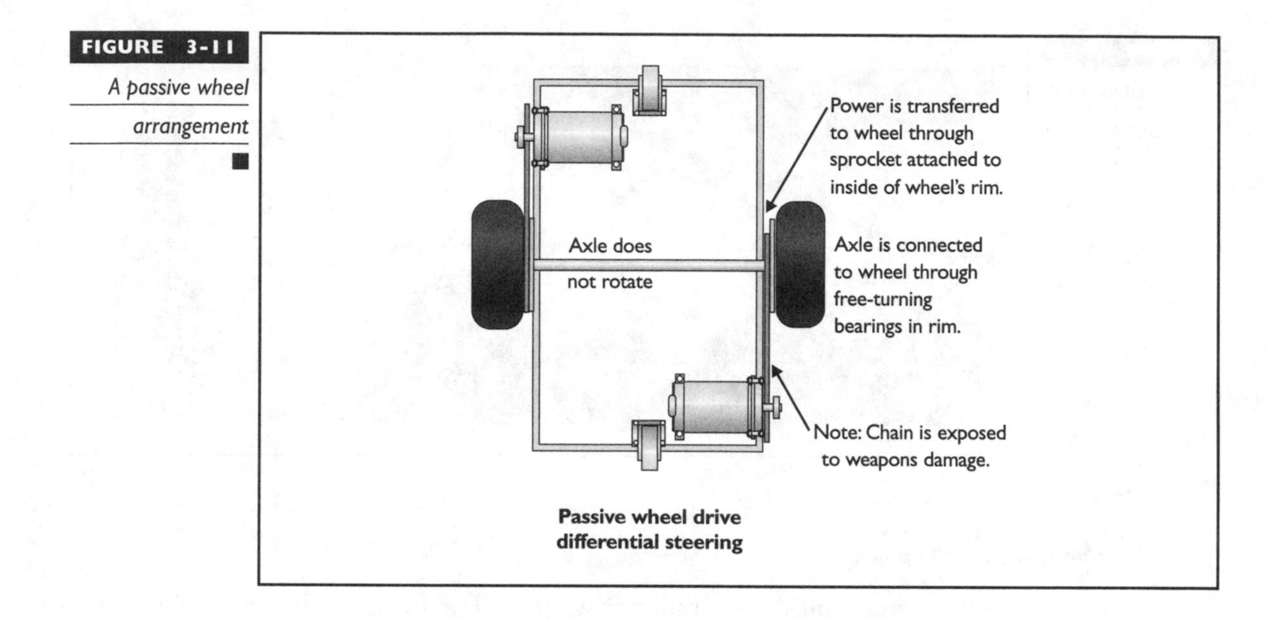

Getting the torque from the shaft to the wheel requires a high-strength hub connection. You should consider using the largest shaft diameter that you can locate and design into your robot. Not only will the larger shaft diameter withstand damage from hazards and weapons much better than a smaller shaft, you will find it easier to machine a slot, a "D" flat, pin holes, or key slots in the larger shaft (see Figure 3-13). With the larger shaft diameter, you will require larger pillow block bearings that will withstand much greater forces. So, larger is better in these cases for greater strength.

FIGURE 3-12 *Placing a hub directly on a gearmotor shaft; the hub can then be directly attached to a wheel.* (courtesy of National Power Chair)

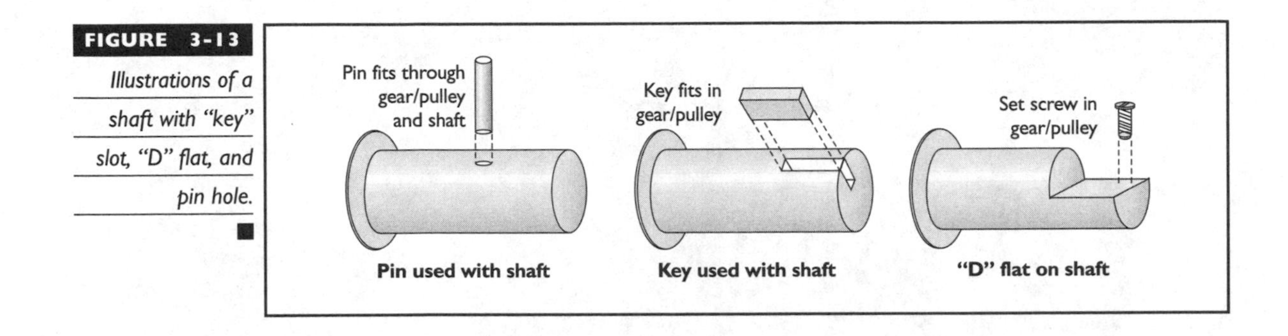

FIGURE 3-13 *Illustrations of a shaft with "key" slot, "D" flat, and pin hole.*

You may be lucky enough to obtain a wheel assembly with a pre-cut slot; then you can cut a corresponding slot in your shaft in which to place a "key" to lock your wheel in place. The wheel is retained on the axle with a nut and washer that allow easy removal. Go-kart and off-road suppliers may be able to furnish you with many wheel/shaft/sprocket assemblies for your robot.

Another way to remove a wheel quickly for fast repairs is to have the wheel permanently mounted to the powered axle. Rather than removing the wheel, you simply flip the robot over and loosen the set screws in your pillow block bearings, remove any retaining shaft collars you may have used and the drive sprocket, and slide the complete wheel/axle assembly out. This obviously has its negative aspects, especially with a heavy robot. It also may create a bit of a problem in reassembly when you have to locate the drive sprocket and chain, and slide the shaft back through. You'll have to locate the flattened part of the shaft you place your set-screws against or the holes through which you must insert pins, and then realign all the bearings and collars before retightening the whole thing.

Protecting Your Robot's Wheels

You might have hard rubber tires with large-diameter axles and heavy rims, but continued pounding by another robot can take its toll on your machine's wheels in nothing flat. An easy way to protect the tires is to have them enclosed within a heavy part of the body's shell, or you can mount a rim around the outside at the tire's most vulnerable parts. You must make this outer shell structure or rim strong so that denting caused by a hazard or opponent's weapon will not cause any part of the metal structure to come in contact with the tire, in which case it could act like a brake or cut the tire.

There are more ways to provide power to wheels than we could ever print in this book. Belt drives have been used successfully, as well as friction drives on the wheels. Canted wheel drives have been used on several robots to provide a wide wheelbase in a smaller-sized robot. Your best approach is to look at what's been done, what bot designs have consistently won over a period of time, and what designs seem to have been problematic. As we mentioned in the beginning, we will never attempt to tell you what is the best design—with a bit of experimenting, you might be able to produce something better than any of today's champion bots.

chapter

4

Motor Selection and Performance

BUILDING a robot requires that you make many decisions—from the type of sensors you'll use to the color you'll paint it. Some of these decisions are trivial, while others will make or break your robot. One decision in the make-or-break category is motors—not just deciding which ones you'll use, but determining how you'll optimize their performance.

Most robots use the same class of motor—the *permanent magnet direct current (PMDC)* motor. These commonly used motors are fairly low in cost and relatively easy to control. Other types of electric motors are available, such as series-wound field DC motors, stepper motors, and alternating current (AC) motors, but this book will discuss only PMDC -type motors. If you want to learn more about other types of motors, consult your local library or the Internet for that information.

Some combat robots use internal combustion motors, but they are more commonly used to power weapons than to drive the robots, largely because the internal combustion engine rotates only in one direction. If you are using an internal combustion engine to drive the robot, your robot will require a transmission that can switch into reverse or use a hydraulic motor drive system. With electric motors, however, the direction of the robot can be reversed without a transmission. Many combat robots combine the two, using electric motors for driving the robot system and internal combustion motors for driving the weapons. Another use for internal combustion engines is to drive a hydraulic pump that drives the robot and/or operates the weapons.

Since most robots use PMDC motors, most of the discussion in this chapter will be focused on electric motors. At the end of this chapter is a short discussion of internal combustion engines.

Electric Motor Basics

Because the robot's speed, pushing capability, and power requirements are directly related to the motor performance, one of the most important things to understand as you design your new robot is how the motors will perform. In most robot designs, the motors place the greatest constraints on the design.

Direct current (DC) motors have two unique characteristics: the motor speed is proportional to the voltage applied to the motor, and the output torque (that is, the force producing rotation) from the motor is proportional to the amount of current the motor is drawing from the batteries. In other words, the more voltage you supply to the motor, the faster it will go; and the more torque you apply to the motor, the more current it will draw.

Equations 1 and 2 show these simple relationships:

$$rpm = K_v Volts$$

$$Torque = K_t Amps$$

The units of K_v are RPM per volt and K_t are oz.-in. per amp (or in.-lb. per amp). Torque is in oz.-in. and RPM is revolutions per minute. K_v is known as the *motor-speed constant*, and K_t is known as the *motor-torque constant*.

These equations apply to the "ideal" motor. In reality, certain inefficiencies exist in all motors that alter these relationships. Equation 1 shows that the motor speed is not affected by the applied torque on the motor. But we all know through experience that the motor speed is affected by the applied motor torque—that is, they slow down. All motors have a unique amount of internal resistance that results in a voltage loss inside the motor. Thus, the net voltage the motor sees from the batteries is proportionally reduced by the current flowing through the motor.

Equation 3 shows the effective voltage that the motor actually uses. Equation 4 shows the effective motor speed.

$$V_{motor} = V_{in} - I_{in} R$$

$$rpm = K_v V_{motor} = K \ (V_{in} - I_{in} R)$$

Where V_{in} is the battery voltage in volts, I_{in} is the current draw from the motor in amps, R is the internal resistance of the motor in ohms, and V_{motor} is the effective motor voltage in volts. It can easily be seen in Equation 4 that as the current increases (by increasing the applied torque), the net voltage decreases, thus decreasing the motor speed. But speed is still proportional to the applied voltage to the motor.

With all motors, a minimum amount of energy is needed just to get the motor to start turning. This energy has to overcome several internal "frictional" losses. A minimum amount of current is required to start the motor turning. Once this threshold is reached, the motor starts spinning and it will rapidly jump up to the maximum speed based on the applied voltage. When nothing is attached to the output shaft, this condition is known as the *no-load speed* and this current is known as the *no-load current*. Equation 5 shows the actual torque as a function of the current draw, where I_0 is the no-load current in amps. Note that the motor delivers no torque at the no-load condition. Another interesting thing to note here is that by looking at Equation 4, the voltage must also exceed the no-load current multiplied by the internal resistance for the motor to start turning.

$$Torque = K_t(I_{in} - I_0)$$

Some motors advertise their no-load speed and not their no-load current. If the motor's specifications list the internal resistance of the motor, the no-load current can be determined from equation 4.

With these equations, as well as the gear ratio, wheel size, and coefficient of friction between wheels and floor, you can determine how fast the robot will move and how much pushing force the robot will have. (How you actually determine this will be explained in Chapter 6.) If you want the robot to go faster, you can either run the motors at a higher voltage or choose a lower gear reduction in the drive system.

Equation 5 is an important equation to know and understand, because it will have a direct effect on the type and size of the batteries that you will need. By rearranging this equation, the current draw requirements from your batteries can be determined. Equation 6 shows this new relationship.

$$I_{in} = I_0 + \frac{Torque}{K_t}$$

For any given torque or pushing force, the battery current requirements can be calculated. For worst-case situations, stalling the motors will draw the maximum current from the batteries. Equation 7 shows how to calculate the stall current, where I_{stall} is the stall current in amps. The batteries should be sized to be able to deliver this amount of current. Batteries that deliver less current will still work, but you won't get the full performance potential of the motors. Some builders purposely undersize the battery to limit the current and help the motors and electronics survive, and others do this simply because they have run out of weight allowance. For some motors, the stall current can be several hundreds of amps.

$$I_{stall} = \frac{V_{in}}{R}$$

Another set of relationships that needs to be considered is the overall power being supplied by the batteries and generated by the motor. The input power, P_{in}, to the motor is shown in equation 8. Note that it is highly dependent on the current draw from the motor. The output power, P_{out}, is shown in mechanical form in equation 9 and in electrical from in equation 10. Motor efficiency is shown in equation 11. The standard unit of power is watts.

$$P_{in} = V_{in} I_{in} = I_{in}^2 R$$

$$P_{out} = Torque \times RPM$$

$$P_{out} = K_t K_v (I_{in} - I_0)(V_{in} - I_{in} R)$$

$$Efficiency = \frac{P_{out}}{P_{in}}$$

The output power is always less than the input power. The difference between the two is the amount of heat that will be generated due to electrical and frictional losses. It is best to design and operate your robot in the highest efficiency range to minimize the motor heating. If the motor is able to handle the heat build-up, it might be best to design the robot (or weapon) to be operated at a higher percentage of the motor's maximum power (to keep the motor as light as possible). For example, a motor that is used to recharge a spring-type weapon might be fine if operated at near-stall load for just a few seconds at a time. The maximum amount of heat is generated when the motor is stalled. A motor can tolerate this kind of heat for short periods of time only, and it will become permanently damaged if it's stalled for too long a period of time. This heat is generated in the armature windings and the brushes, components that are hard to cool by conduction.

Figure 4-1 shows a typical motor performance chart. These charts are usually obtained from the motor manufacturer, or a similar chart can be created if you know the motor constants. The motor shown in Figure 4-1 is an 18-volt Johnson Electric motor model HC785LP-C07/8, which can be found in some cordless drills. The constants for this motor are shown in Table 4-1. This motor is discussed here as an example motor to describe how all of the motor constants relate to each other and how they affect the motor performance.

Figure 4-1 graphically displays how the motor speed decreases as the motor torque increases and how the motor current increases as the applied torque on the motor increases. For this particular motor, maximum efficiency is approximately

FIGURE 4-1 *Typical motor performance curves.*

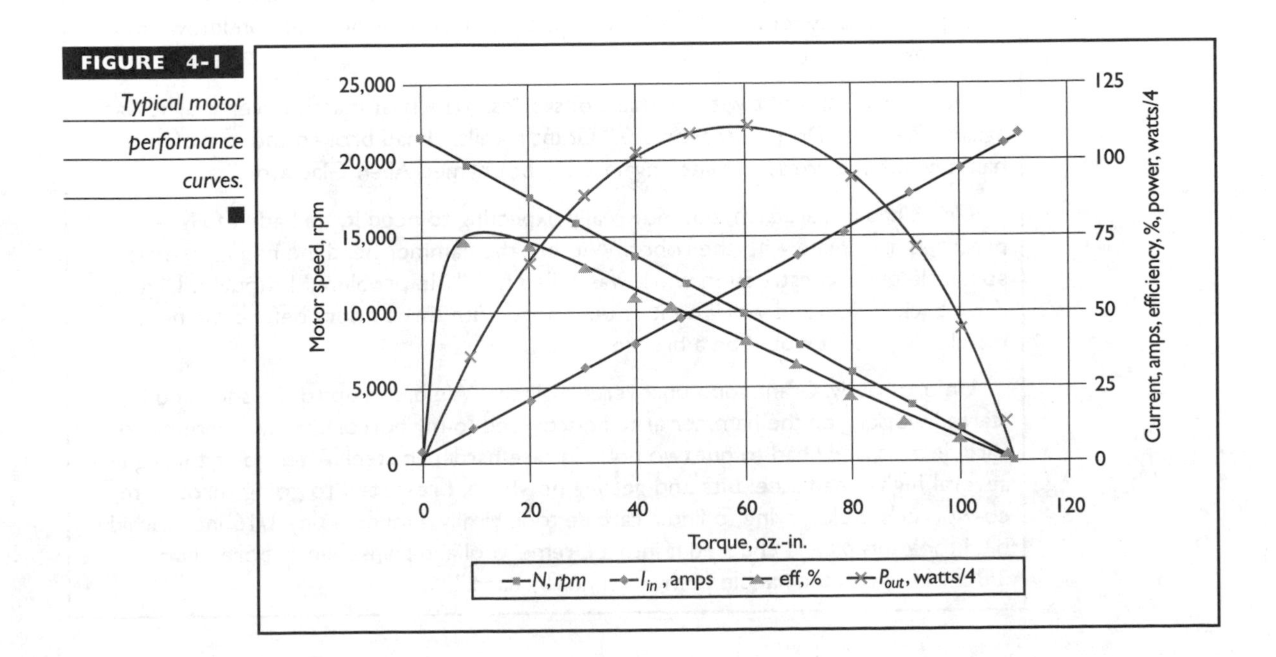

I_0	1.934 amps
R	0.174 ohms
K_v	1,234.6 rpm/volt
K_t	1.097 oz-in/amp

TABLE 4-1 *Motor Constants for Figure 4-1*

75 percent and it occurs when the motor is spinning at approximately 19,000 RPM. Maximum output power from this motor occurs when the motor speed decreases to about 50 percent of its maximum speed and the current is approximately 50 percent of the stall current. For all permanent magnet motors, maximum power occurs when 50 percent of the stall current is reached. Motor manufacturers recommend that motors be run at maximum efficiency; otherwise, motors will overheat faster.

True Story: Grant Imahara and Deadblow

Grant Imahara started his career in robotics as a kid by drawing pictures of robots from movies and television. Later, his designs evolved into LEGOs, and then cardboard and wood. "Only recently," he laments, "have I had the tools and equipment to build them out of metal."

Though Grant got his start as part the Industrial Light and Magic team at Robot Wars in 1996 (he's an animatronics engineer and model maker for George Lucas' ILM special effects company), he is perhaps best known for his creation known as Deadblow.

Deadblow is a robot with its share of stories. "The best match I ever fought was against Pressure Drop in season 1.0," Grant recalls. "I had broken the end of my hammer off in a previous match against a robot named Alien Gladiator."

Grant had a spare arm, but, not really expecting to need it, he hadn't fully prepared it to mate with the robot. Without the hammer head, he had no weapon, so a little quick construction work was called for. "'No problem,' I thought. I'll just drive back to ILM and work on it at our shop. With three hours before the next match, I figured it would be a breeze."

Unfortunately, Grant soon uncovered a glitch. "We drove up to the shop and I started working on the hammer arm. I discovered to my horror that we were out of carbide mills, and I had to put two holes in case-hardened steel. After going through several high-speed steel bits and getting nowhere, I resorted to going through my co-worker's desks, trying to find a carbide tool. Finally, I found a tiny 1/16-inch carbide bit. I took this bit and chucked it into a Dremel tool and painstakingly bored two 3/8-inch holes in the handle of my hammer by hand."

Grant Imahara and Deadblow (continued)

With only an hour left and a 20-minute drive to get back to the competition, Grant still wasn't overly concerned. "But then we hit Sunday evening traffic back into San Francisco. We were going to be late. Forty-five minutes later, I ran into Fort Mason with the new hammer in hand. And we threw it into the robot." As the announcer called Team Deadblow to line up for the fight, they were still screwing the armor back onto the robot. "If you look carefully," Grant says, "you can see that my normally put-together look had become severely disheveled. I was out of breath and about to pass out and the match hadn't even started yet! I had a 'go for broke' attitude for that match, and the adrenaline was pumping. Deadblow went in and pummeled Pressure Drop with a record number of hits. By the end, I could barely feel my hands because they were tingling so much."

Determining the Motor Constants

To use the equations, the motor constants, K_v, K_t, I_0, and R must be known. The best way to determine the motor constants is to obtain them directly from the motor manufacturer. But since some of us get our motors from surplus stores or pull them out of some other motorized contraption, these constants are usually unknown. Fortunately, this is not a showstopper, because these values can be easily measured through a few experiments.

You'll need a voltmeter and a tachometer before you start. To determine the motor speed constant, K_v, run the motor at a constant speed of a few thousand RPMs. Measure the voltage and the motor speed, and record these values. Repeat the test with the motor running a different speed, and record the second values. The motor speed constant is determined by dividing the measured difference in the motor speeds and the difference between the two measured voltages:

$$K_V = \frac{RPM_2 - RPM_1}{Volts_2 - Volts_1}$$

All permanent magnet DC motors have this physical property, wherein the product of the motor speed constant and the motor torque constant is 1352. With this knowledge, the motor torque constant can be calculated by dividing the motor speed constant by 1352. The units for this constant is *(RPM / Volts)* × *(oz.-in. / amps)*. Equation 13 shows this relationship.

$$K_t = \frac{K_V}{1352}$$

The next step is to measure the internal resistance. This cannot be done using only an ohmmeter—it must be calculated. Clamp the motor and output shaft so that they will not spin. (Remember that large motors can generate a lot of torque and draw a lot of current, so you need to make sure your clamps will be strong enough to hold the output shaft still.) Apply a very low DC voltage to the motor—a much lower voltage than what the motor will be run at. If you do not have a variable regulated DC power supply, one or two D-cell alkaline batteries should work.

Now measure both the voltage and current going through the motor at the same time. The best accuracy occurs when you are measuring several hundred milliamps to several amps. The internal resistance, R, can be calculated by dividing the measured voltage, V_{in}, by the measured current, I_{in}:

$$R = \frac{V_{in}}{I_{in}}$$

It is best to take a few measurements and average the results.

To determine the no-load current, run the motor at its nominal operating voltage (remember to release the output shaft from the clamps, and have nothing else attached to the shaft). Then measure the current going to the motor. This is the no-load current. The ideal way to do this is to use a variable DC power supply. Increase the voltage until the current remains relatively constant. At this point, you have the no-load current value. The no-load current value you use should be the actual value for the motor running at the voltage you intend to use in your robot.

After conducting these experiments, you will now have all of the motor constant parameters to calculate how the motor will perform in your robot.

Power and Heat

When selecting a motor, you should first have a good idea of how much power that your robot will require. A motor's power is rated in either watts or horsepower (746 watts equal 1 horsepower). Small fractional horsepower motors of the type that are usually found in many toys are fine for a line-following or a cat-annoying robot. But, if your plan is to dominate the heavyweight class at *BattleBots*, you will require heavyweight motors. This larger class of motors can be as much as 1,000 times more powerful than the smaller motors.

A small toy motor might operate at 3 volts and draw at most 2 amps, for an input requirement of 6 watts (*volts × amps = watts*). If the motor is 50-percent efficient, it will produce 3 watts of power. At the other end of the spectrum are the robot combat class motors. One of these might operate at 24 or 48 volts and draw hundreds of amps, for a peak power output of perhaps 5 horsepower (3,700 watts) or more. Two of these motors can accelerate a 200-pound robot warrior to 15-plus mph in just a few feet, with tires screaming. One 1997 heavyweight (*Kill-O-Amp*) had motors that could extract 1,000 amps from its high-output batteries! The power that your robot will require is probably somewhere between these two extremes.

Your bot's power requirements are affected by factors like operating surface. For example, much more power is required to roll on sand than on a hard surface. Likewise, going uphill will increase your machine's power needs. Soft tires that you might use for greater friction have more rolling resistance than hard tires, which will increase the power requirements. Do you have an efficient drive train, or are you using power-robbing worm gears? How fast do you want to go?

An internal combustion engine produces its peak horsepower at about 90 percent of its maximum RPM, and peak torque is produced at about 50 percent of maximum RPM. The higher the RPM, the more energy it consumes. Compare this to the PMDC motor, which consumes the most energy and develops its peak torque at zero RPM. It consumes little energy at maximum RPM, and it produces its peak horsepower at 50 percent of its unloaded speed.

At 50 percent of maximum speed, the PMDC motor will draw half of its maximum stalled current, as seen earlier in Figure 4-1. Unfortunately, much of the current going into the motor at this high power level is turned into heat. Figure 4-2 shows how much heat is generated in the example motor used to create the statistics in Figure 4-1.

It is obvious to see that the minimum amount of heating occurs when running the motor near its maximum speed and efficiency. It can also be seen in Figure 4-2 that as the motor torque increases, a near exponential increase in motor heat results. Motors can tolerate this amount of heat only for short periods of time. Continuously running a motor above the maximum power output level will seriously damage or destroy it, depending on how conservatively the manufacturer rated the motor.

Many motors are rated to operate continuously at a certain voltage. You can increase the power of your motor by increasing the voltage. Figure 4-3 shows how a motor's speed, torque, and current draw are affected by increasing the input voltage to the motor. In Figure 4-3, you can see that the motor speed is doubled

FIGURE 4-2 *Heat generated in an electric motor.*

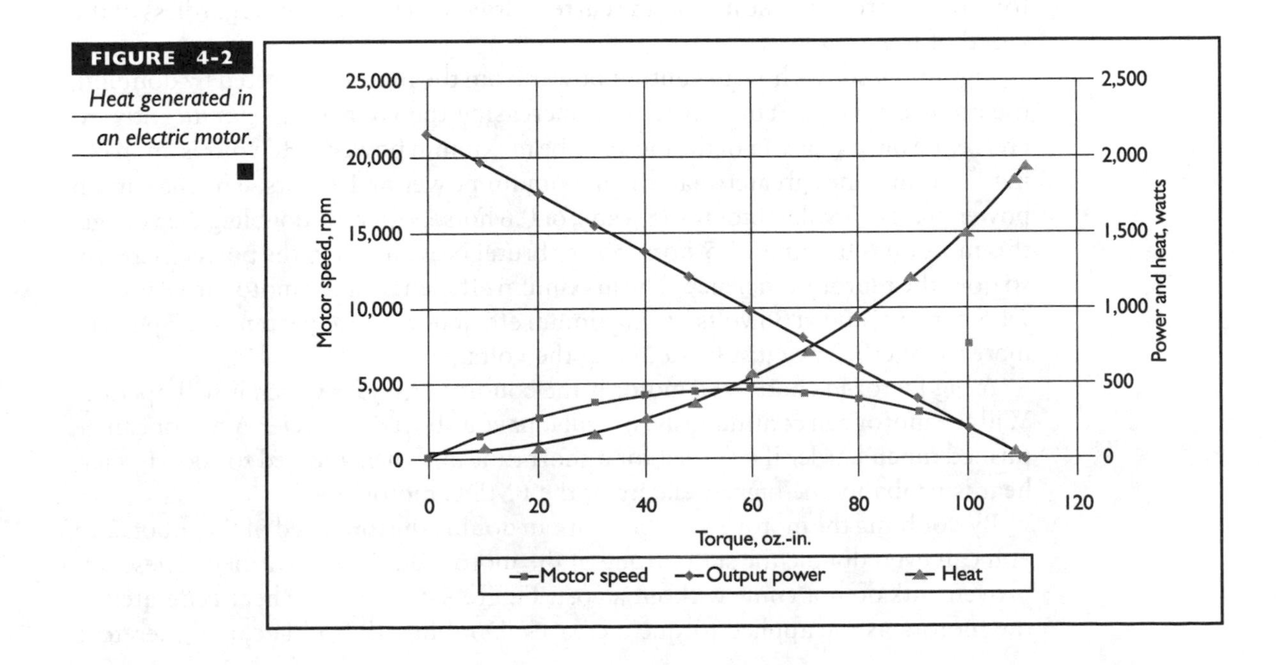

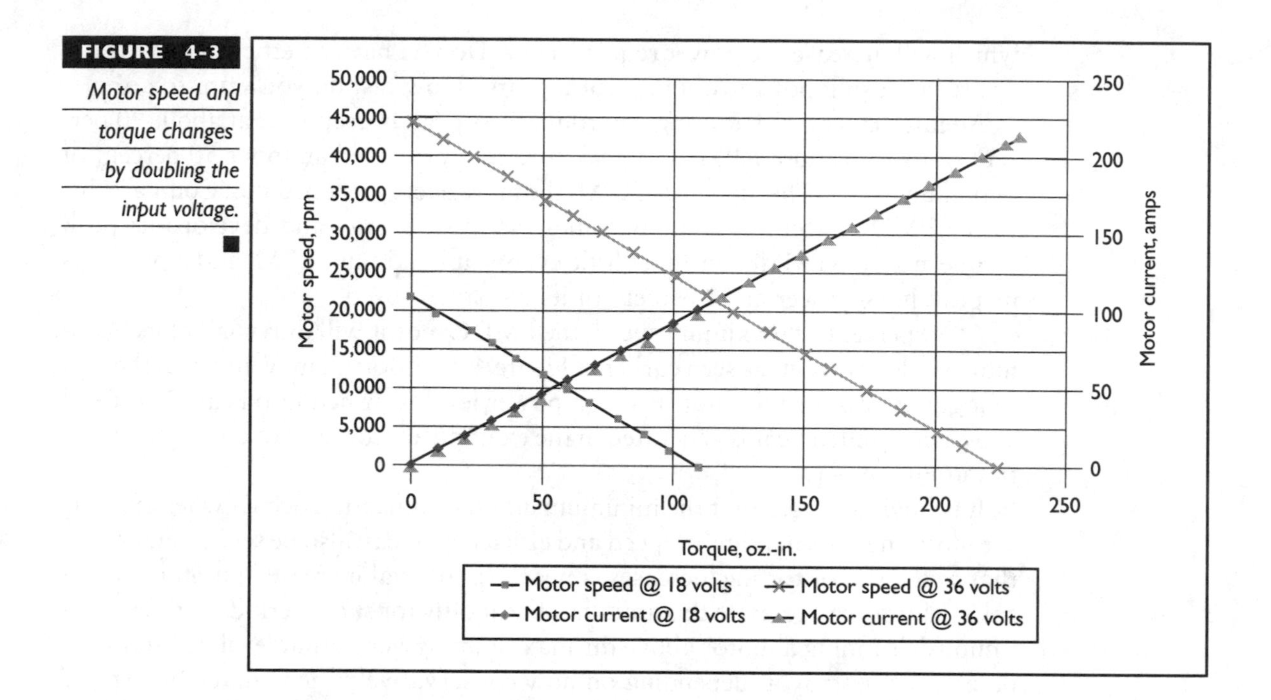

FIGURE 4-3 Motor speed and torque changes by doubling the input voltage.

and the maximum stall torque is doubled when the input voltage is doubled. Recall from equation 4 that the motor's speed is proportional to the applied voltage. In Figure 4-3, you will notice that the current draw line from the 18-volt and 36-volt cases are on top of one another. Remember that the current draw is only a function of the applied torque on the motors, and it is not related to the voltage. So for a fixed torque on the motor, the current draw will be the same regardless of the speed of the motor.

Figure 4-4 shows how the output power from the motor is affected by doubling the applied voltage. You can see that increasing the voltage can significantly increase the output power of the motor. The maximum power at 36 volts is approximately four times greater than the maximum power at 18 volts. The maximum power of this 18-volt motor is 448 watts, or 0.6 horsepower. By doubling the voltage, this motor has become a 2.5-horsepower brute! Not only does the power increase, so does the motor's efficiency. The maximum efficiency of the motor at 18 volts is 74.5 percent, and at 36 volts the maximum efficiency is 81.6 percent—a 7 percent increase in efficiency just by doubling the voltage!

A big factor in choosing a motor is the conditions under which it will operate. Will the motor run continuously, or will it have a short duty cycle? A motor can be pushed much harder if it is used for a short time and then allowed to cool. In fact, heat is probably the biggest enemy of the PMDC motor.

By doubling the motor's voltage, you can double the top speed of the robot, and you can even double the stall torque of the motor. But be forewarned: These improvements do not come without a cost. Figure 4-5 shows the heat generated in the motors as the applied torque increases. Doubling the voltage, and therefore

FIGURE 4-4 *Motor power changes by doubling the input voltage.*

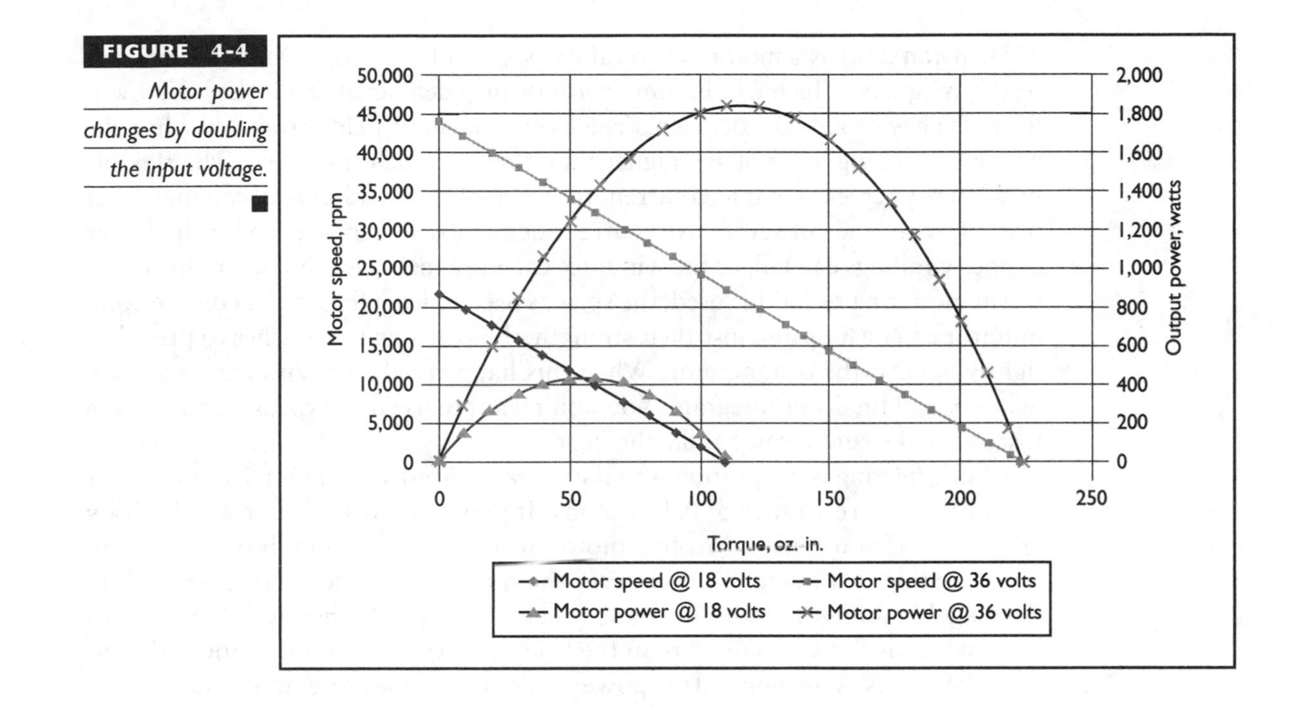

the current, increases the heat by a factor of four! Stalling the motor will cause the motors to overheat and be seriously damaged in a short period of time. Nothing is free in the world of physics.

FIGURE 4-5 *Heat generated by doubling the applied voltage.*

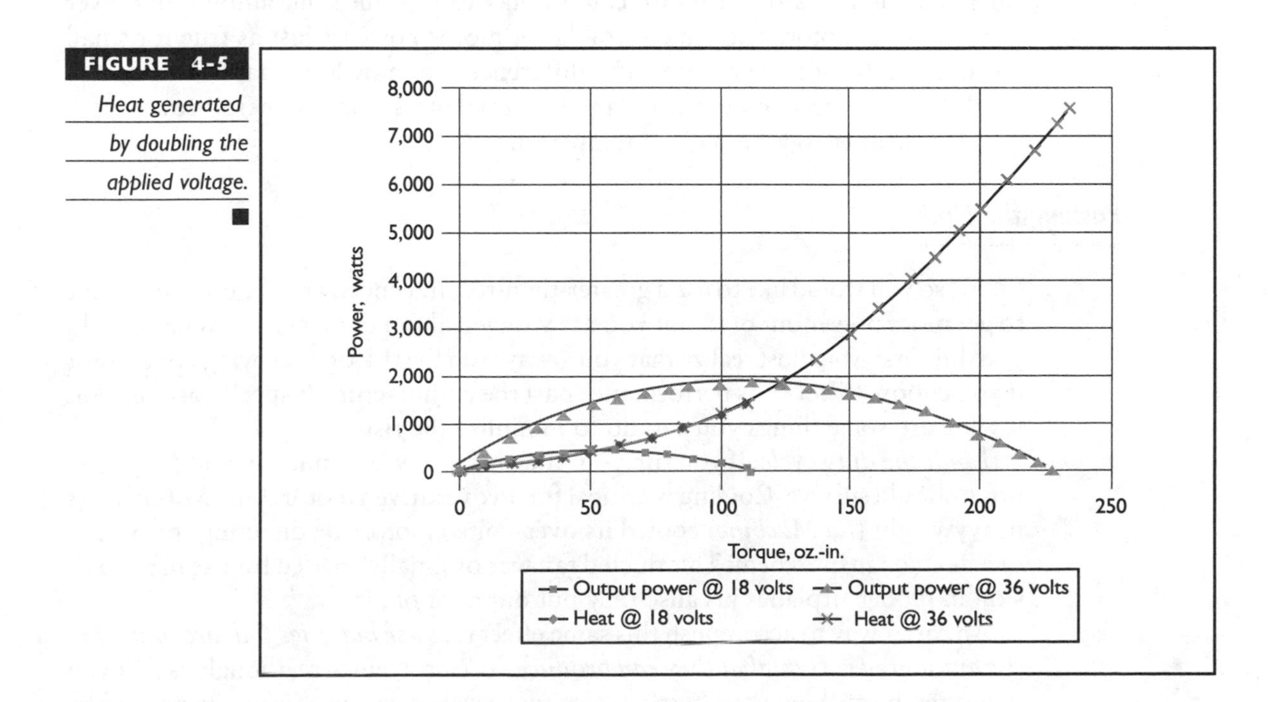

Heat can destroy a motor in several ways. Most lower-cost PMDC motors use ferrite magnets, which can become permanently demagnetized if they are overheated. They can also be demagnetized by the magnetic fields produced when the motor is running at a voltage higher than that at which it is rated. The flexible braided copper leads that feed current to the brushes (called *shunts*) can melt after just a few seconds of severe over-current demands. The insulation on the heavy copper windings can fail, or the windings can even melt. Depending on the motor brush mounting technique used, the springs used to keep the brushes on the commutator can heat up and lose their strength, thus causing the brushes to press less tightly against the commutator. When this happens, the brushes can arc more, heat up, and finally disintegrate. You don't want to use that expensive motor as a fuse, so make sure it can handle the heat.

Motor heating is proportional to the $current^2 \times resistance$. Our 18-volt motor example has a resistance of 0.174 ohms. If you were to stall it, it would draw 103 amps. If you stalled the same motor at 36 volts, it would draw 207 amps. Since heating is a function of $current^2$, the motor would get four times as hot. Pushing 207 amps through a resistance of .174 ohms will generate 7,455 watts of heat, which is five times more than the heating output of a typical home electric space heater. Now imagine all the power of your portable heater multiplied by five and concentrated into a lump of metal that weighs just a few pounds. You can see why survival time is limited.

The physical size of the motor that would best fit your robotic needs is in large part determined by the amount of heat that will be generated. Some people find it surprising that a 12-ounce motor can produce exactly the same amount of power as a 5-pound motor. The same formula for motor power is just as true for small motors as it is for large motors. The difference is in how long that power can be produced. The larger motor has a larger thermal mass, and can therefore absorb a lot more heat energy for a given temperature rise.

Pushing the Limits

Okay, so you would like to use a greater-than-recommended voltage on your motor to get more power out of it, but you are worried about damaging it. What should you do? First, you must realize that you always run the risk of destroying your motor if you choose to boost its performance past the manufacturer's specifications. Following are some things you can do to minimize the risk.

Limit the duty cycle. If you run your motor for, say, 1 minute on and 5 minutes off, it should survive. Cooling is critical for an overdriven motor. One *Robot Wars* heavyweight (*La Machine*) cooled its over-volted motors by directing the output of a ducted fan into them. This ducted fan was originally created for use in propulsion in model airplanes because they put out a lot of air.

An easier way to accomplish this same effect is to *use batteries that are limited in the amount of current that they can produce*. The problem here, though, is that you will often be pushing your battery to output levels that will shorten its useful life. Even the sealed lead-acid batteries can sometimes boil and leak under heavy loads.

Another method that can be used to help control the heat buildup in the motors is to *use an electronic speed controller (ESC)*. The ESC is a device that meters the flow of current to your motor. It does this by rapidly switching the current on and off, several hundred to several thousand times per second. One way in which controllers from different companies differ is in the frequency at which they chop the current to the motor. The motor takes a time average of the amount of time the current is on versus the time between each cycle. As a result, the motor will see a lower "average" current and voltage than it would if it were on continuously. Hence, the motors will see less heating.

As stated before, nothing happens for free in the world of physics. Electronic controllers get hot and require heat sinking. They also can generate radio frequency interference, which might cause problems in a radio-controlled robot. Chapter 7 will provide a more detailed discussion on electronic speed controllers.

High-Performance Motors

If you are still not satisfied with the performance of your motor (and money is no object), you might want to purchase a high-performance motor. High-performance motors have one major difference (and several minor ones) from regular motors—in a word, *efficiency*. We have been discussing motors with 50- to 75-percent efficiencies. That is the range for fair to very good ferrite magnet motors. When we step up to rare-earth magnets, we get into a whole new realm of performance. The efficiency figures for small rare-earth magnet motors range from about 80 to 90 percent.

Rare-earth magnets are made from either cobalt or neodymium alloys. The magnetic fields are so powerful that they are actually dangerous to handle. A moment's inattention may result in a nasty crush as your finger is caught between them and a stray piece of metal. The added bonus with cobalt alloy magnets is that they are resistant to demagnetization. A motor with cobalt alloy magnets is virtually immune to demagnetization, no matter how much voltage you pump into it or how hot it gets. Motors with rare-earth magnets run much cooler than ferrite motors. While running under ideal operating conditions, a ferrite motor turns about 33 percent of the power it consumes into heat, whereas the rare-earth motor wastes only about 10 to 20 percent of the electricity you feed it.

Another class of high-performance motor is the brushless PMDC motor. The brushes in an ordinary motor can be the source of several problems: they spark and cause radio interference, they are a source of friction, and they wear out. The brushless motors have sensors that detect the position of the rotor relative to the windings. This information is sent thousands of times a second to a special controller that energizes the windings at the optimum moment on each revolution of the motor. In a brushless motor, the windings are stationary and the magnets spin—exactly the opposite of a conventional motor. This configuration is capable of much higher speeds. You can get motors that spin at 50,000 RPM or more. The major drawback to the high-performance motors is that they are significantly more expensive then regular motors.

Motor Sources

You can acquire electric motors in two ways: you can purchase them from a motor manufacturer or retail store, or you can salvage them from other pieces of equipment. Many robot builders use salvaged motors because they usually cost less than 20 percent of the original cost of buying a brand new motor. Appendix B in this book lists sources for obtaining robot motors.

Robotics companies are starting to sell motors that are specifically designed for combat robots. For example the 3.9-horsepower Magmotor sold by *http://www.RobotBooks.com* has become the standard motor used in several champion *BattleBots*. Figure 4-6 shows a photograph of the motor.

Because electric motors are so common, they can be found easily. Some of the best places to get good electric motors are from electric bicycles, electric scooters and mopeds, electric children's cars where the kids ride and drive, electric model cars and planes, trolling motors, windshield wiper motors, power window motors, power door locks, and even powered automobile seat motors can be used. Some people have even used automotive and motorcycle starter motors and electric winches from the front of a pickup truck or from a boat trailer.

Probably the two best places to get electric motors are from electric wheelchairs and high-powered cordless drill/drivers. The advantages to the electric wheelchair motors are that they already come with a high-quality gearbox, and the output shaft has a good set of support bearings. Depending on which type of motor you get, you could directly attach the wheels of the robot to the output shaft of these motors. Several companies sell refurbished wheelchair motors. One of the best places to get these motors is from National Power Chair (*http:// www.npcinc.com*). Figure 4-7 shows a wheelchair motor.

FIGURE 4-6 24-volt, 3.9-horsepower electric motor. (courtesy of Carlo Bertocchini)

FIGURE 4-7 *24-volt, 185 rpm, 896 in-lb. stall torque wheelchair motor.* (courtesy of National Power Chair)

Cordless drill motors are excellent motors for driving small- to medium-sized robots. Some heavyweight robots have successfully used cordless drill motors, which are small and compact, and can deliver a lot of torque and speed for their size by using planetary gears. One of the other advantages to using cordless drill motors is that they already come with a set of high-capacity batteries and battery chargers. This almost becomes an all-in-one package for building combat robots. The drawbacks to using cordless drill motors are that there is no simple way to mount the motors in the robot; they don't have output shaft bearings to support side loads; and the output shaft is threaded, which makes it difficult to attach anything to it. The best way to use them is to make a coupling and pin it directly to the threaded output shaft. The coupling then attaches directly to a bearing-supported shaft or axle. Figure 4-8 shows the electric motor, gearbox, and clutch from a Bosch 18-volt cordless drill reconfigured into a robot gearbox to drive two sprockets.

FIGURE 4-8 *Bosch 18-volt cordless drill motor converted into a robot drive motor.*

Internal Combustion Engines

Not all robots use electric motors to drive and power the weapons. Some robots use internal combustion engines to perform this important task. These engines are much smaller than those found in automobiles and are usually obtained from gasoline-powered lawnmowers, rototillers, or even weed whackers. The energy density of gasoline is about 100 times greater than that of batteries, and this makes gasoline an attractive source for powering large combat robots. Conversely, gasoline is also the main factor in not selecting this method of power—it is flammable and dangerous. Figure 4-9 shows a 119 cc air-cooled, two-cycle, gasoline-powered cut-off saw by Partner Industrial Products. This saw, equipped with a 14-inch diameter saw blade, was used as the primary weapon in Coolrobots super heavyweight champion *Minion*.

Because most combat robots use electric motors, this book will not go into details of how to use internal combustion engines in combat robots. By reading the rules and regulations of the *BattleBots* competition, you will get a good understanding of what is allowed and not allowed with gasoline engines. The key elements for a gasoline-powered robot is to be able to control the engine if it is upside down, making sure that the fuel does not leak and that fuel flow remains constant in the rough jarring environment, and that you can throttle the speed up and down as you need to. A lot of the gasoline safety and performance schematics will be similar to those of high-powered gasoline-powered model aircraft. Good candidate gasoline engines for combat robots are chainsaw engines, because they have a carburetor that can operate in all positions.

Since internal combustion engines operate in one direction only, a transmission that has a reverse gear must be used if the gas-powered engines are used to drive

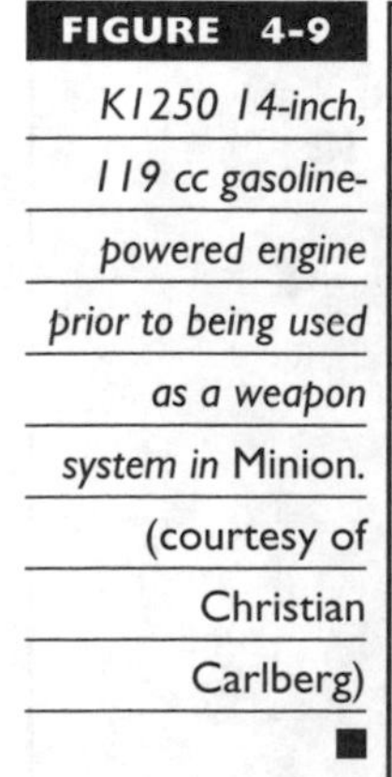
FIGURE 4-9 *K1250 14-inch, 119 cc gasoline-powered engine prior to being used as a weapon system in* Minion. (courtesy of Christian Carlberg)

the robot. If the engine is used to drive a hydraulic pump, the pump needs to have a solenoid valve to reverse the direction of the hydraulic fluid. Probably the most common use for gasoline engines is to power spinning weapons because these weapons spin in only one direction.

For more information on how to use an internal combustion engine in a combat robot, talk with other robot builders that have used them and read up on how to use large engines in model aircraft.

Conclusion

The motors are the muscles of your robot. By understanding how the motors work and how to push them to their limits, you will be able to determine the appropriate motors, the types of batteries, and the appropriate-sized electronic speed controllers for your robot. When building your combat robot, the motors are usually the first major component that is selected. Sometimes the motors are selected based on performance goals, and other times the robots are built around a set of motors that you already have. Both are acceptable ways to build competitive combat robots.

Understanding how current works in the motors will help you determine what type of battery you will need. Chapter 5 will cover how to determine the appropriate size of battery you will need for a robot. Understanding how fast a motor turns and how much torque the motors can generate will help you determine what type of speed reduction/transmission the robot will need to meet your desired goals. Chapter 6 covers this topic. By understanding how the voltage and current relate to one another, determining the right type of speed controller can be accomplished. Chapter 7 will discuss how to select the appropriate-sized electronic speed controller. Understanding how heat can destroy the motors will help you avoid accidental meltdowns.

Before selecting a motor, you should understand how the subjects presented in Chapters 3 through 7 relate to one another. Now, this isn't required—in fact, many robot builders simply pick a motor and build a robot around it. If they're lucky, everything works out just fine. However, most robot builders learn the hard way, as things break because they inadvertently pushed components past their capabilities. How you choose to build your robot is totally up to you.

chapter 5

It's All About Power

ELECTRICALLY powered competition robots are quite demanding on their batteries, which must weigh as little as possible yet supply a lot of current. such requirements push the batteries to their limits. the high current demands can have some surprising results on battery performance, and you need to consider this when selecting the type of battery to use.

This chapter discusses how to determine battery requirements, how these requirements affect battery performance, and how to estimate battery life. At the end of this chapter is a discussion on the various pros and cons of different battery types that can be used in combat robots. Understanding how well the batteries perform is crucial to your ability to build a winning competition robot.

Battery Power Requirements

The batteries' primary purpose is to keep your robot powered during the competition. These competitions can last up to 5 minutes, so the battery must supply all the power to the robot during that time. Selecting an appropriately sized battery that will confidently run your robot throughout the entire match can be a significant competitive advantage. The lightest battery will allow the robot to use the weight savings for other things, such as weapons and armor. A properly selected battery will have enough capacity to supply full running current continuously to your robot's motors; and it will be able to supply the peak currents that will allow your robot's motors to deliver the maximum torque, when needed.

Measuring Current Draw from the Battery

You can find out from the motor specification sheets exactly what current draw to expect when running the motor. Adding up all the currents from the various motors on your robot will tell you the maximum and typical motor running currents to expect.

Because many of us use motors that come without data sheets, we have to measure the running currents ourselves. To do this, you need to have a good battery from which to draw the current. You might ask yourself, "Do I really need to buy a battery to test what size battery I need?" Yes, you do, if you want to be able to measure the current draw. The battery voltage of this test battery must not droop while testing for the current draw. In other words, the voltage must remain constant throughout the tests. The advantage of using a large lead acid battery for the current draw tests is that, because it will provide a long run time, you can use this battery during the initial testing phases of the robot. After you have selected the appropriate batteries for your robot, you can use them for all of the final test phases.

In most cases, fighting robots will draw a lot of current—much more than the maximum current rating of most multimeters. The best tool to use to measure the current draw is a high-current ammeter capable of measuring more that 100 amps.

Using Ohm's Law to Measure Current Draw

You can also measure the resistance of the motor and calculate the current draw from this measurement using Ohm's Law. The formula to do this is *current = voltage / resistance*. This formula doesn't necessarily provide a reliable measure, however, because, first, the resistances are very low for competition motors and most ohm meters are not accurate at such low resistance levels. Second, if this measurement is made accurately, it must be made considering the resistances of the complete wiring harness, motor drivers, and motor. Last, even if the measurement is done accurately, the calculated current will be much higher than actual due to frictional and heat losses.

In all fairness, if measured accurately, the peak motor currents can be determined using an ohm meter and this formula:

$$V = IR \text{ or } I = \frac{V}{R}$$

Here, the current, I, is in amps; the voltage, V, is in volts; and the resistance, R, is in ohms. To use this method, place a high-power, small-resistance-value resistor in series with your robot's battery supply. Then, using a voltmeter, measure the voltage across this resistor.

Suitable Resistor and Measurement Basics

If you have access to a low-value, high-wattage resistor, you should use it to perform your measurements—but resistance, high-wattage resistors are hard to find. The resistance should be less than 0.01 ohms. If your motor's expected peak current draw is 100 amps, you will need at least a 100-watt resistor. If you don't have access to such a resistor, a 0.01-ohm resistor can be made with 6.2 feet of readily available #12 copper wire. The wire needs to be slightly longer than 6.2 feet, but you can connect the voltmeter at the place on the wire that is 6.2 feet from the battery. In addition, it is a good idea to keep the insulation on the wire and to coil up the wire so that it is easy to handle. Temperature causes the resistance to change, so use the wire at room temperature and don't use it so long that it heats up.

1. Place the resistor in series with your robot's battery.
2. Measure the voltage across the resistor (a 6.2-foot-long coil of #12 wire, or the high-wattage resistor) with the robot running in normal battle-like conditions. When measuring this voltage, the value will likely be variable and may appear unstable. Take the maximum reading, and then take a reading that appears to be the nominal or average value. The robot's motors must be loaded to simulate those of a real battle, or else you will measure a value that is much too low—up to five to ten times too low than battle-use values.
3. When you have gathered these voltage values, calculate the current by placing the voltage readings into the formula *current* = *voltage* / 0.01 ohms. The 0.01 ohms is the resistance of the 6.2-foot-long wire. If you are using a high-wattage resistor, then substitute the 0.01 ohms for the resistance of your resistor. For example, suppose that when running the experiment, you noted a maximum voltage of 1.2 volts and an average of 0.5 volts. Plugging these values into the formula yields a maximum current value of 120 amps (120 amps = 1.2 volts / 0.01 ohms) and a typical current of 50 amps (50 amps = 0.5 volts / 0.01 ohms). After you have found the maximum current value and the typical current value, you have the information that you need to choose the correct battery for your robot.

Blowing Fuses on Purpose?

An alternative method for measuring current draw is one of the easiest methods and is fairly accurate. You can use the fuse holder that is in-line with your robot's battery to measure draw. Fuses are commonly used for testing, but few people use fuses during an actual competition. It is usually better to risk an electrical fire than to blow a fuse and be a sitting duck for your opponent to destroy your bot with impunity. A blown fuse in battle also means an automatic loss!

To use this method, you'll need a handful of fuses of various amperages. Start with a fast blow fuse, and select values that you think it will survive. Install this fuse and test run your robot in battle-like conditions.

note *It is important that you test your robot in battle-like conditions, or else the measurement will yield a current draw that is lower than what will occur in the robot arena.*

Keep changing the fuse values until you find the fuse value that will survive and the highest fuse value that fails. Between these two values is your robot's maximum

> **Potting the Battery...NOT!**
>
> *Potting* is encasing the battery in epoxy or some other compound. At first, this might seem like a good idea because it will protect the battery. *Don't do it!* All batteries have internal gas vents. If you were to pot the batteries and then overcharge one or more of them, the buildup of internal pressure inside the battery would cause the battery to explode! If you want to encase the battery, put it in a well-vented but protected place in your robot.

current draw. Using this method, you can find the maximum running current of your robot to within 5 amps. Next, switch to slow-blow fuses. You want to find the fuse that lasts for about 1 minute while running your robot in battle-like conditions. This fuse value will yield your typical running current.

After you have found the maximum current value and the typical current value, you have the information that you need to size your battery.

Battery Capacity Basics

Batteries come in several varieties:

- Sealed Lead Acid (SLA)
- Nickel Cadmium (NiCad)
- Nickel Metal Hydride (NiMH)
- Alkaline
- Lithium Ion

Each of these will be discussed later in the chapter in the section "Battery Types."

The *amp-hour* (Ahr) rating of a battery specifies its capacity to hold energy. In simple terms, it can be viewed as the number of amps that the battery will supply during a 1-hour period. Even so, all batteries' amp-hour ratings are specified at the place where that particular battery technology will be the most efficient, anywhere from dozens of hours for alkaline batteries to 1 hour for NiCads and NiMH. In addition, some battery types are specified at various run-time capacities. Because competition matches only last for 2 to 5 minutes (at *BattleBots,* the preliminary elimination rounds are 2 minutes, finals are 3 minutes, and rumbles are 5 minutes), the results for how the various battery types compare may surprise you. One surprise is alkaline batteries. Although they are considered to have the highest energy density of almost any common battery type, they end up dead last when evaluated for high-current, short-run applications.

When purchasing batteries, always check their Ahr ratings because many name-brand battery manufacturers are selling subcapacity cells. For example, a

D-cell NiCad should always have a capacity greater than 4Ahr, yet many name-brand D-cell NiCads can be found with Ahr capacities of less than 2.5Ahr.

Virtually all brand-new rechargeable batteries will have a higher energy capacity after going through a few charge/discharge cycles. The minimum recommended break in period is three cycles, although capacity will increase during the first ten charge/discharge cycles.

For all battery types, if you want to increase voltage, just add the batteries together in series. From any battery type, you can build up as high a voltage as needed. All the batteries in series must be the exact same type of battery in voltage and capacity. If you want to increase for current capacity, add battery packs with the equal voltage and current capacity together in parallel.

Figure 5-1 shows two battery packs wired together to increase the voltage or current. When connecting batteries together in series, the voltage is added together and the current capacity is the same as a single battery pack. When the batteries are wired together in parallel, the voltage remains constant but the current capacity is added together.

caution *Remember that each battery pack must have identical total voltage and current capacity or you will damage the batteries.*

Preventing Early Battery Death

With proper care, most combat rechargeable batteries can run through 200 to 1,000 charge cycles. Under battle conditions and extreme current draws, the actual figure will be closer to 200 than 1,000, though. If you do a lot of practice driving, you should consider getting new batteries after two or three competitions. To get the maximum amount of charge cycles, you must pay attention to the following areas.

First, follow the proper care and charging guidelines for your particular rechargeable battery. All rechargeable batteries require about 5 to 50 percent more charge placed into them than is taken out of them. Improper charging by either overcharging or undercharging is probably the biggest killer of rechargeable batteries. An automatic charger specifically designed for your particular battery type is the best defense against harming the battery by improper charging.

Second, rechargeable batteries can become severely damaged by being deeply discharged. While the battery is in hard use, and whenever the battery charge is below 80 percent of the rated charge, it is possible that some of the cells within the

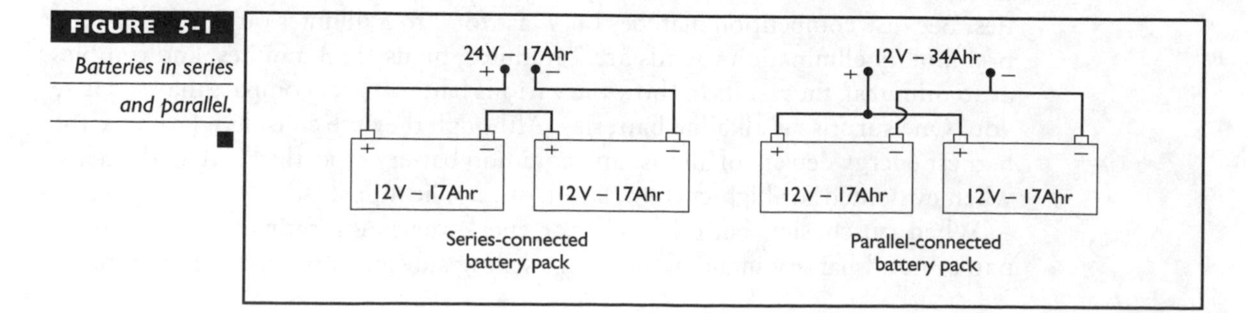

FIGURE 5-1 *Batteries in series and parallel.*

battery will switch polarity. Cell reversal can cause permanent damage to the battery, which will greatly reduce the charge cycles. Most lead acids will recover well from a deep discharge (to about 1.5 volts per cell), as long as the discharge was rapid. Deep discharging a lead acid over a period of days is likely to damage it. NiCads require an occasional deep discharge (to about .9 volts per cell) to maintain their full capacity, but going deeper than this risks polarity reversal on the weaker cells.

The third major killer of rechargeable batteries is shelf life. Even if you follow all of the appropriate care instructions, most combat robot batteries will require replacement long before the maximum number of charge cycles is reached. The shelf life of a typical rechargeable battery is five years when stored at 25° C. If the battery is stored 10 degrees cooler (15° C), shelf life will increase to 10 years; and if the battery is stored in a typical refrigerator (5° C), the shelf life will increase to 20 years! Conversely, if a battery is stored in a hot Arizona garage (average 40° C), shelf life can be reduced to less than two years. In addition, don't store below 0° C. Within reason, store your batteries in the coolest place possible.

Sizing for a 6-Minute Run Time

Choosing to compare battery types at 6-minute run times has many benefits. First, 6 minutes provides some measure of run-time safety margin because generally the longest fighting competitions can last up to 5 minutes in duration. Sizing to 6 minutes prevents the deep discharge. In addition, the 6-minute run time is $1/10^{th}$ of an hour, which makes it easy to calculate the current that the battery can supply for the 6-minute period. To yield the average current that the battery can supply for 6 minutes, multiply the 6-minute amp hour rating by 10. (Ideally, it makes more sense to size the battery for the particular competition. For example, *BattleBots* matches never run more than 3 minutes and the majority of the matches only last 2 minutes. The rumbles last 5 minutes, but only a small fraction of the robots make it to the rumble. In this case, to be a little more aggressive, you could size the battery for 4 minutes and just plan to skip the rumble.)

Except for the NiCad battery type, limited information is available on what happens when the battery is discharged in a short period of time. Because NiCad batteries are often used in the hobby radio control market, a lot of information is available on how they perform for these short run times.

note *The information presented here has been gathered from many manufacturers' data sheets and application notes. From the data sheets and experiments, a special conversion factor was derived for each battery type. This conversion factor is used to convert the nominal Ahr rating of each battery type to the 6-minute run-time period (see Table 5-1, later in this chapter). This allows easy comparison of one battery type to the other for battery capacity. These factors should be considered "rules of thumb"; for best accuracy, individual battery data sheets should be consulted and actual experiments with the batteries should be conducted.*

Comparing SLA, NiCad, and NiMH Run-Time Capacities

In this chapter, a comparison between 4 different battery types that have 6-minute run-time capacities between 4 and 6Ahr. With these batteries, you can draw 40 to 60 amps for 6 minutes. All are 12-volt batteries or 12-volt battery packs. This is a common motor voltage and eliminates having to scale the readings to make the comparisons here. For this comparison only, the selected batteries/ packs are listed here:

- PowerSonic, part number PS-12180, SLA 17.5Ahr
- PowerSonic, part number PS-12120, SLA 12Ahr
- Panasonic, part number HHR650D NiMH 6.5Ahr, pack of 10 D-cells
- Panasonic, part number P440D NiCad 4.4Ahr, pack of 10 D-cells

Comparing Amp Hour Capacity

First, let's compare the Ahr capacity verses run time of these batteries. Figure 5-2 shows what happens to the capacity of the battery if you change the rate at which you

True Story: Jim Smentowski and Nightmare

Jim Smentowski guesses that he's invested well over $30,000 into his robots, though it's hard to pin him down to an exact figure. "I stopped counting," he admits. "Then again, this is an obsession, so you aren't supposed to keep track."

Although Jim has always been mechanically minded, he didn't have an easy start with robotics after seeing *Robot Wars* for the first time in 1996. "I got into it because the concept of fighting robots fascinated me. I had no idea how to make it happen, I just knew, somehow deep inside, that this was something I had to do. I just started doing research. On the web, talking to other builders, talking to manufacturers of parts, picking up all the info I could from anywhere I could. It took a lot of time, and nobody ever just handed me the info I needed, I had to spend a lot of time and make a lot of mistakes before I got to where I am."

But where he is is a good place, indeed. The man behind such renowned robots as *Nightmare, Backlash,* and *Hercules,* he's a top-rated competitor on *BattleBots.*

Nonetheless, when asked to recall one of his most exciting moments under the lights, Jim chose an early competition that, as he explains, was "an exciting moment that was not a win at all."

"Back in 1997," he explains, "I had the chance, as a rusty rookie builder, to face one of the top robots in the sport, *Biohazard,* in the rumble. He beat me, of course, but I was the last to fall of all the other bots in the rumble, and *Biohazard* had to work hard to defeat me. It was then that I knew that I might have what it takes to actually build a machine capable of winning. I've been on that quest ever since."

Jim adds, wistfully, "Oh, and I still haven't defeated *Biohazard*... But I'm getting closer."

FIGURE 5-2 *12-volt battery types compared to capacity versus run time.*

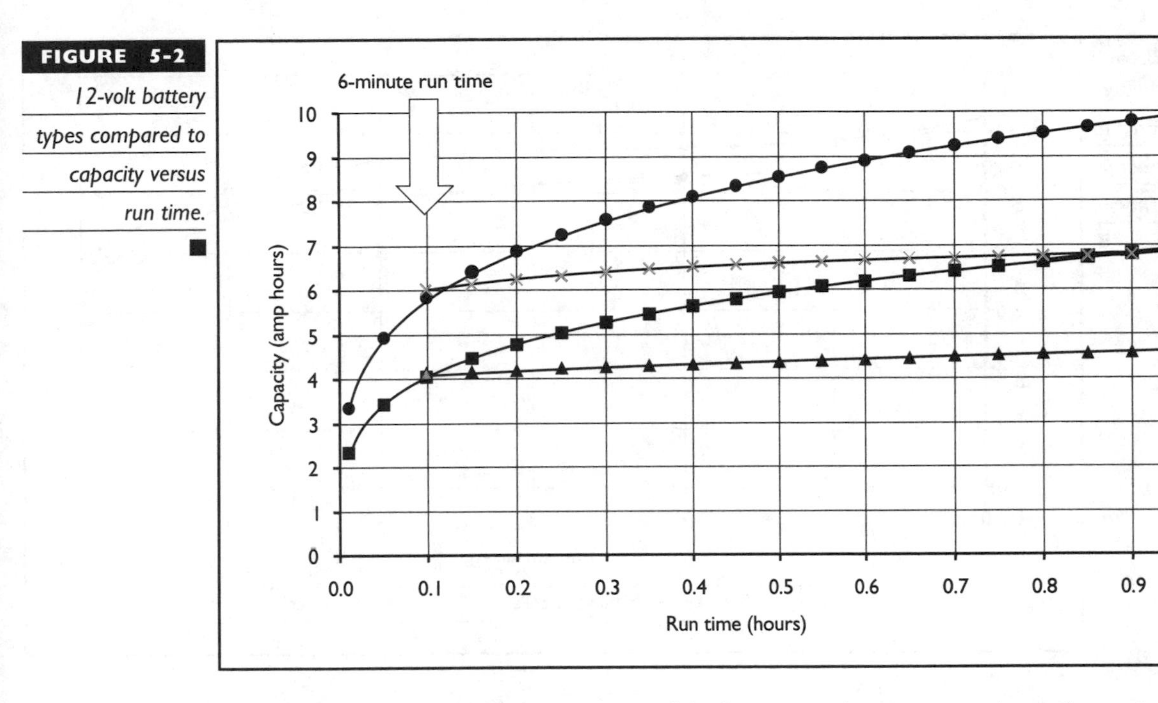

draw most of the current out of the battery in the given run times shown in the figure. Notice the fairly steep slope for both of the SLA batteries and how both the NiCad and NiMH are almost flat. The physics of each battery technology will determine the shape of these curves. This curve is repeatable between the various battery manufacturers, so capacity can be predicted for various run times. This is true even for the steep slope of the SLA batteries.

To determine the 6-minute run-time capacity for a battery, look at the 1/10th hour (or 6-minute) run time data in Figure 5-2. The average current the battery can deliver for 6 continuous minutes will be 10 times this 6-minute run-time value. In Figure 5-2, you will see that the discharge rates for the 17Ahr SLA and the 6.5Ahr NiMH batteries are nearly identical at approximately 6Ahr; thus, the average current these batteries can deliver during the 6 minutes is 10 times this value, or 60 amps. For the 12Ahr SLA and the 4.4Ahr NiCad, the 6-minute run-time capacity is about 4Ahr, so these two batteries can deliver on average about 40 amps for 6 minutes.

Voltage Stability

Figure 5-3 shows the voltage supplied by the various battery types for the 6-minute run time. This graph assumes that your robot will try to drain the battery in 6 minutes. Only three curves are shown, as these graphs are normalized for the 6-minute run times. Both of the 17Ahr and 12Ahr batteries will see nearly the same type of voltage change when both are discharged to the same level in 6 minutes. Notice how stable the voltage is out to 5 minutes and that the voltage starts to drop off rapidly after 5 minutes for all battery types. The NiMH voltage discharge is flat

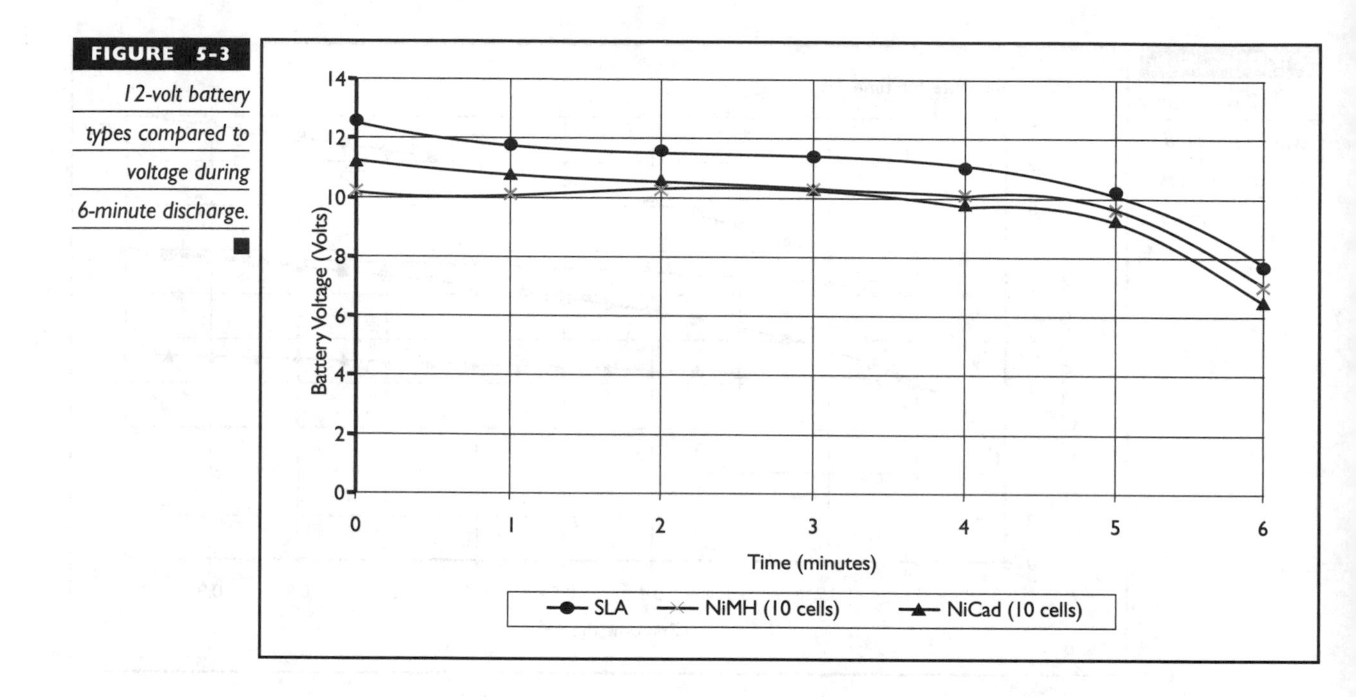

FIGURE 5-3 *12-volt battery types compared to voltage during 6-minute discharge.*

and even increases slightly as it warms during discharge. Both the SLA and the NiCad have slowly fading voltage curves.

If you are familiar with traditional battery literature, you would not expect the NiCad's voltage to fade. This is due to the high discharge rate and the increase of series resistance in the cells. Of particular interest is the fact that throughout the discharge, the SLA battery voltage is above the other two battery types. Why would this be the case? The reason for this is that all the SLA battery types have the lowest internal resistance, hence they have the lowest internal voltage drop.

Voltage Stability for Peak Currents The preceding section brings up a good point. What happens to the battery voltage when one tries to draw various amounts of current from the battery? Figure 5-4 shows how the internal voltage losses increase as the current demand increases.

The voltages shown on these graphs use Ohms Law. The formula is

voltage loss = (*internal resistance of the battery*) × (*current draw*)

---

note *Remember that for the NiCad and NiMH packs, the internal resistance of each cell is added together. For 10 cells, then, the total internal resistance is 10 times the internal resistance of 1 cell.*

Figure 5-4 should provide an intuitive feel for what is happening inside the batteries. It shows the relationships for the various battery types using batteries of similar 6-minute capacities. Notice the voltage loss in the NiCad pack when trying to

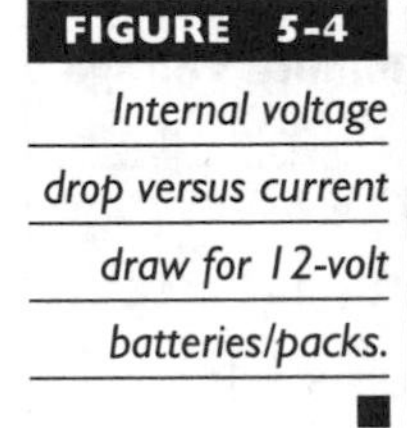

FIGURE 5-4 *Internal voltage drop versus current draw for 12-volt batteries/packs.*

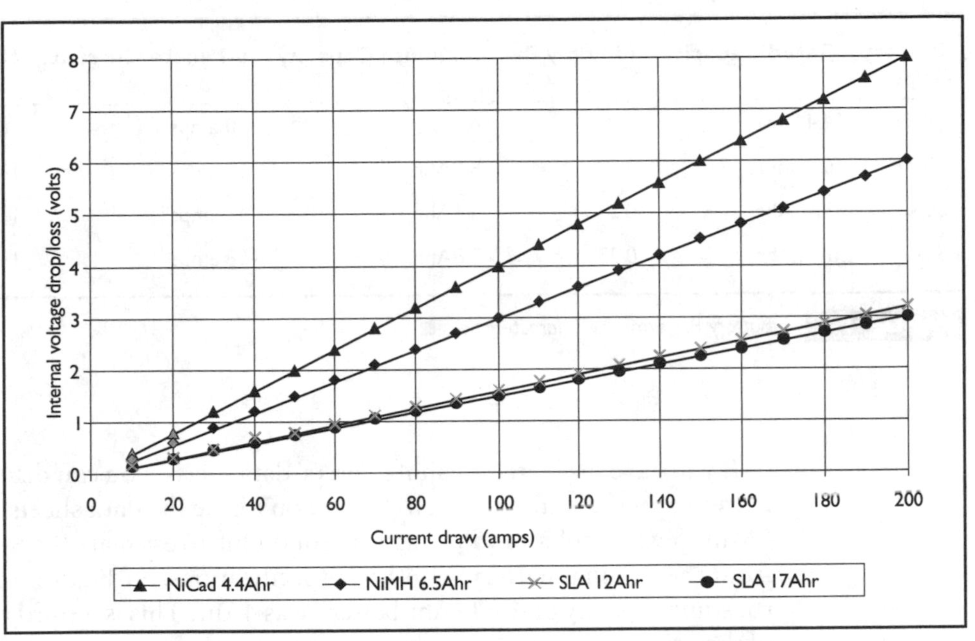

draw 200 amps. The 8 volts lost inside the batteries is turned into heat, and the batteries get very hot—in this case, 1,600 watts of heat. On the other hand, the motors will receive only 4 volts and will run much cooler. Of course, the motors will run much slower and also deliver a great deal less torque. Clever battery/motor designs might use this fact to raise the normal running voltage to the motors, knowing that under high load, the voltage will drop and prevent the motors from burning up. If you do this, remember that many motors and batteries have burned out using this method during the competitions. On the other side of the spectrum, the SLA battery will hold the output voltage at 9 volts while delivering 200 amps. This increases the speed and torque to the motors compared to the other battery types.

When comparing batteries with similar 6-minute capacities, the series resistance of a particular battery characteristic must be looked at carefully. This data will be included in the battery data sheets from the manufacturer. For example, only the high-quality NiCads and NiMHs will have a series resistance as low as what is shown in Figure 5-4. Another point to consider is that, in general, the larger the Ahr capacity of the battery, the lower the internal series resistance.

Wrapping Up the Comparison

It is sometimes easier to see the battery comparison in table form. Note the additional information—the weight of each battery, the rated maximum current, and the 6-minute power density. Tables 5-1 and 5-2 show the battery performance characteristics used to create Figures 5-2, 5-3, and 5-4.

Battery Type	Rated Capacity	Multiply By	6-Minute Capacity	6-Minute Current	6-Minute Voltage
NiCad	4.4Ahr	0.90	4.0Ahr	40 amps	10.3 volts
NiMH	6.5Ahr	0.92	6.0Ahr	60 amps	10.3 volts
SLA	12.0Ahr	0.33	4.1Ahr	41 amps	11.5 volts
SLA	17.5Ahr	0.33	5.8Ahr	58 amps	11.5 volts

TABLE 5-1 *Battery Performance Characteristics* ■

If you have the battery manufacturers' data sheets, you can determine the actual battery-specific values yourself. If you don't have the data sheets, you can use the "Multiply By" column values as a rule of thumb to estimate the 6-minute capacity from the original battery amp hour specifications. In Figure 5-2, the 6-minute run-time capacity of the 12Ahr battery was 4Ahr. This is a third of the Ahr rating of the original battery. Also, from Figure 5-2, you can see where the 0.9 and 0.92, 6-minute conversion factors for the NiCad and NiMH batteries come from.

The following two equations show how to estimate the 6-minute and the peak current capacity of a battery:

$$C_{6min} = 10K_f C_{batt}$$

$$C_{peak} = 10C_{batt}$$

The first equation addresses the 6-minute capacity, where C_{6min} is the average current the battery can deliver for 6 minutes, in amps; K_f is the 6-minute conversion factor, as seen in Table 5.1; and C_{batt} is the original battery Ahr specification. In the second equation, the peak current capacity is C_{peak}.

Battery Type	Rated Max. Current	Voltage @ Max. Rated Current Volts	Weight	Rated Power Density	6-Minute Power Density
NiCad	100 amps	7.90 volts	1.4Kg	3.1Ahr/Kg	2.9Ahr/Kg
NiMH	100 amps	10.30 volts	1.8Kg	3.6Ahr/Kg	3.3Ahr/Kg
SLA	120 amps	10.22 volts	4.1Kg	2.9Ahr/Kg	1.0Ahr/Kg
SLA	175 amps	9.78 volts	5.9Kg	3.0Ahr/Kg	1.0Ahr/Kg

TABLE 5-2 *Battery Performance Characteristics* ■

note *These equations are rule-of-thumb–type equations for estimating current capacity in a battery. To obtain the exact values, consult the battery manufacturer's data sheets. Some high-performance batteries have a much higher peak current capacity, while other batteries' peak current capacity is measured in millisecond time frames. These questions provide a good starting point for estimating the life of a battery.*

Electrical Wiring Requirements

Another part of the battery selection process is selecting the proper wire sizes between the batteries and the motors. The electrical wires must withstand the current requirements without overheating. The wire's current rating is determined by the gauge of the wire and the type and thickness of the insulation around the wires. If the wire size is too small for the amount of current passing through it, the wire will heat up to the point where the insulation will melt—and in the worse case, the wire may melt. Table 5-3 shows the conservative American Wire Gauge (AWG) values for various maximum currents through copper wire. This table is a good starting point for selecting the appropriate wire sizes for your robot.

The figures in Table 5-3 are conservative and considered safe for normal home use. But some robot builders use #12 wires for 200-plus amps, #10 for 350-plus amps, #8 for 500-plus amps, and #4 for 1,000-plus amps. (These are peak amp draws; average amp draws are much lower.) The key is to use the high-temperature insulation.

Current	Minimum AWG
13 amps	#20
18 amps	#18
20 amps	#16
28 amps	#14
38 amps	#12
53 amps	#10
78 amps	#8
105 amps	#6
142 amps	#4
196 amps	#2
266 amps	#0

TABLE 5-3 *American Wire Gauge Copper Wire Minimum Current Ratings* ■

You should use only multi-stranded wires—the more strands, the better. Do not use solid core wires because they have the tendency to break due to the vibrations and impacts within the robot. Most wires use PVC for the insulation; but for higher temperature handling capability and flexibility, use wires with Tefzel, Kapton, Teflon, or Silicone insulation.

Battery Types

Sealed lead acid (SLA), nickel cadmium (NiCad), and nickel metal hydride (NiMH) batteries can be successfully used for competition. Two other battery types worth mentioning are the Lithium Ion and the Alkaline types. Although not recommended, these two battery types are common enough that some people might consider using them in their robots.

In most competition robot contests, the regular lead acid batteries that are used on automobiles, boats, and motorcycles are prohibited because these batteries allow access to the internal liquids, and they can leak acid if they are turned upside down or if they become damaged—which can also damage the arena and pose a safety hazard. The lead acid batteries that are allowed in these events are called *sealed* lead acid batteries, because they have no ports for checking the internal fluids and they can be operated in any orientation (see Figure 5-5). These batteries are often called Gel-Cells, immobilized lead acid batteries, or glass-mat lead acid batteries.

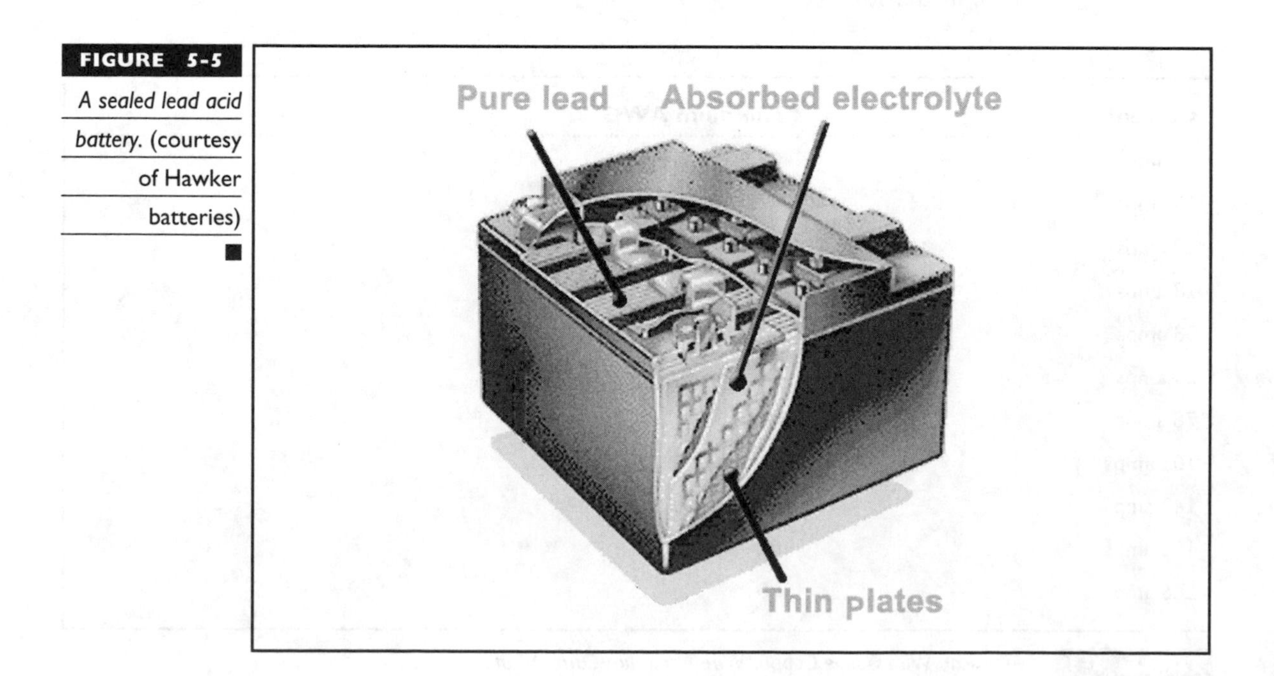

FIGURE 5-5 *A sealed lead acid battery.* (courtesy of Hawker batteries)

Sizing the Battery

If your robot draws an average current of less than 40 amps and has a peak current less than 100 amps, you can select from SLA, NiCad, or NiMH batteries with ease. Just size your battery to make sure that the 6-minute rating and peak current rating is higher than your robot requires.

If your robot will draw an average of more than 40 amps or more than 100 amps peak, use SLA batteries or parallel packs of NiCad or NiMH batteries. The SLA is easier, but not necessarily better. Remember, do not mix different types and sizes of batteries together.

Sealed Lead Acid

The rugged construction of SLA batteries is well suited for combat robot use. SLA batteries do not leak and they are a mature battery technology. Figure 5-6 shows various SLA batteries.

In general, the Ahr rating of the SLA is specified at the 20-hour rate. Multiply by 0.33 (see Table 5-1 for the 0.33 conversion factor) to convert this 20-hour rate to the 6-minute rate. For example, an SLA battery with a capacity of 12Ahr has a usable 6-minute capacity of 4.1Ahr (4.1Ahr = 0.33 × 12Ahr) and will provide an average current of 41 amps (41 = 10 × 4.1Ahr) for the 6-minute duration. Typical SLA batteries have a peak current delivery capacity of 10 times its 20-hour capacity. In this example, the battery can supply a peak current of 120 amps (120 = 10 × 12Ahr).

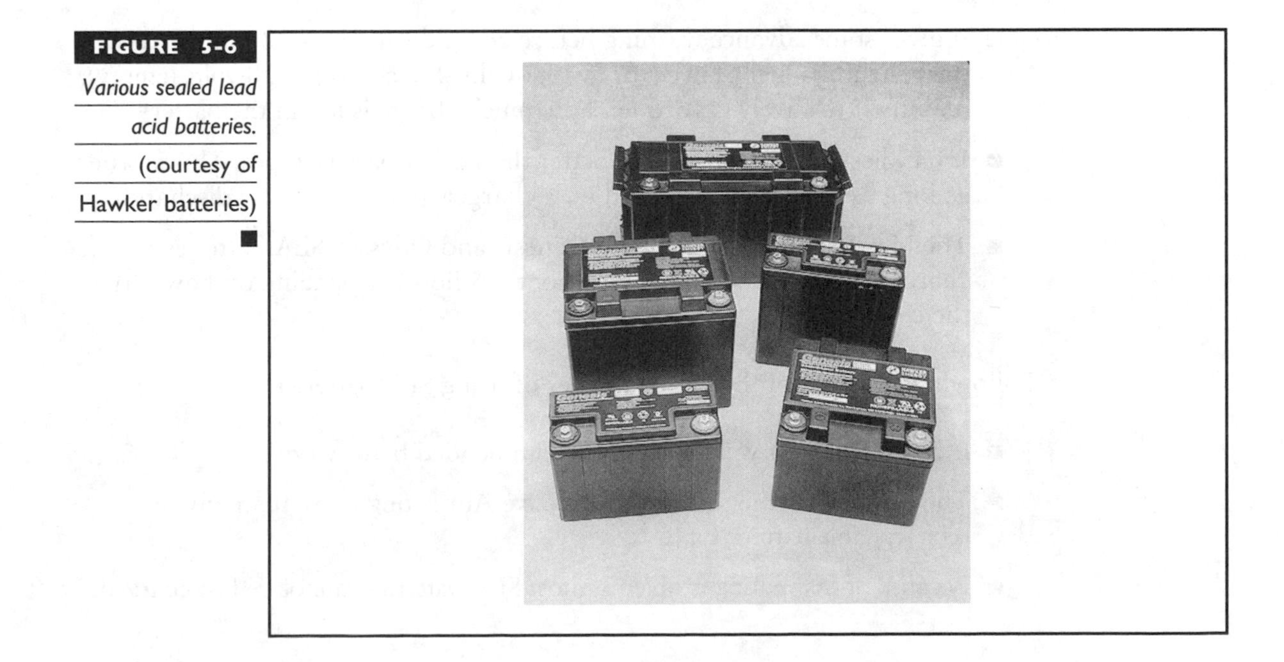

FIGURE 5-6 *Various sealed lead acid batteries.* (courtesy of Hawker batteries)

Hawker brand SLA batteries (www.hepi.com) have peak current delivery up to 40 times the 20-hour capacity. For example, the Hawker 16Ahr battery can source 680 amps for 5 seconds—or about 42 times the 20-hour capacity.

Charging is accomplished by applying 2.45 volts per cell and limiting the current to the battery manufacturer's recommended charging current. The exceptions to this rule are the Hawker batteries, which do not require current limiting. A 12-volt SLA battery has six cells, so the charging voltage is 14.7 volts (14.7 = 6 × 2.45 volts). If you are leaving the battery on the charger for an extended period of time, the charging voltage should be reduced to 2.27 volts to 2.35 volts per cell. When storing SLA batteries, you should be sure to charge them fully every six months. A good automatic automotive charger will work well for fast charging; however, it is important to use a battery that can handle fast charging and to use a charger that does not force charge into the battery after it is fully charged.

Following are some of the advantages of using SLA batteries:

- It's the least expensive rechargeable battery type, so it's easier to purchase more than one battery at a time.
- Up to 300 charge/discharge cycles to 80-percent capacity are possible.
- When stored at 25° C, it loses less than 1 percent of its charge per day.
- It can supply the highest current of any battery type.
- The wide range of battery capacities makes it easy to size the battery to the job.
- It gives some advance warning before going dead. For a 12-volt battery, the voltage gradually lowers from 13.2 volts (full charge) to 10 volts (empty), making it relatively easy to tell how much charge is left in the battery.
- It handles fast deep-discharge better than other battery types. This is true as long as the battery is placed on a charger quickly after the discharge.
- The Hawker brand Cyclon and Genesis and Odyssey SLA batteries can be charged in about 30 minutes to about 1.5 hours depending on how large the charger is.

Following are some of the disadvantages of using SLA batteries:

- It has the highest weight of any recommended battery type.
- The 6-minute rating drops the effective Ahr rating more than any rechargeable battery type.
- Because of gas venting problems, most SLA batteries cannot be fast charged.

- Because the acid in the SLA battery will attack the plates of the battery when discharged, it must always be stored in a charged state and must be periodically recharged when in storage. If stored uncharged for an extended period of time, the battery will die.

Which SLA Manufacturer Is the Best?

Most SLA batteries have similar capacity performance. Even so, the Hawker brand (formerly Gates) stands out as the best SLA battery manufacturer. Cyclon, Genesis, and Odyssey batteries can be 1.5-hour fast charged (or faster), can be repeatedly fully drained with little battery degradation (down to 9 volts), have the lowest shelf leakage of the SLA lineup, can supply three times more peak current than other batteries with similar Ahr ratings, and have good shelf life.

The SLA battery manufacturer to avoid is Panasonic. Many of the Panasonic brand SLA batteries have built-in thermal cutoff switches (a safety feature), making fast, high-current discharge impossible. The Power Sonic brand seems to have a good price/performance value. For the largest robots, the Optima battery brand is great. Optima is a good battery, but the 12-volt Optima weighs almost 40 pounds.

Are SLA batteries too heavy to have a competitive advantage? Not at all. Electric wheelchairs, golf carts, even electric racing go-karts and boats use SLA batteries. If your robot requires high sustain currents or high peak currents, the SLA battery may have the best performance.

Nickel Cadmium (NiCad)

The rugged construction of NiCad batteries is well suited for combat robot use. Though NiCads are a mature battery technology, they are still seeing incremental improvements in price and performance. Fast-charge/fast-discharge NiCads are required for competition applications.

The Ahr rating for this battery type is specified at the 1-hour discharge rate. To determine the 6-minute, run-time capacity, multiply the 1-hour capacity rating by 0.9 (see Table 5-1). Sometimes, even with a fast-discharge NiCad, this 6-minute discharge rate is higher than a NiCad's datasheets will allow. For example, a D-cell NiCad battery pack with a capacity of 5Ahr has a usable 6-minute capacity of 4.5Ahr ($4.5 = 0.9 \times 5$Ahr) and will provide an average current of 45 amps ($45 = 10 \times 4.5$A) for the 6-minute duration. Even so, a typical fast-charge/fast-discharge C-cell or D-cell NiCad datasheet will show only an average drain of 35 to 40 amps, with short duration (less than 100 milliseconds) peak currents of 100 to 130 amps. For higher current draw, you need to parallel multiple battery packs together or run outside the manufacturer's recommendations.

Fast charging is accomplished by applying the current equal to the Ahr rating of the battery for about 1.5 hours. Charge must be terminated when the battery starts to heat up, when the battery voltage begins to decline, or some combination

FIGURE 5-7 *Various NiCad Batteries (courtesy of Panasonic)*

of the two. Generally, a charger designed for this purpose must be used. Excellent fast chargers for NiCads are readily available.

Slow charging can be accomplished by sending a current equal to 1/10th of the Ahr rating of the battery for 15 hours. It is important that you not allow the battery to remain on this type of charger for long periods (longer than 24 to 48 hours) or else the NiCad will suffer from voltage depression (about .1 to .2 volts per cell). Figure 5-7 shows various NiCad batteries.

Following are the advantages of NiCad:

- It has an excellent cost verses performance ranking.
- For long-term use and with proper care, the NiCad can be less expensive in the end—even less than the SLA.
- With proper care and storage, NiCads can last through more than 1,000 charge cycles—though a chance to run this many charge cycles is not likely to happen in the harsh world of a combat robot.
- NiCad packs are small, so they can be stored in your refrigerator for long periods of time.
- The NiCad battery is moderately priced, so you can purchase more than one battery pack.
- The energy density is good—three times that of SLA—and in this application surpassed only by NiMH.

- NiCads can be stored with or without a charge, without damaging effects. However, it is usually safe to store the batteries in the discharged state.
- NiCads have no memory effects when used for this application. Because they are fully discharged during a combat match, this avoids memory effects.

Following are some disadvantages of NiCad

- When stored at 25° C, the NiCad battery loses 1 percent of its charge per day.
- When fully charged, the NiCad will self-discharge to an 80-percent charge in about three weeks.
- Occasional cycling to 80-percent voltage is required to keep the internal resistance of the battery low. If the robot is noticeably slower, you know the battery has reached this 80-percent level. It is best to do this every 20 charge cycles or so. During the testing phase, usually the batteries are repeatedly drained.
- NiCads are high-maintenance batteries, requiring careful monitoring, charging, cycling, and low temperature storage to yield long life.
- NiCads have cadmium; and although safely housed in the battery, cadmium is a toxic element and must be disposed of properly.

Nickel Metal Hydride (NiMH)

The rugged construction of NiMH batteries is well suited for combat robot use. This is an emerging battery technology that is still seeing constant improvement. Fast-charge/fast-discharge NiMH packs are required.

The Ahr rating of this battery type is also specified at a 1-hour rate. Multiply by 0.92 (see Table 5-1) to convert this 1-hour rate to the 6-minute Ahr rate. For example, a D-cell NiMH battery pack with a capacity of 6.5Ahr, has a usable 6-minute capacity of 6Ahr ($6 = 0.92 \times 6.5$Ahr) and can provide a calculated average current of 60 amps ($60 = 10 \times 6$Ahr) for the 6-minute duration. Even so, the specification data sheets show that for the fast-charge/fast-discharge C-cell or D-cell batteries, the maximum average current is only about 40 amps, with the peak current limit of about 100 amps. For higher current draw requirements, it is necessary to parallel the batteries.

For fast charging, use only a charger designed for NiMH. Using a charger designed only for NiCads, for example, will usually destroy NiMH batteries. Because this technology is relatively new, chargers for this type of battery are harder to come by than NiCad or SLA chargers.

Following are some advantages of NiMH:

- The NiMH energy density is the best of all the usable battery types currently available.
- With proper care and storage, NiMHs will last through more than 300 charge cycles.
- Because NiMH packs are small, it is easy to keep them in the refrigerator for long-term storage.
- The voltage output remains constant until almost fully discharged. This provides full power to your robot for the duration of the match.
- They can be stored without a charge without damaging effects.
- They have no memory effects when used for this application.
- They have no cadmium, so they don't have the related health problems.

The emerging NiMH battery technology will see improvements. In time, expect a lower cost, a higher number of charge cycles, lower internal resistance resulting in a higher maximum current rating, and lower self-discharge rates.

Following are some disadvantages of NiMH:

- It is the most expensive rechargeable battery technology.
- It has the lowest life of the rechargeable battery technologies. After 300 charge/discharge cycles, the battery capacity measurably degrades while the internal resistance increases.
- When stored at 25° C, the NiMH battery can lose up to 5 percent of its charge per day. When fully charged, the NiMH can self-discharge to an 80-percent charge within five days!
- Occasional cycling to 80-percent voltage is required to keep the internal resistance of the battery low. If the robot is noticeably slower, you know the battery has reached this level. It is best to do this every 20 charge cycles or so.
- NiMH are high-maintenance batteries, requiring careful monitoring, charging, cycling, and low-temperature storage to yield long life.

Alkaline

The alkaline battery is the most common primary battery in America. It is used to power most products from radios to flashlights. Small robot kits often will use them as the power source. Alkaline batteries cannot handle a high rate discharge, so they don't work well for combat robots.

The alkaline battery works best when powering low-current devices. When used to power high-current devices, the performance is dismal. Even so, many robot kits use AA alkaline batteries to drive servos and the onboard electronics. When stalled, these servos can try to draw 1 ampere, bringing short order to the AA alkaline batteries. Usually, these robots will see a performance increase if the alkaline batteries are changed over to standard NiMH or NiCad cells.

Following are some advantages of alkaline:

- They are readily available.
- They have the least expensive startup cost.
- It is easy to replace the battery with a known fresh battery.
- They are low maintenance—you can throw away the old ones.

Here are the disadvantages of alkaline:

- In the long run, they are the most expensive battery type.
- They have the poorest 6-minute energy density of all the batteries.

Lithium Ion

Lithium ion batteries are common rechargeable batteries used in computing applications. They have high-energy density when current is pulled out at a moderate rate. However, the voltage drops when pulling current out at a high rate. In addition, the battery can fail when pulling out current at a higher than moderate rate. Therefore, lithium ion batteries do not work well for combat robots. Another negative factor is that the typical shelf life of the lithium ion rechargeable battery is only two years if stored at 25° C.

Combining Drill Motors, Batteries, and Battery Chargers

Many cordless power drills come with rechargeable batteries and fast chargers. Many competition winners have used these drill/battery/charger combinations to have a complete solution to the problems of supplying motors for the robot, getting good batteries, and getting fast chargers. In addition, spare batteries for these motors are readily available. Four-wheel-drive robots using four power drill motors have had good success in the combat arena. If you go this route, use good-quality cordless power drills with NiCad or NiMH battery types.

Installing the Batteries: Accessible vs. Nonaccessible

It is best to install your robot's batteries where they can be easily accessed for replacement. Due to the relatively short time period between matches, and because it can be difficult if not impossible to put a full charge on the batteries if they remain in the robot, the best idea is to replace the batteries with freshly charged batteries between matches. To do this quickly, batteries need to be placed in the robot in such a way that allows for quick and easy replacement.

If the battery is not accessible, so that the builder or operator cannot replace the batteries between matches, you need to come up with another recharge scheme. If you're using a nonaccessible battery, the robot could be fast charged between matches while still in the robot. Even so, as a competitor, you can count on incidents of no time to top off the battery charge between matches. In such cases, the battery must have the capacity to be able to run the robot through two or maybe even three matches before requiring a recharge. Of course, you need to account for this when selecting the battery capacity and when installing the battery in your bot.

Now you probably know more about batteries than you ever knew you would know. The batteries are the heart and blood of your robot. You take care of your batteries, and they will take care of your robot. The 6-minute run time estimates are the minimum your robot will need to survive in the competition arena. You should always have spare batteries when you go to any competition. The last thing you want to see happen is to watch your winning robot stop dead because the batteries went dead. If your robot can handle the weight and size of larger batteries, then consider using them to get a little more assurance that your robot will survive all the way through a tough match.

chapter 6

Power Transmission: Getting Power to Your Wheels

ONE of the most important considerations in the design of your robot is locomotion. You can use a propeller, or even a jet engine, to "blow" your machine along, but these sorts of propellants are not allowed in most competitions and would prove to be quite ineffective anyway. Moving parts that actually touch the floor are the preferred method of providing controlled movement to your robot, with wheels being the most chosen method.

The following are some definitions used in this book:

- **Speed reduction** Transforming high RPM and low torque power into low RPM, high torque power.
- **Speed reducer** The device that does the speed reduction.
- **Gear reduction** Speed reduction using gears.
- **Power transmission** Every device and component that transmits power from the motor to the wheels (including the speed reducer).
- **Transmission** A speed reducer with more than one reduction ratio. Note that a transmission is only one component in the "power transmission." The two terms are not interchangeable.

The purpose of the power transmission is to transmit the rotational energy from the motors to the wheels of the robot, and many different ways can be employed to do this. The simplest way is to use a *direct drive method*. With this method, the wheel's hub, or axle, is directly connected to the motor—either directly on the output shaft of the motor or the output shaft of a gearmotor.

A *gearmotor* is a single unit with a gearbox and a motor combined into one convenient package. The gearbox is used to decrease the rotational speed of the motor to a more usable output shaft speed. Many electric motors' rotational speeds range between 3000 to 20000 RPM. This speed is too fast for directly driving a robot's wheels—unless you want your robot to move at warp speed. The gearbox also increases the actual torque of the electric motor to a much higher value on the output shaft. The higher torque will give your robot more pushing power.

Although many robot builders use the gearmotor approach, some have used non-gearmotors to power the wheels directly. For example, the middle and heavyweight entries from team Whyachi used direct-drive Magmotors in their robots.

Early in the robot design process, you usually decide that you want your robot to move at a certain speed and have the ability to push a certain amount of weight. These specifications can help you select an appropriately sized motor.

Ideally, you will be able to find a prepackaged gearmotor that will meet your specifications. If you cannot find the perfect gearmotor, you will have to settle for whatever you can find and live with a different robot speed and strength—or you can build your own speed reducer.

The type of power transmission you'll need for your robot is a simple speed-reduction setup, not the type of power transmission commonly found in automobiles or motorcycles. In some cases, you may want to *increase* the speed of a gearmotor; but in most cases, you will be *reducing* the speed of the motor. This type of power transmission usually consists of a set of chains and sprockets, timing belts, V-belts, gears, or even a secondary gear box.

The power transmission is also often used to transmit the power of the motor to two or more robot wheels. In most cases, two separate axles are driven at the same time through chains and sprockets, timing belts, and V-belts.

True Story: Grant Imahara and Deadblow

"The most spectacular failure I had was in Las Vegas, during season 2.0," says Grant Imahara, the renowned builder behind *Deadblow*. "I was waiting to fight a robot named *Kegger* built by a team called 'Poor College Kids.' It was probably going to be a pretty easy match, but BattleBots teaches you not to be overconfident, because anything can happen."

Indeed, Grant has seen just about everything. He was there at the birth of the sport, since Marc Thorpe, an Industrial Light and Magic co-worker, created *Robot Wars* in 1995 and gave Grant tickets to attend. "I was captivated, and knew that I had to build a robot of my own."

Deadblow was the result of that obsession; and at this particular event, Grant found himself charging the onboard air tanks—essential to power the weapons—in preparation for competition. "I was filling my two onboard air tanks from an external SCUBA tank, which was a pretty standard thing for me to do. I had done it a million times. But this time I heard a loud 'pop,' followed by a rush of high pressure air coming out of the robot."

Grant describes how the nearby mass of people backed away uneasily at the ominous sound of rushing air. "That pop meant that I had ruptured one of my air lines and the weapon— *Deadblow's* only weapon—couldn't work without air. I knew that if the robot couldn't fight at its designated time, I would have to forfeit my match."

Grant Imahara and Deadblow (continued)

Fortunately, the Washburn family was nearby in the contestant stands. Shane Washburn, Grant explains, was a co-worker at ILM and he had fought against Grant with his bot *Red Scorpion* in previous years. Moreover, Shane's father, Ray, was a welder and hydraulics expert, and his brother, Jon, was an emergency medical technician. "They heard the air line rupture and were immediately at my side. While I was desperately trying to turn off the SCUBA tank, the Washburns and my crew were taking the screws out of the top cover. We wheeled my robot out of the way and the *BattleBots* people and Team Poor College Kids graciously allowed another match to go before us. This bought a little time, but not much. Ray ran all the way back to the pits and grabbed all the air fittings from my toolbox. We fixed it there on the spot in about five minutes—I couldn't have done it without their help."

Despite the catastrophic failure, Grant adds that he went on to beat *Kegger* with just a single onboard air tank. But there's a lesson in the story: "Always inspect all of your equipment for wear and damage, even if you don't think you had any."

Power Transmission Basics

As stated, the purpose of the power transmission is to reduce the speed of the motor to some usable speed for the robot and to transmit the power to the wheels. The speed of a robot is a function of the rotational speed of the wheels and the diameter of the wheels. Equation 1 shows this relationship, where v is the velocity of the robot, D is the diameter of the driven wheels, and N is the rotational speed of the wheel. So, to determine the required rotational speed of the wheel, Equation 1 is solved for N, which is shown in Equation 2.

$$v = \pi D N$$

$$N = \frac{v}{\pi D}$$

If your robot has 10-inch-diameter wheels and the rotational speed of the robot is 300 RPM, the speed of the robot will be 9,425 inches per minute, or about 8.9 miles per hour (MPH). If you want your robot to move 20 MPH, this same wheel will have to spin at 673 RPM. This is one fast robot.

After you have an idea of the wheel speed you want, you need to determine how much of a speed reduction in the power transmission you will need to convert the motor speed to the wheel speed. This is done by using a combination of different sprocket diameters, pulley diameters, or gear diameters. The speed ratios of a gear train are just a ratio of the gear diameters.

Figure 6-1 shows a sketch of the same type of speed reduction. The leftmost sketch shows two gears in mesh, and the sketch on the right shows a belt/chain gear reduction. One thing to note here is that with the gear reduction, the direction of the driven gear is opposite of that of the driving gear. With the belt/chain system, the directions of both pulleys/sprockets are the same.

FIGURE 6-1 Simple speed reduction schematic

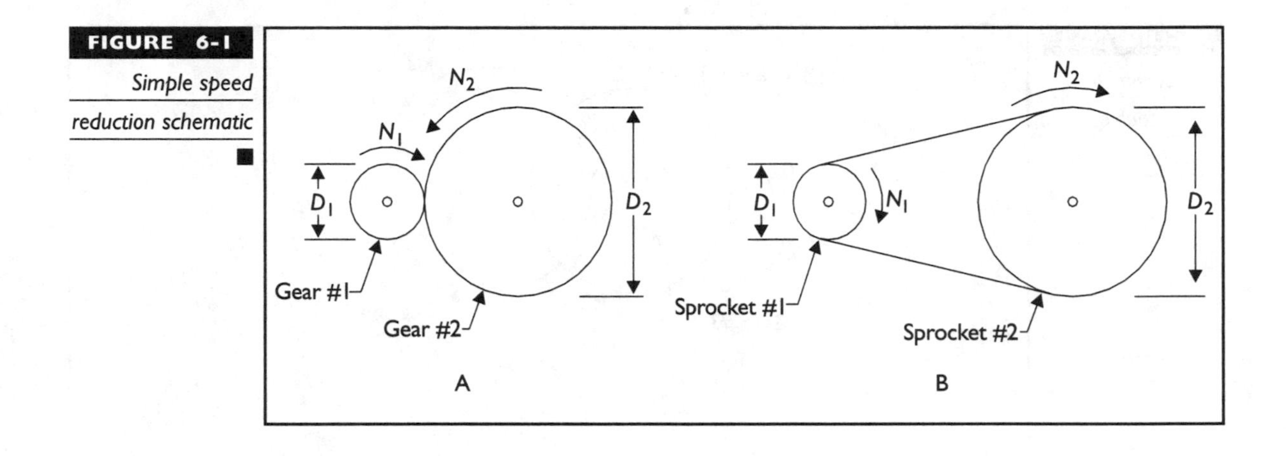

Equation 3 shows how the speed of the output gear relates to the speed of the input gear.

$$N_2 = \frac{D_1}{D_2} N_1$$

In Equation 3, D_1 and N_1 are the diameter and rotational speed of the driving gear, and D_2 and N_2 are the diameter and rotational speed of the driven (output) gear. When D_1 is greater than D_2, the output gear will spin faster than the driving gear; when D_1 is less than D_2, the output gear will spin slower (gear reduction) than the driving gear. When driving two shafts together, such as a front and rear axle being driven with only one motor, the gear/sprocket diameters between the two axles must be the same or the wheels will spin at different speeds.

If you have a 3000 RPM motor and you want a wheel speed of 300 RPM, you will have to reduce the speed of the motor by a factor of 10. By looking at equation 3, you can see that the output gear, D_2, will have to be 10 times bigger than the input gear, D_1. This is a pretty big gear reduction with only two gears. If you were using a 1.5-inch-diameter gear on the motor shaft, you would have to use a 15-inch-diameter gear on the wheel. If the wheel is only 10 inches in diameter, the gear's diameter will cause the gear to strike the ground, since it is larger than the wheel. When this type of situation occurs, three or more gears/pulleys/sprockets must be used together.

Figure 6-2 shows a more complex speed reduction.

Though the configuration shown in Figure 6-2 seems complicated, it can be simplified by looking at it as two separate two-gear systems. In this example, the speed of gear number 2 is the same as what is shown in Equation 3. The speed of gear number 4, N_4, is first shown in Equation 4 that follows. It has the same exact form as what is seen in Equation 3. Since gears numbers 2 and 3 are physically attached to the same shaft, they will both spin at the same speed, which is shown in Equation 5. Because of this, you can substitute Equation 3 into Equation 4 to de termine the final speed of the output shaft. Equation 6 shows the speed reduction for

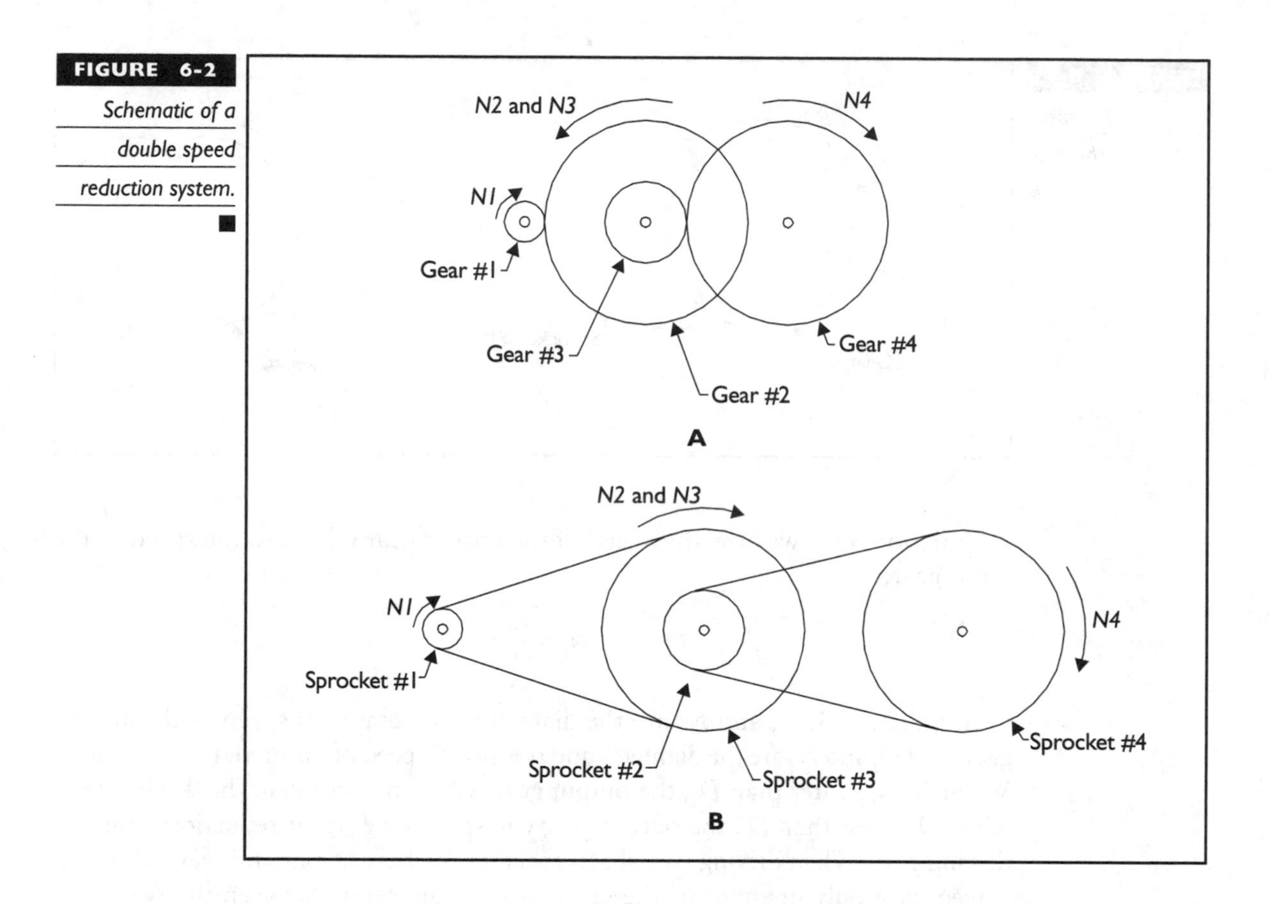

FIGURE 6-2 *Schematic of a double speed reduction system.*

the gear reduction shown in Figure 6-2. The D_1 / D_2 is the first gear reduction ratio, and D_3 / D_4 is the second gear reduction ratio.

$$N_4 = \frac{D_3}{D_4} N_3$$

$$N_3 = N_2$$

$$N_4 = \frac{D_3}{D_4} \frac{D_1}{D_2} N_1$$

In the previous example, you looked at a speed reduction of 10. With the double-speed reduction system, you have a lot of options for choosing gear diameters. The product of the first and second stages in the speed reducer must be 10. For example, you can choose the first gear reduction to be 4 and the second gear reduction to be 2.5. In this case, you can use the same 1.5-inch-diameter gear on the motor shaft, and then the second gear should be 6 inches in diameter. This is smaller than the 10-inch-diameter wheels used in this example. The third gear could be a 2-inch-diameter gear, which would mean that the last gear should be 2.5 times larger or 5 inches in diameter. These gear sizes are much more manageable than trying to do this entire gear reduction in one step.

As a general rule, the greater the gear reduction, the more gears you will have to use to achieve the gear reduction. In the real world, you may not find the exact gear and sprocket diameters you want. This may be because the actual sizes do not exist. For example, if you are using sprockets instead of gears, it is rare to be able to find a sprocket that has a diameter 10 times greater than the driving sprocket. You will usually have to choose components that are close to the values you want. Thus, the speed reduction will be a little lower or higher than what you want.

Torque

The output torque is also a function of the gear ratios, but the torque and gear ratios have an inverse relationship. When the speed is reduced, the torque on the output shaft is increased. Conversely, when the speed is increased, the output torque is reduced. Equation 7 shows the torque relationships from Figure 6-1. The direction in which the torque is being applied is identical to the rotational directions.

$$T_2 = \frac{D_2}{D_1} T_1$$

T_1 and D_1 are the torque and the diameter of gear 1, and T_2 and D_2 are the torque and diameter of gear 2. If D_2 is greater than D_1, the output torque is increased. From Figure 6-2, the output torque is shown in equation 8.

$$T_4 = \frac{D_4}{D_3}\frac{D_2}{D_1} T_1$$

In the previous example, where we were looking for a 10-to-1 speed reduction, this will increase the output torque by a factor of 10.

During the robot design process, the power transmission must be considered at the same time while you're selecting the motors. The number of gears, sprockets, and pulleys and their sizes can have a significant impact on the overall structural design of the robot. To simplify the overall power transmission design, you should choose a motor that has the lowest RPM so that the number of components in the power transmission (or speed reducer) can be minimized.

Force

The robot's pushing force is a function of the robot's wheel diameters and the output torque on the wheel, and the coefficient of friction between wheels and floor. By definition, torque is equal to the force applied to some object multiplied by the distance between where the force is applied and the center of rotation. In the case of a gear, the torque is equal to the force being applied to the gear teeth multiplied by the radius of the gear. Equation 9 shows this relationship, where T is the torque, F is the applied force, and r is the distance from the center of rotation and where the force is being exerted. Equation 10 shows how the force is related to the applied torque.

$$T = Fr$$

$$F = \frac{T}{r}$$

Using this relationship, you might think that your 500 in.-lb. torque motors and your 10-inch-diameter wheeled robot would have a pushing force of 100 pounds (100 pounds = 500 in.-ibs. / 5-inch radius). But this isn't the case. Wheel friction becomes part of the equation. Without friction, powered wheels will never move a vehicle, and turning the vehicle would be virtually impossible. In most mechanical devices, friction is undesirable; but for wheels, friction is good. For combat robots, the more friction you can get the better your robot can push. The frictional force to move an object across a horizontal floor is equal to the product of the coefficient of friction between the floor and the object's surface and the weight of the object. Equation 11 shows you how it works:

$$F_f = \mu F_w$$

where F_f is the frictional force, μ is the coefficient of friction, and F_w is the weight of the object.

Figure 6-3 shows a schematic of the various forces acting on a wheel. F_w is the weight force acting on this wheel. For a really rough approximation, this value could be estimated by dividing the robot's total weight by the number of its wheels. This applies only a rough estimate to the weight of a wheel, and it is true only if the robot's center of gravity is at the geometrical center between the wheels. Computer-aided design (CAD) software can help provide the actual values for the wheels, or they can be directly measured by putting a scale under each wheel.

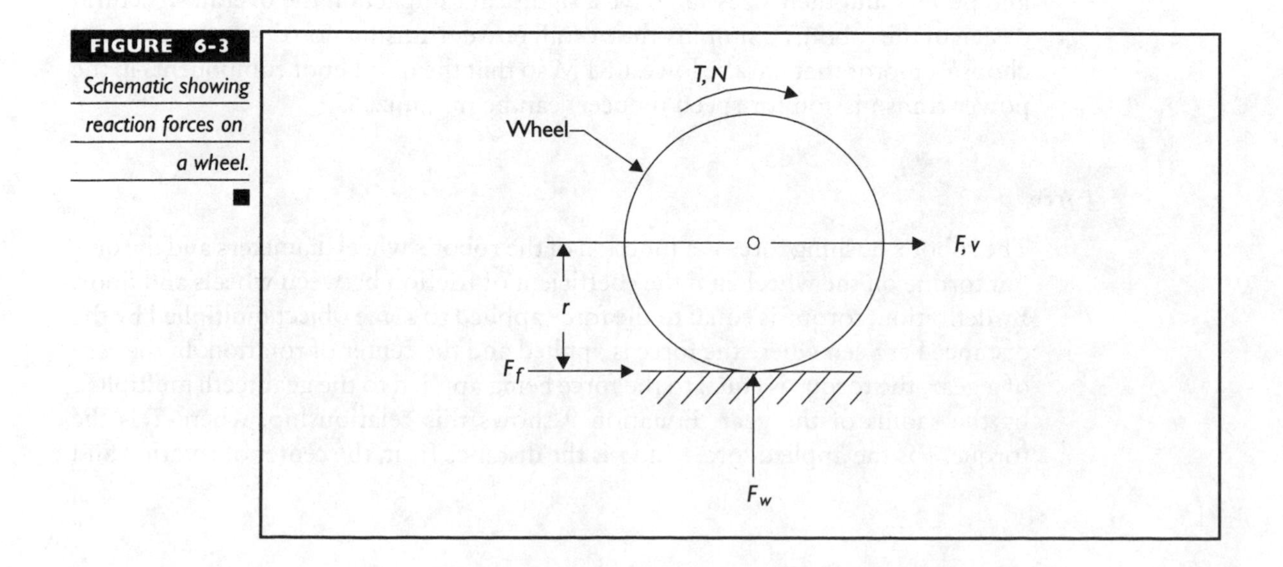

FIGURE 6-3 *Schematic showing reaction forces on a wheel.*

So how much can a robot push? The maximum pushing force will be equal to the sum of all the frictional forces, F_f, for all of the wheels. When the reaction forces of an immovable object, such as a wall or a bigger robot, exceeds the total frictional forces, your robot will stop moving—and, in this case, your robot could actually be pushed backward! By combining Equations 9 and 11, the torque required to produce the maximum pushing force will be as shown in Equation 12.

$$T_f = F_f r \text{ or } T_f = \mu F_w r$$

For a robot with all identical wheels and motors that can deliver all the torque it could need, the total maximum pushing force, F_{max}, will become the product of the weight of the robot and the coefficient of friction. Equation 13 shows this.

$$F_{max} = \mu W_{robot}$$

If the motor torque can produce a force greater than the frictional force, the wheels will spin. If the maximum torque of the motors cannot produce forces greater than the frictional forces, your robot's motors will stall when you run up against another robot or a wall. In Chapter 4, you learned that stalling a motor is not a good idea, so it is a better idea to have the wheels spin rather than being stalled. Equation 14 shows the stall torque relationship for each wheel. This information can be used to help you determine the speed reduction in the power transmission and help you pick the right-sized motors. Equation 13 is a rather interesting equation. This maximum force is the maximum force your robot can exert, or it is the force another robot needs to exert on your robot to push it around. This force is a function of two things: weight of the robot and the coefficient of friction between the robot's wheels and the ground. So, this tells you that increasing your robot's weight can give you a competitive advantage.

$$T_{stall} = \mu F_w r$$

One of the difficult tasks in determining the pushing force is determining the coefficient of friction. The coefficient of friction between rubber and dry metal surfaces can range from 0.5 to 3.0. In your high school science classes, you probably learned that the coefficient of friction cannot be greater than 1.0. This is true for hard, solid objects; but with soft rubber materials, other physics are involved. It is not uncommon to find soft, gummy rubber that has coefficients of friction greater the 1.0, and some materials have a coefficient of friction as high as 3.0. For all practical purposes, the coefficient of friction for common rubber tires and steel surfaces is between 0.5 and 1.0.

The other factor that affects the coefficient of friction is how much dirt is on the surface. A dirty surface will reduce the overall coefficient of friction. This is why off-road tires have knobby treads to help improve the friction, or traction.

As a worse-case situation, assume that the coefficient of friction is equal to 1 and size all your components so you will not stall the motors in these conditions. This will give most robots a small safety margin. If you want to be more conservative, use a coefficient of friction greater than 1.

Location of the Locomotion Components

Most combat robots are fairly simple, internally. They consist of a power source, a set of batteries; motors for the wheels; a radio-controlled (R/C) system receiver; controllers to take an R/C signal from the receiver and send power to the motors; and a weapon system's actuators, if they're required in your design. Other components appear in various robot designs, such as microcontrollers to process incoming data to pulse width modulation signals, DC to DC converters, fans to cool controllers, and so on, but these are generally smaller and can be placed in tight-fitting places.

The location of the main drive motors is the most critical concern in the placement of large robot subsystems. Usually, these motors are quite large. The other large subsystems, such as batteries and controllers, can be located "wherever possible." Motors have to be close to the wheels, and their position and orientation is critical. Quite often they are mounted in the lowest part of the robot. The motors must be positioned accurately, especially if a series of gears are used to transmit the power to the wheels, and the chains or gears need to be aligned in the same plane as the wheel system.

Mounting the Motors

Mounting of the motors in any application is important, but combat robots present another magnitude of problems for their motors. The motors are trying to wrench themselves out of their mounts from extreme torque conditions. At the same time, their mounts are being shaken so intensely that the mounting screws can be sheared in half. So you must design your robot to handle such extremes.

Quite often, a DC motor you might find in a surplus catalog has several threaded holes in the front face where the output shaft is located. Using these mounting holes for screwing the motor to a plate is okay for the types of applications for which the motor was originally designed, but using these holes may not suit an extreme situation in combat robots. To determine whether these holes are suitable, you may need to subject the motor / mounting brackets to a shock test. The large inertial mass of the motor may just shear off the screws as you slam the assembly into your garage floor. Unfortunately, you might have to use an "easy-out" to remove the remaining portion of the screws. Use your judgment here.

You're in a far better situation if your motors have a flange mount around the front face of the motor. If you need more strength, you can drill out the threaded holes and make larger holes for through-hole, high-strength bolts. A flanged base mount can be found on some older motors. Flange-based motors offer a higher strength method of mounting compared with the threaded face hole method.

Another method to use for mounting motors is to secure the face with the existing mounting holes to a motor bracket you've fabricated, and then secure the back part of the motor with several high-strength clamps and a machined block in which to rest the motor. Use high-strength hose clamps that have a machined screw—not the "pot metal" types found in some hose clamps. This back clamping will prevent the heavy motor from moving. See Figure 6-4.

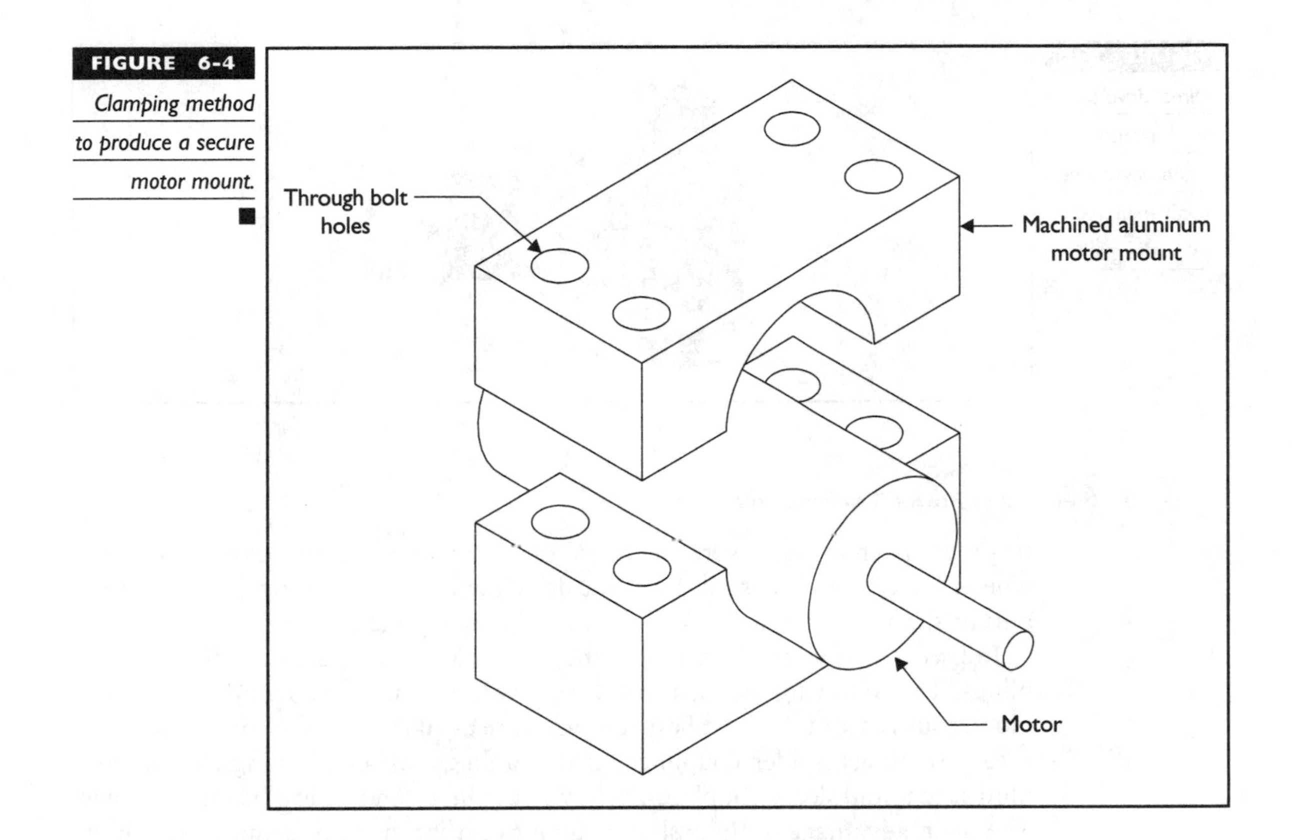

FIGURE 6-4 Clamping method to produce a secure motor mount.

Thermal Considerations for the Motor

One of the drawbacks of using a higher-than-recommended voltage on a DC motor is the possibility of overheating. Even though combat matches generally last only a few minutes, intense heat built up in a motor can destroy it. This is not a power transmission issue, but it certainly is a mounting consideration. Some motors use a fan at one end to draw in air for cooling, but the intermittent action of the motor may mean that the motor is cooking in its built-up heat while it is off. You must also remember that the windings that heat up are in the armature, which is the rotating component that is isolated from the case, so heat sinks are not as effective as one might think. If the armature heats up too much, it can begin to disintegrate, slinging wire pieces all over the inside of the motor. If that happens, you're in for a bad day.

How can you keep these motors cool? If you've run the motor on your bench while under load and you've noticed that the case gets extremely hot, you may want to mount it in a machined aluminum block to absorb and conduct the excess heat away from the motor. Some competitors have also used a small blower to force air through the motor to augment the fan. Have the fan run even when the motor is off to continue the cooling process as much as possible.

FIGURE 6-5 *Direct-drive power transmission showing a wheel directly mounted to the gearbox.*

Methods of Power Transmission

In previous chapters, several methods of interconnecting the motors with the wheels have been discussed. In direct-drive methods, the motor or gearmotor's output shaft is connected directly to the wheels (see Figure 6-5).

Indirect-drive methods include a chain, belt, and even a series of flexible couplings. The following sections will discuss various chain and belt drive systems. Numerous types of flexible shaft couplings are available, such as universal joints, shear couplings, spider couplings, grid couplings, offset couplings, chain couplings, gear and sleeve couplings, bellows couplings, and helical beam couplings. The main advantage of these shaft couplings is that they can connect two shafts that are slightly misaligned. Figure 6-6 shows a Lovejoy flexible coupling. A *Lovejoy* coupling is a spider coupling. They come with three different parts: two bodies and a spider. The shaft bodies come with different bore diameters so that different shaft diameters can be coupled together. The spider's material is made out of urethane, Hytrel, or rubber. The selection of the spider material is based on the applications the coupling is going to be used for.

For high-powered robots, careful design of the components and mounting locations will be needed to minimize shaft misalignment.

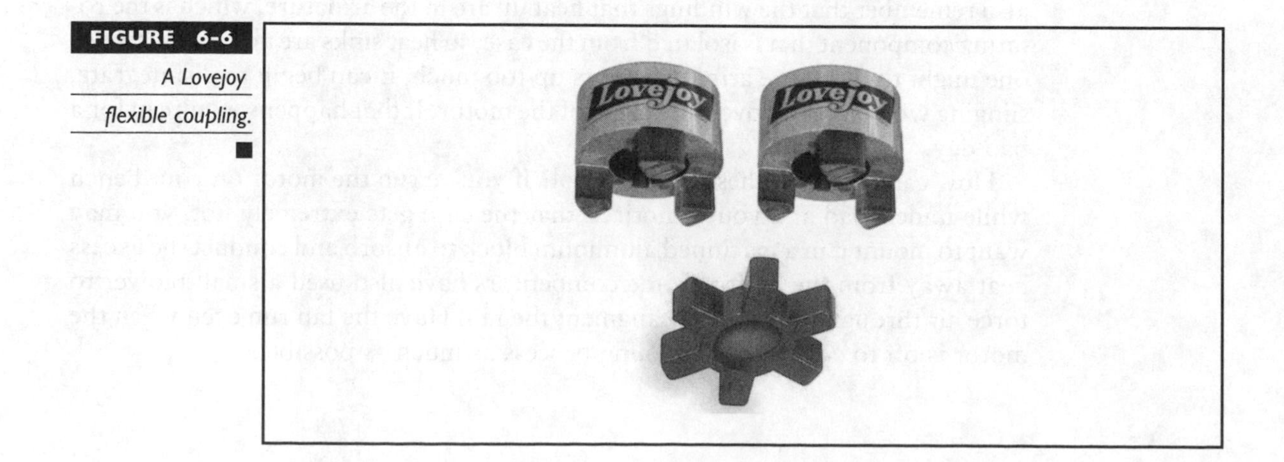

FIGURE 6-6 *A Lovejoy flexible coupling.*

Chain Drive Systems

Rather than starting with some more exotic designs that use a flexible shaft or even an articulated shaft fitted with swivel joints, let's instead jump right to the method that is used the most—a chain drive. This type of interconnection between the wheels and motors offers a lot of pluses. If the proper chain is used, it has the capacity to transfer a lot of power to the wheels. It also has the ability to take up "slop" in the system without requiring precise spacing between the motor and wheel/axle sprocket.

Buying the Chain

What is the proper chain for your robot? You might be tempted to use a bicycle chain. Hey, you can pedal hard, even stand on the pedals when going uphill, and still not break the chain. The quality of mass-marketed bicycle chains is not up to industrial standards, however. Invest a few bucks in some good roller chain. It will be money well spent and can save you from a few headaches in the long run.

The proper term for this type of chain is *single strand roller chain*. Generally, the pitch on these types of chains ranges from 1/4 inch to 3/4 inch. A 1/2 inch pitch means that the spacing of the sprocket's teeth are 1/2 inch apart (or the chain's rollers are 1/2 inch apart). The industrial roller chain is specified with an ANSI number, generally 25 to 80. See Table 6-1 for a list of some of the common chains.

A typical ANSI #40 industrial roller chain, for example, will have a 1/2-inch pitch and a 5/16-inch roller width; it will have a maximum allowable load of 810 pounds; and the chain will break when the load gets up to 4,300 pounds. The maximum allowable load is based on continuous operation. Exceeding the maximum allowable load will shorten the life of the chain. If you exceed the average tensile strength, the chain will break.

Some builders have ganged up two sprockets on each end to double the strength. In actuality, the strength is not quite doubled due to slight differences in

ANSI No.	Pitch, in Inches	Roller Width, in Inches	Chain Width, in Inches	Max Working Load, in Pounds	Average Tensile Strength, in Pounds
25	1/4	1/8	0.31	140	1,050
35	3/8	3/16	0.47	480	2,400
40	1/2	5/16	0.65	810	4,300
50	5/8	3/8	0.79	1,400	7,200
60	3/4	1/2	0.98	1,950	10,000
80	1	5/8	1.28	3,300	17,700

TABLE 6-1 Standard Chain Size and Load Specifications

chain-link spacing and subsequent uneven loading on one of the two chains, but we won't cover the dynamics and physics of this scenario. This is still an acceptable method of applying redundancy for safety. When one of the chains fails, you still have another to carry most of the load. Double-strand roller chain is the best way to increase load capacity, and the cost of this type of chain is only about twice that of single-strand chain.

Most supply houses will supply the chain as a random-length loop or as long pieces of various lengths. Cutting the chain may require that you punch or drill out the rivet on one part of a link. You can buy a set of chain maintenance tools for in-the-field chain repairs; these would include a roller chain breaking tool, which is far easier to use than a hammer and a punch. Also available are chain pin extracting tools and a unique roller chain puller that allows you to tighten the chain before inserting a master link connector. For maximum chain strength, a chain can be custom ordered from the manufacturer in the exact length you need. If you choose to go this route, you will not need a master link.

The *master link* is a separately purchased connector link that allows you to create a continuous loop of chain. You should also buy several extra master link connectors to fasten the chain together at the length you'll want. This fastener consists of a side piece of a link with two pins that fit in the roller parts of the two ends of the chain, and a figure-8 side piece to fit over the pins on the other side. A clip snaps over the slotted ends of the pins, locking the master link in place. Figure 6-7 shows a typical chain.

Chain Sprockets

The sprockets used with roller chains look a little bit like gears, but they have more rounded teeth and are not meant to mesh with each other like a "standard" gear. For combat robots, you should buy only steel sprockets for their strength. These sprockets are specified by an ANSI number (sprockets and chains must have the same ANSI number, or they will not mesh together because the pitch lengths will not be the same), the number of teeth on the sprocket, and the shaft bore size. Most sprockets you will find include a keyway to lock them to a shaft with a similar

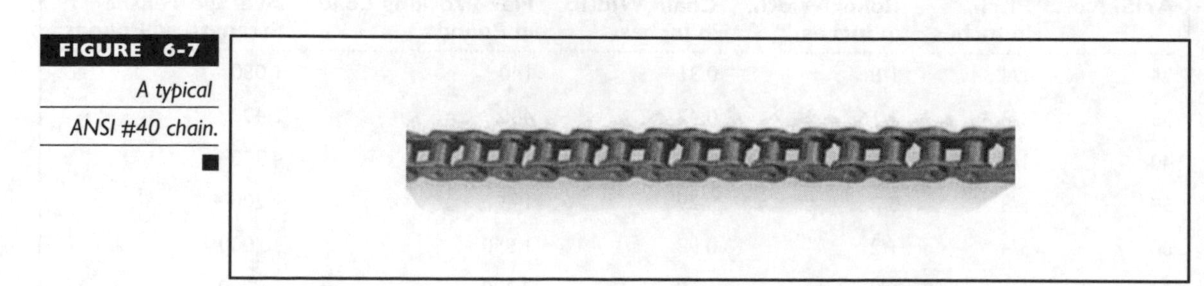

FIGURE 6-7 *A typical ANSI #40 chain.*

keyway. Some of the smaller diameter sprockets may have one or two set screws in the place of a keyway. These will work adequately with a flattened area on the shaft for lower torque applications, such as for small hobby robots. For combat robots, use keyways on all sprockets, gears, and pulleys. Doing so is a battle-proven method to secure components to shafts.

You might also want to apply one or more idler sprockets to take up slack in the chain. Quite often you place your motor(s) and wheel(s) in set locations and then apply the chain. More than likely, you'll find that the chain is too loose (or maybe too tight). Having a bit of slack in the chain and using a sprocket idler on a small spring-loaded lever arm will keep the chain at a specified tightness and will prevent the chain from flying outward with centrifugal force under high speeds.

When implementing a sprocket and chain system, all of the sprockets must have the same pitch as the chain to which they are connected. When calculating the speed and torque ratios, you should use the number of teeth instead of using the actual diameter. If you use the sprocket diameter, use the specified pitch diameter, not the outside diameter of the sprocket. The pitch diameter is the actual diameter in which the chain will wrap around the sprocket.

To locate the sprockets on the robot, you can determine the distance between the sprockets in two ways. The proper method would be to calculate the center distances and then design the robot to accommodate the dimensions. Appendix C shows the calculations for determining the center distances. The other method, which is used by many beginners, is to place the two sprockets wherever you want them and then take a long length of chain and wrap it around both sprockets, holding the two ends in your hand. Then you cut the chain at the appropriate place, apply the master link, and possibly use an idler sprocket to take up the slack. Figure 6-8 shows a sprocket.

FIGURE 6-8 *A typical 12-tooth ANSI #40 sprocket.*

Belt Drive Systems

In addition to chain drive systems, a belt drive system can be used to transmit power from the motor to other devices such as wheels and weapons. Many different types of belt drive systems are available, but the three most common are flat belt, synchronous belt, and V-belt systems.

Flat Belts

Flat belts are commonly used for applications that need high belt speeds, small pulley diameters, and low amounts of noise. Flat belts are in common use when one large motor drives several different pieces of machinery. They cannot be used for applications in which absolute synchronization between two pulleys is required. This is because these belts require friction to maintain motion, and slippage or creepage can occur. Flat belts must be kept under tension to transmit power from one pulley to another. Because of this, a belt tensioning device is required.

One advantage of this type of system is that a flat belt could be wrapped directly between the motor shaft and larger diameter pulley attached directly to the robot wheel. A similar application is commonly seen inside small electronic equipment such as tape recorders and videocassette recorders, and you can find them turning the rotary brushes in vacuum cleaners.

The drawback to these types of systems is that the two pulley surfaces must be perfectly parallel. If they are not, the belts will run off the pulleys. To prevent this from happening, flanges need to be placed on the sides of the pulleys to constrain the belts in place.

For combat robotic applications, these types of belts can be used for spinning weapon systems. If the weapon gets stalled, the motor will slip under the belt, which helps to protect the motor from stalling and burning out. These types of belts also offer little power transmission ability due to the small frictional area at each pulley.

Synchronous Belts

Synchronous belts are more commonly known as timing belts. The name *timing belt* is derived from their popular use in car engines, where they're placed between the cam and crankshaft and are used to synchronize the cams inside the engine. Timing belts are similar to flat belts in their operation. The physical difference between these two belts is that the timing belts have teeth on one or both sides of the belt. This allows timing belts to synchronize the speeds between all the pulleys that are being driven by the belt. Figure 6-9 shows a timing belt.

Because the teeth on the belts are used to drive the pulleys, similar to the chain drive systems, the belt tension requirements are much less for synchronous belts

FIGURE 6-9 *Timing belt used to drive a spinning weapon.* (courtesy of Andrew Lindsey)

than regular flat belts. Timing belts can transmit significantly more torque than regular flat belts. They provide a much more quiet operation than chain drive systems. They have no backlash (they don't slop when changing directions), so they are ideal in precise positioning systems such as automated and robotic machine tools.

For combat robots, synchronous belts can be used to convert a two-wheel-drive robot into a four-wheel-drive robot, and they can be used for speed reductions. The drawbacks to timing belts are that the costs for the belts and pulleys are fairly high compared to belt systems and chain drive systems, and they require the pulleys to be precisely aligned and in the same plane with each other.

Table 6-2 shows a list of the traditional belt sizes. Table 6-3 shows a list of high-performance belt sizes.

Belt Type	Pitch, Inches
MXL	.080 inch
XL	.200 inch
L	.375 inch
H	.500 inch
XH	.875 inch
XXH	1.250 inch

TABLE 6-2 *Traditional Belt Size Designations*

Belt Type	Pitch, mm	Pitch, in.
2 mm GT	2.0	0.079
3 mm GT	3.0	0.118
5 mm GT	5.0	0.197
3mm HTD	3.0	0.118
5 mm HTD	5.0	0.197

TABLE 6-3 *High-Performance Belt Size Designations*

For a similar belt pitch, the high-performance belts are significantly stronger than the standard belts. Consult belt manufacturers such as Gates Rubber Company or Stock Drive Products to obtain actual belt specifications for your speed and torque requirements. As with chain drive systems, different pitches have different belt designations. A timing belt's ability to transmit torque is based on the belt's power rating (*torque* × *speed*) and the belt's width factor. The baseline width factor for timing belts is 1 inch.

To determine the load-carrying capability of a timing belt, you multiply the power rating of the belt by the belt width factor and divide the result by the rotational speed of the smallest pulley diameter. With timing belts, general relationships can be used to describe the load-carrying capabilities of the belts. You can obtain this information directly from the belt manufacturer, who should also provide a belt design datasheet that will explain how to compute these values directly.

The pulley centerline distances are computed in a similar manner to how the centerline distances are computed with chain drive systems. The calculations are shown in Appendix C. They require more work to implement because the center distances have to be determined after the selection of the timing belt is made. Timing belts are available only in fixed lengths.

V-Belts

V-belts have more of a trapezoidal cross-section, and the pulleys have a *V*-like shape to them. The proper name for a V-belt pulley is a *sheave*. V-belts are the most commonly used type of belt drives. They are seen in virtually every type of machinery where synchronization is not required. Virtually every automobile on the road has at least one V-belt on the engine. They can transmit more power than traditional flat belts because V-belts have two frictional contact surfaces.

V-belts come in two general classifications: standard and high capacity. Five standard sizes are called A, B, C, D, and E, and they range from 1/2-inch wide to 1.5-inches wide. For the high-capacity classifications, the three different sizes are 3V, 5V, and 8V. Their widths range from 3/8- to 1-inch wide. As with timing belts, V-belts come in fixed lengths.

The power transmitting capability of a V-belt is dependent on the belt tension and the angle of wrap around the sheave. The greater the belt tension, the greater the torque transmitting capability. As with synchronous belts, V-belts are available only in fixed lengths. To determine which size of V-belt to use, you should consult the belt specification datasheets from the belt manufacturer.

For combat robots, V-belts could be used for drive belts in the power transmission and for speed reduction applications. But the most common use for V-belts is for driving weapons. As with flat belts, using V-belts in this way will allow the belt to slip if the weapon is stalled. With V-belts, more torque can be transmitted from the motor to the weapons, thus making them more effective than regular flat belts. The belt slippage when the weapon has stalled may be desirable in this situation because the drive motors are protected from complete stall and possible burnout.

Gearboxes

The compact form of a power transmission is to use a gearbox between your motors and wheels. Earlier, we talked about using gearmotors for robots. A gearmotor consists of a gearbox mounted to an electric motor. Inside the gearbox are gears, shafts, bearings, oil/grease, and a rigid case. A gearbox consists of precisely designed components. Within a gearbox there are various configurations of gears to obtain the speed reduction. The common methods consists of spurs gears, planetary gears, helical gears, worm gears (shown in Figure 6-10), or some combination of these gears.

FIGURE 6-10 *A worm gearbox attached to an electric motor. Note the screw-type gear in the center of the gearbox.*

Mounting Gear Assemblies

Now that we've covered gear assemblies and methods of gear reduction, we should mention the relative difficulty of constructing a gear reduction power transmission using off-the-shelf gears. The most difficult part of the process is the extreme precision required in the placement of two adjacent gears. If they are placed too close together, the gears will bind and not turn freely. If the two gears are too far apart, "gear slop" will occur and actual gear slippage might occur. To place the gears at a proper spacing, you must calculate the center distances and make the exact distance measurements of the two shaft's centers on the gearbox, and then carefully drill and bore the holes using a milling machine.

It is best to bolt the two sides of your gearbox together before drilling to ensure that all holes on one side align with those on the other side. You must use ball bearings or bronze bushings to support the gear shafts, and they need to have accurate holes bored to allow the bearings to be pressed in firmly. Remember that when you drill those first holes and later bore them out for your bearing assemblies, any mistakes in placement will mean that you will have to start from scratch, which will mean two new sides for your gearbox. Plan accordingly and measure carefully.

Securing Gears to Shafts

The second difficult part in building your own gearbox is fastening two gears of different diameters on a single shaft. If the gears are to rotate freely like an idler gear, you don't have to fasten them securely to transfer torque between gears. However, if you intend to construct an assembly like the one shown in Figure 6-2, gear number 2 and gear number 3 or sprocket number 2 and sprocket number 3 must be securely attached to the shaft to transmit torque between them. This will more than likely require a hardened steel pin protruding through the gear's hub, through the steel shaft, and through the other side of the hub. If the hub extends on both sides of the gear, a second pin is recommended. You should always use a hardened steel pin—never use a cotter pin.

You must also be careful to align the gears with each other within the gearbox. Securing the shafts against side-to-side slop can be accomplished using collets fastened on the shaft inside the bearings. If this all sounds a bit complicated, you're right. It really is complicated for the first-time machinist. A better way to go to achieve speed reduction is to use a chain and sprockets. The distance between sprocket shaft centers can be a lot less precise to accomplish the same ratio of speed on the sprockets. If your robot needs a gearbox, you should use a gearbox that has already been designed and manufactured.

The most common term for these gearboxes is *speed reducers*. A wide variety of speed reducers are in use, including parallel shaft speed reducers, where the input shaft and the output shaft are parallel. The other general class of speed reducers are called *right-angle drives*. In this type of system, the output shaft is at a right angle with respect to the motor shaft. Many wheelchair motors, windshield wiper motors, and power window motors are right-angle drives.

Most commercial speed reducers use standard mounting methods defined by NEMA (National Electrical Manufacturers Association). Two NEMA general classifications should be considered: the NEMA Frame size, and the NEMA Face size. The NEMA Frame size defines the standard dimensions for the motor mounting holes to secure the motor frame and defines the height of the motor shaft from the mounting surface. The NEMA Face size defines the bolt hole pattern on the front face of the motor. Two NEMA speed reducers are shown in Figure 6-11.

With NEMA face mounting, the speed reducer is bolted directly to the face of the motor. With NEMA frame mounting, both the speed reducer and motor are mounted to the same base plate. The NEMA specifications also specify the motor shaft diameters, shaft length, and the type of keyway. A motor and a speed reducer with the same NEMA specification will fit together like a glove. If you are trying to fit together a motor and a speed reducer that do not have the same NEMA specification, then you will have to build an adapter to mate these two components together.

Right-angle drives are usually made up of worm gears, which can provide high speed reductions. The right-angle drives that use bevel gears usually have low gear reductions.

Speed reducers using spur gears are the lowest cost speed reducers when compared to helical gear speed reducers and planetary gear speed reducers. Helical speed reducers can transmit higher torque than spur gear speed reducers, and they run quieter. But helical speed reducers are sometimes less efficient than regular spur gear speed reducers. Planetary gearboxes offer a high gear reduction in a small package, and Harmonic speed reducers are the most compact speed reducer.

FIGURE 6-11 *Two typical right-angle drive speed reducers using the 56C NEMA standard face size.*

Planetary and Harmonic speed reducers are the most expensive forms of a gearbox. For low- to medium-power robots, one of the most cost-effective methods of obtaining a planetary gearhead is to pull one out of a cordless drill. But when doing this, you will have to build a special mount for the gearmotor because these motors are not designed to be stand-alone gearmotors.

When building combat robots, it is generally a good idea to start with motors or gearmotors and use a chain or belt drive system to increase or decrease the output shaft speed of the motor/gearmotor to drive the robot's wheels, than to design a custom gearbox. If you plan to use a commercially available high-powered electric motor, look for electric motors that use NEMA face and frame mounting methods so that standard gearboxes can be used with them. With a good power transmission, your robot should have all the speed and pushing force it should need in a contest.

chapter 7

Controlling Your Motors

IF batteries are the source of power for a robot, and motors are the source of movement and locomotion, you might consider the electronic speed controller (ESC) the "ringmaster" of all robot systems. The ESC is the device that controls the amount of voltage that goes to the motors and the direction in which the motors turn in your robot. Without an ESC, you cannot control your robot.

The ESC is probably the most critical component in the entire robot, so you must select it carefully. An improperly selected controller will usually result in a short life for your robot and can damage the motors or the batteries. If the ESC fails during a competition, you can pretty much count on losing the match.

This chapter will explain several different approaches to implementing electronic speed and direction controls, including simple relay controls and solid-state electronic variable speed controllers. Each approach has its advantages and disadvantages and should be selected according to the application.

Relay Control

A *relay* is an electric device used to switch a high-powered electric circuit with a low-powered signal. Inside a relay is an electromagnetic coil and a set of movable electric contacts. When power is sent through the relay coil, it creates a magnetic field inside the relay case. The magnetic field then pulls a piece of metal connected to a set of movable electrical contacts into contact with stationary set of contact points—thus making an electric circuit and allowing power to flow to the load. When the power to the coil is interrupted, the magnetic field disappears and a spring pushes the movable contacts back into their original position, breaking the circuit. Figure 7-1 shows a schematic of a typical single-pole double-throw (SPDT) relay (see the next section for a definition of relay types).

Poles and Throws

Relays contain one or more circuits. The number of circuits in a relay are referred to as *poles*. A relay with one circuit is called a *single-pole (SP)* relay. A relay with two circuits is called a *double-pole (DP)* relay.

FIGURE 7-1 *Typical automotive surplus SPDT relay.*

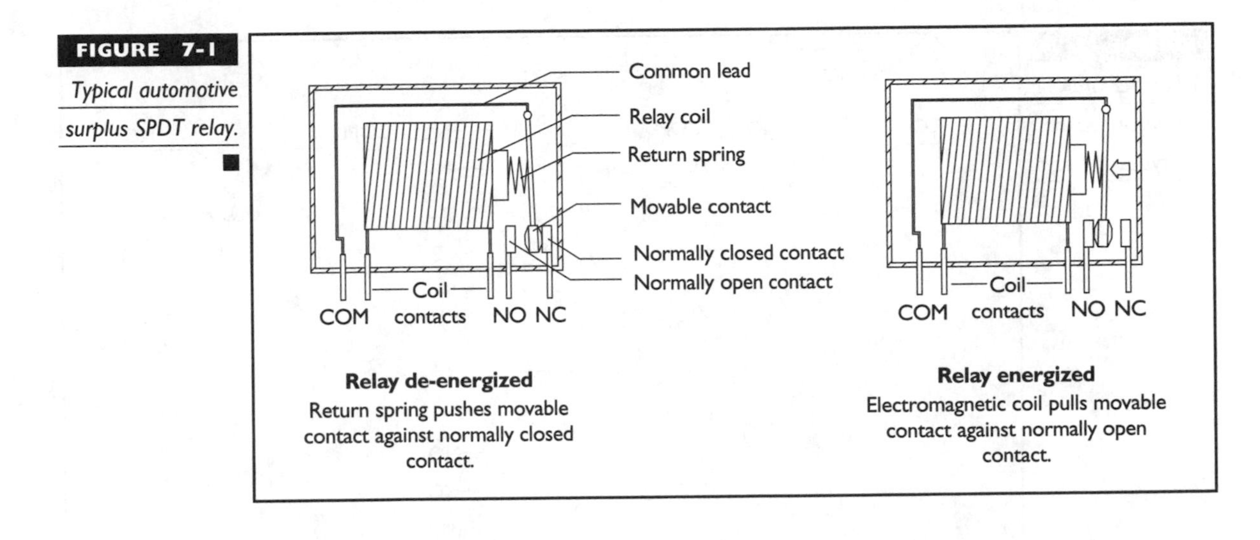

Relays also comprise two kinds of contacts: *normally open (NO)* and *normally closed (NC)* contacts. Normally open contacts (also known as Type A contacts) do not allow power to flow until the relay coil is energized. Normally closed contacts (also known as Type B contacts) allow power to flow when the relay is de-energized, but they break the connection when the relay is energized. Both of these types of relays are called *single-throw (ST)* relays. Many relays contain an NO and an NC contact with one common wire (known as the *COM contact*) between them so that the relay will make one contact and break another when it is energized. This is known as a *double-throw (DT)* relay (also known as a Type C contact).

Most relays are either single- or double-pole relays, and each of these can be either single- or double-throw relays. So relays are usually given a four-letter designation—the first two letters are the number of poles, and the second two are the number of throws. The SPDT relay shown in Figure 7-1 is a single-pole (circuit) double-throw relay.

Figure 7-2 shows the schematic drawings of SPST, SPDT, and DPDT relays. The dashed line between the two contacts in the DPDT relay shows that both contacts move together, but they are not electrically connected to each other.

Current Ratings

When choosing relays to use in your robot, you should first look at and compare each relay's current and coil voltage rating. Relays will have a rating for the amount of current their contacts are designed to switch. The current holding capacity of a relay is much greater than its current switching capacity, and manufacturers usually don't bother giving a rating for the relay's holding capacity.

When a relay breaks the circuit with a significant current flowing, a momentary electrical arc will result between the relay contacts as they separate. The relay contacts

FIGURE 7-2

Schematic drawing of three common types of relays.

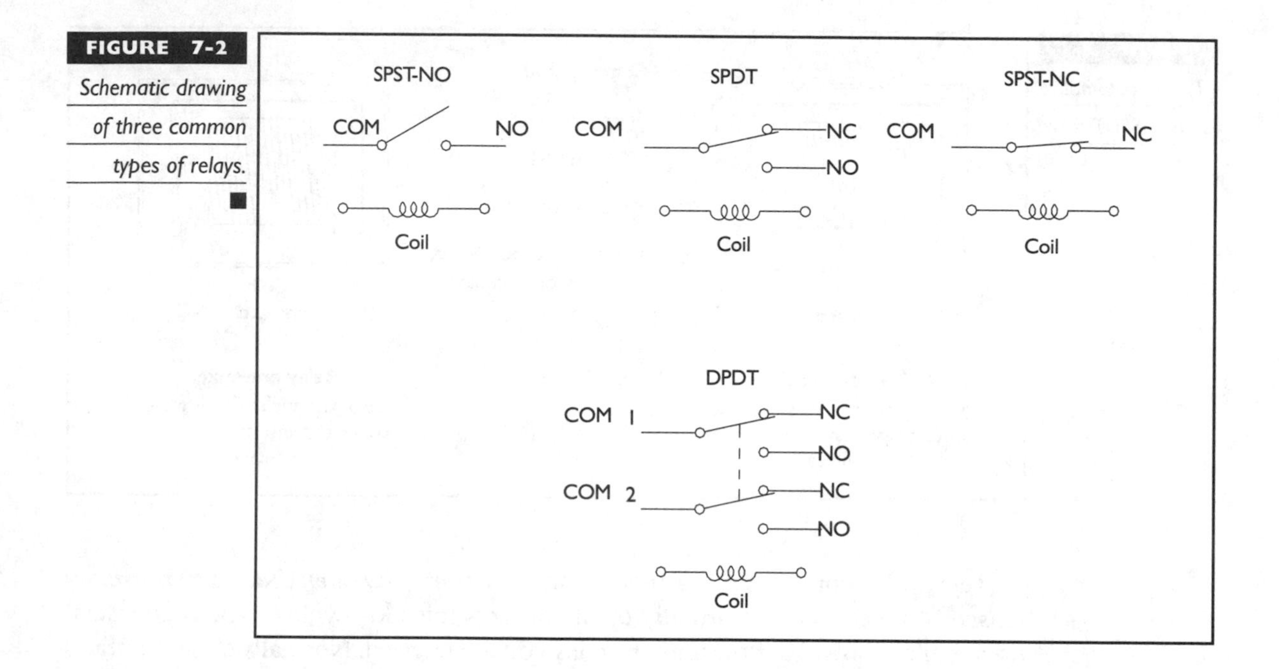

are designed to survive a certain amount of arcing. If you switch a relay while it's carrying more current than it's designed for, the arc can pit and erode the relay contacts. Once damaged, the relay contacts have less effective switching area, making them more likely to be damaged by arcing on the next disconnect. Arcing also occurs when the relay contacts meet; and if the contacts are sufficiently damaged or running current too far over their rating, the contacts can actually weld together on contact. The relay's return spring won't be sufficient to break the contacts apart, and the relay will remain stuck on even when the coil is de-energized.

Welded relay contacts can result in dangerous situations in which a robot fails to shut off—or its weapons won't shut off—and the machine runs wild. To avoid this situation, properly sized relays must be used. The chances of welding the relay contacts greatly increase when the relays are switched on and off rapidly in a short period of time, allowing them to "chatter." Switching generates heat in the relay contacts, and switching the relay contacts repeatedly without letting them cool off makes it more likely that they will weld together.

Motors for combat duty can draw from a few amps (for weak motors) to several hundred amps for major weapon or drive motors in the larger weight classes. Relays for your robot should have high-current capacity, and they should be compact, durable, and easily available. Many relays used in robotic combat are the automotive surplus type, which typically have 12-volt DC or 24-volt DC coils and contacts rated for from 10 to 60 amps. These relays usually have both NO and NC contacts (making them double throw) and should be able to handle most small-sized motor needs.

A relay designed to handle higher current demands is known as a *solenoid* relay. Shown in Figure 7-3, this type of relay uses a solenoid—an electromagnetic

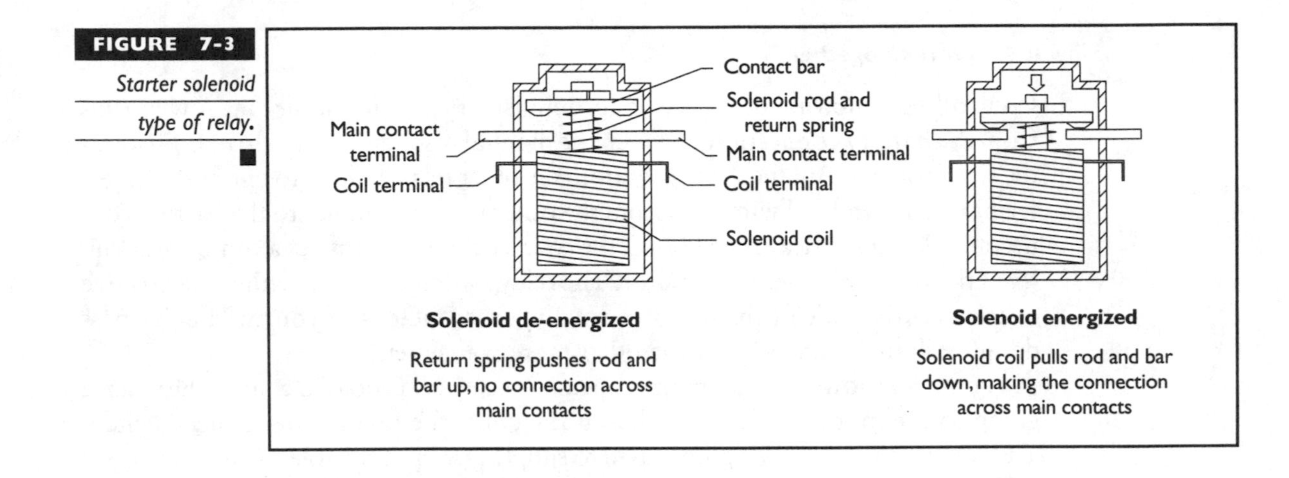

FIGURE 7-3 Starter solenoid type of relay.

coil with a movable metal rod down the middle—to pull a shorting bar across a pair of contacts. Commonly known as *starter solenoids*, they are used for high-current, intermittent-duty applications such as running the starter motor on an internal combustion engine. Industrial starter solenoids are available for power levels of up to 400 amps. Some solenoids have one side of the coil internally connected to one of the internal contacts. These are designed for automotive use, in which the motor circuit and the coil circuit have a common return line to the battery. These solenoids can be used for robot combat applications, provided that the common line is taken into account when designing the electrical system.

One thing you cannot do is connect multiple relays in parallel to get a higher current capacity. The closing of a relay contact is a slow event, as compared to the time it takes for current to start flowing through the motor. Because of manufacturing differences, all the relays would not close at the same time, so the first relay to make contact—or the last relay to break contact when opening the circuit—would take the entire motor load by itself. So a bank of relays wired in parallel can still safely switch only as much current as any single relay acting alone could.

The coil of a relay should be operated at the voltage for which it was designed. Running the coil of a relay on less than its design voltage can result in insufficient pressure on the contacts, reducing the area of metal through which current is flowing and increasing the chances of welding. Running a relay coil on more than its intended voltage can result in the coil burning up and overheating, especially on relays designed for intermittent use. Running the relay coil on more than its intended voltage doesn't offer any advantage in reliability or performance, although it may make the robot's wiring simpler if the motors are being run off a different voltage than the relay was designed for. For the duration of a typical combat match, most relays can survive twice their intended operating voltage, although this should be tested prior to a match. The voltage polarity applied to the relay coil itself usually doesn't matter, but some relays have diodes internally wired across the coil connections and must be connected with the appropriate polarity.

How It All Works Together

Controlling a motor with a relay is accomplished via a simple circuit. A wire runs from the battery connection, through the manual disconnect switch, to one side of the relay contact. Another wire goes from the other relay contact to one of the motor terminals, and a final wire runs from the other motor terminal to the battery connection. When the relay is energized, the contact closes and makes a complete circuit from the battery through the motor. The relay switch can be on either the positive or the negative side of the motor—usually, other factors of your wiring harness design will make one way or the other more convenient.

Figure 7-4 shows a simple wiring schematic using a solenoid to control the voltage going to the motor. This figure does not include the manual disconnect switch. The control switch in the figure is used to supply power to the solenoid's coil to open and close the circuit.

A manual disconnect switch physically disconnects the batteries from the rest of the robot. For safety purposes, a disconnect switch should be placed in all combat robots. You do not want the robot to be accidentally turned on by you or another person while you're working on the robot; and sometimes a short can occur during maintenance, which will cause a motor to turn.

Many robot contests, such as *BattleBots*, *Robot Wars*, and *Robotica*, require that a manual disconnect switch (sometimes called a *kill* switch) be installed in all

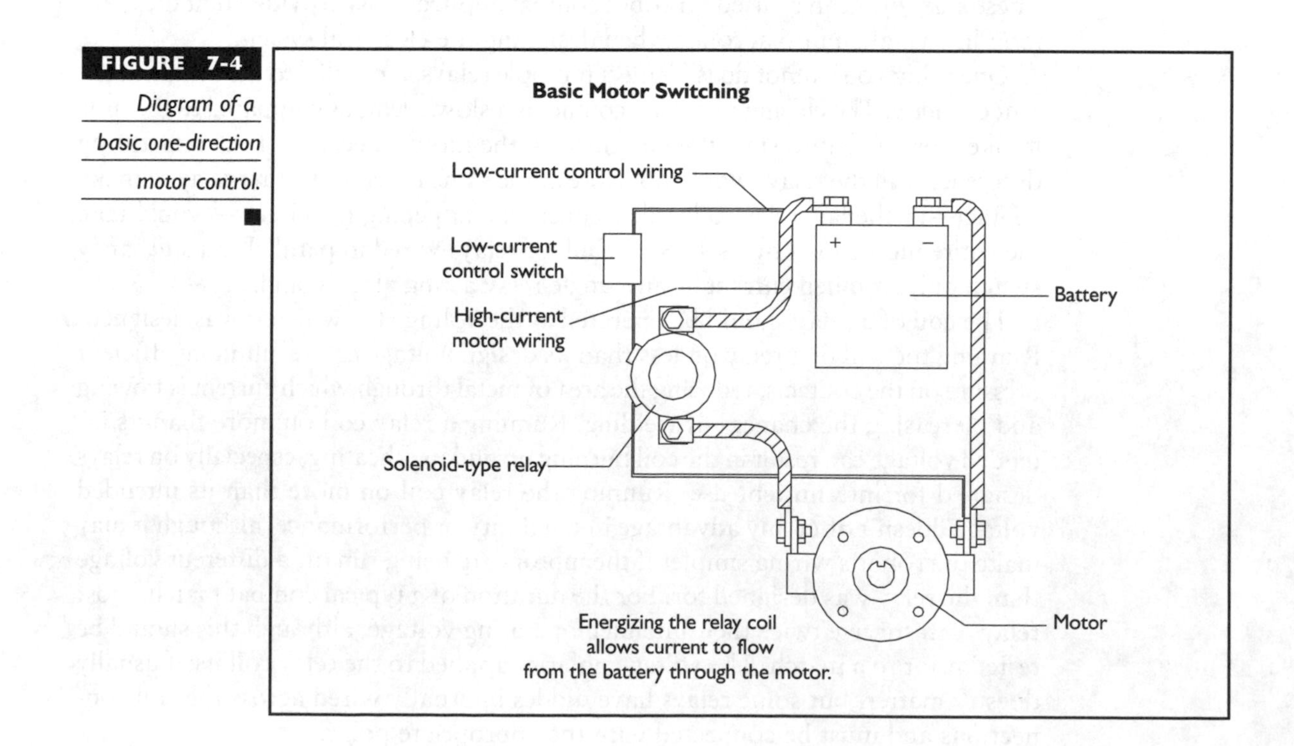

FIGURE 7-4 *Diagram of a basic one-direction motor control.*

competing robots between the batteries and drive and weapons motors. The manual switch must be rated to safely handle the current that will pass through the switch, which can be more than 100 amps. Team Delta (*www.temadelta.com*) sells several types of manual disconnect switches, in addition to a device called a *removable link*, which is a physical wire connected through a plug that can be physically pulled out of the receptacle to break the electrical connection.

Some weapon designs will require that you actively stop the weapon when it is not running. Large spinning weapons, for example, may need to be actively braked to spin down fast enough to be compliant with competition rules. A permanent magnet DC motor will act as a brake if its leads are shorted together. To get this effect on your combat robot, you will need to add a second relay wired to short the motor's leads together when you want the weapon to stop. Figure 7-5 illustrates how to implement braking on an electric motor.

caution *Tuke great care with wiring so that the braking relay and the motor-run relay can never be energized at the same time. This will result in a dead short across the battery that could result in fire, smoke, and a dead robot.*

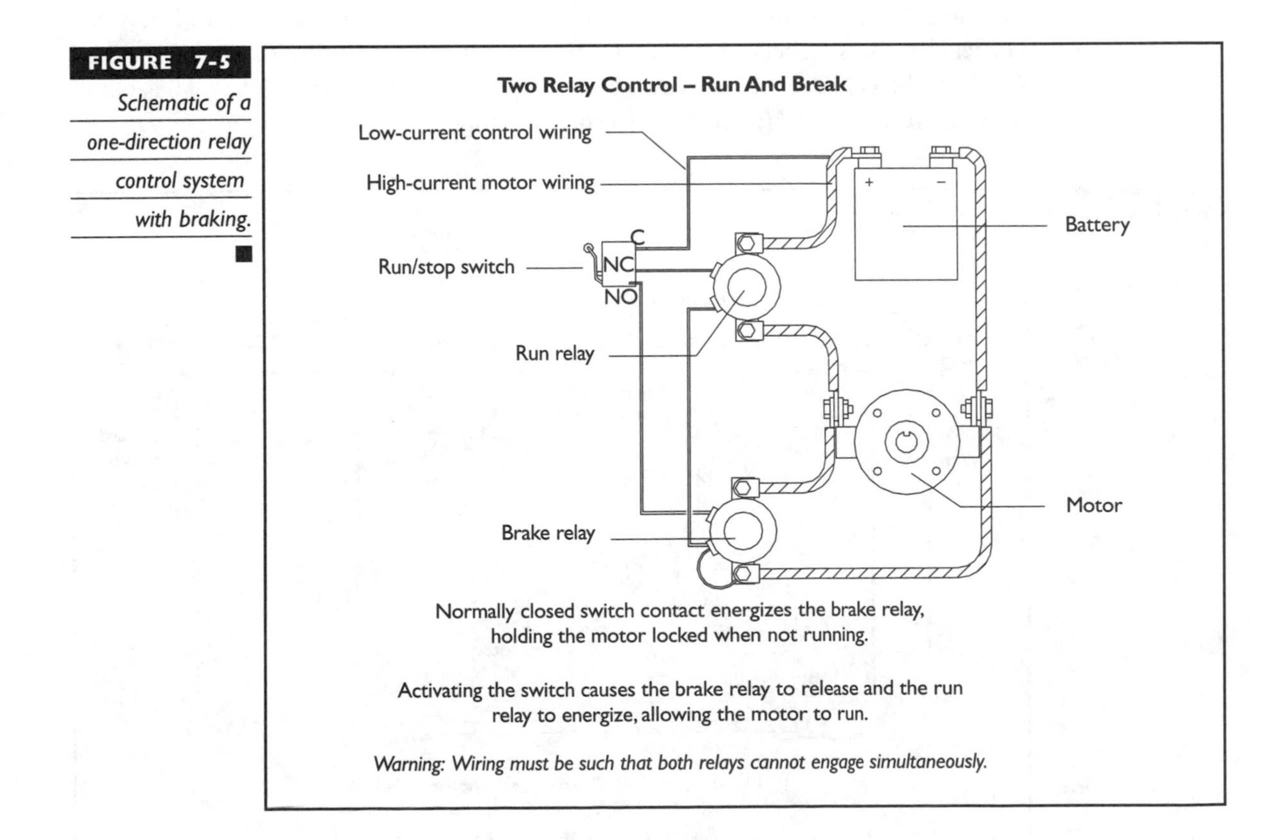

FIGURE 7-5 *Schematic of a one-direction relay control system with braking.*

Driving with an H-Bridge

Relay control gives you only two speeds—full speed or stopped. Some weapon systems require that you reverse the direction of the motor, and the motors of your robot's drive train will also need to be reversible. Running a motor in both directions will necessitate that you switch both sides of the motor between the plus and minus sides of the battery. The circuit for doing this is called an *H-bridge*. An H-bridge gives you the ability to reverse direction, but you'll still be going full speed in whichever direction you choose. When can you get away with this?

Most weapons don't need more than simple on/off control. A saw or spinner weapon usually needs a single relay to switch it on or off. Large high-inertia spinners may need a second relay for braking purposes. Hammer and lifting arm weapons will need an H-bridge arrangement for reversing direction, but they usually do not need to run at variable speeds. An H-bridge using solenoids for motor control is shown in Figure 7-6.

An H-bridge uses four relays, one from each motor terminal to each battery terminal. In Figure 7-6, relays A and B connect one motor terminal to the positive and negative sides of the battery, respectively, and relays C and D connect the other side of the motor to the positive and negative sides of the battery. When you look at Figure 7-6, imagine a vertical line passing between relays A and B, and a vertical line passing between relays C and D. Then imagine a horizontal line passing through the center of the motor, connecting to the two vertical lines. These lines now form the letter *H*; hence the term, H-bridge.

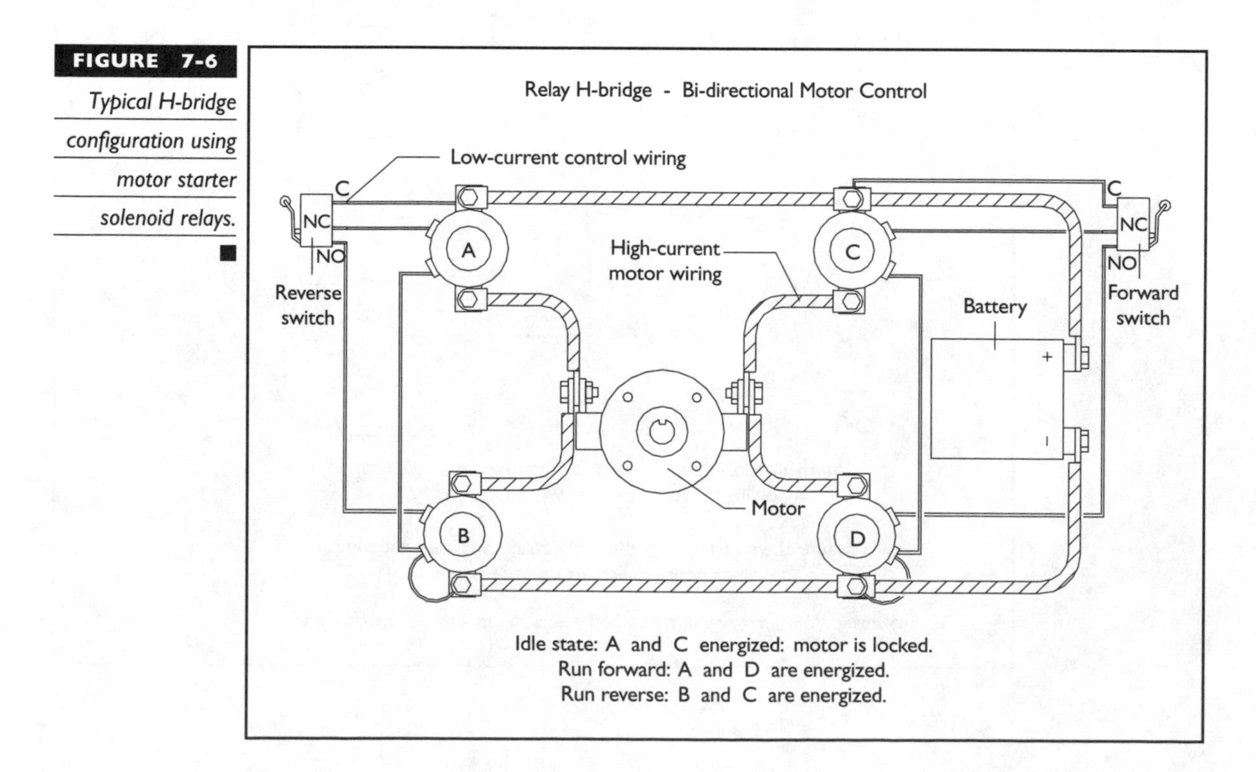

FIGURE 7-6 *Typical H-bridge configuration using motor starter solenoid relays.*

In Figure 7-6, we assume two low-current, SPDT control switches to drive the relay coils, although smaller relays with Type C (or SPDT) contacts can be used in place of the switches. Note that just like the relays described in the previous section, the control switches have NO, NC, and common terminals. In the resting state, the NC legs on the forward and reverse switches result in relays A and C being energized and relays B and D being de-energized. No battery current can flow through the motor because no path exists from the motor to the negative terminal of the battery, and the motor terminals are shorted to each other through relays A and C. The motor is stationary and locked in place.

To run the motor forward, the forward switch is activated, which causes relay C to de-energize and relay D becomes energized. The motor now has one terminal connected to the positive side of the battery through relay A, and the other terminal connected to the negative side of the battery through relay D. This makes a complete circuit and causes the motor to run. To run the motor in reverse, the reverse switch is activated, causing a current flow from the battery, through relay C into the motor and out through relay B into the other side of the motor.

note *If both the forward-going and reverse-going switches are activated, the circuit path will be broken and the motor terminals will be shorted together.*

A significant danger of relay control is the possibility of contacts bouncing on severe impact that a combat robot will receive during a battle. A severe shock impact in a direction relative to the relay orientation can be sufficient to overcome the force of the return spring holding the contact bar out, thus causing a momentary connection across the relay's contacts. Having a weapon motor switch on for a moment might not be a catastrophic event, but it can be dangerous if people are nearby and a weapon starts to move. If a momentary short occurs within the motor braking relay while the motor is running, or if one of the nonactive relays in the H-bridge is shorted while the other side of the H-bridge is active, a dead short across the main motor batteries will result. In the relay circuit shown Figure 7-6, this can happen even when the motor is not running—because half the relays in the circuit are always energized, a momentary contact bounce of any of the non-energized relays will cause a catastrophic short. The dead-short battery current will inevitably weld the contacts together, resulting in the entire wiring harness going up in smoke and one dead robot.

Turning Switches On and Off

In a remote controlled robot, you will need a way to turn switches on and off remotely. This can be done either electronically or mechanically. The electronic approach will be discussed in the solid-state logic section. A mechanical approach will require some form of an actuator to turn the switch on and off physically. One of the cheapest and easiest ways to mechanically actuate a switch is to simply use a standard hobby radio-controlled (R/C) servo to throw a switch.

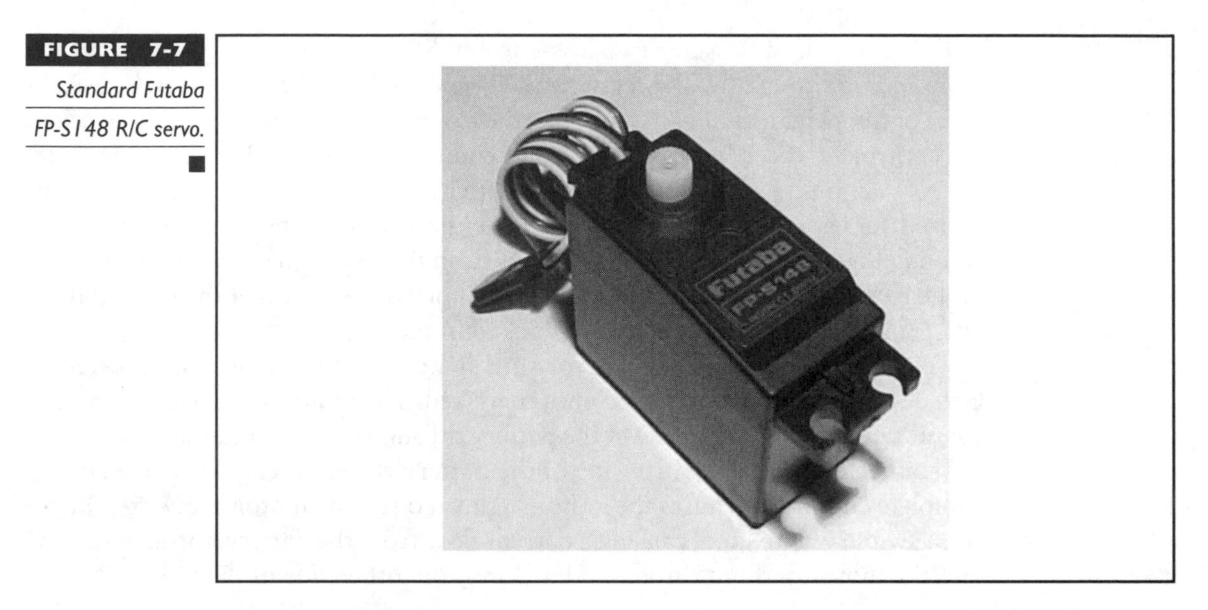

FIGURE 7-7 *Standard Futaba FP-S148 R/C servo.*

Radio-Controlled Servos The R/C servo discussed here is the same type of servo that is commonly found in R/C model airplanes. Figure 7-7 shows a photograph of one of these servos.

The servo will respond to the signal from the radio transmitter by rotating its output shaft to various commanded positions. A servo arm (commonly called a servo *horn*) attached to the output shaft can be used to move a switch to an on or off position, which can supply power to the coils of the relays. The most reliable way to do this is to use a roller-type lever switch and a round servo horn manually cut into an egg shape. By doing this, the servo horn is being converted into a cam. Two lever switches positioned on opposite sides of the servo can be used to trigger two different motor circuits, or to drive a single motor in forward or reverse direction. The basic R/C servo configuration is shown in Figure 7-8. Microswitches can be used to drive small motors or to switch relays for driving larger motors.

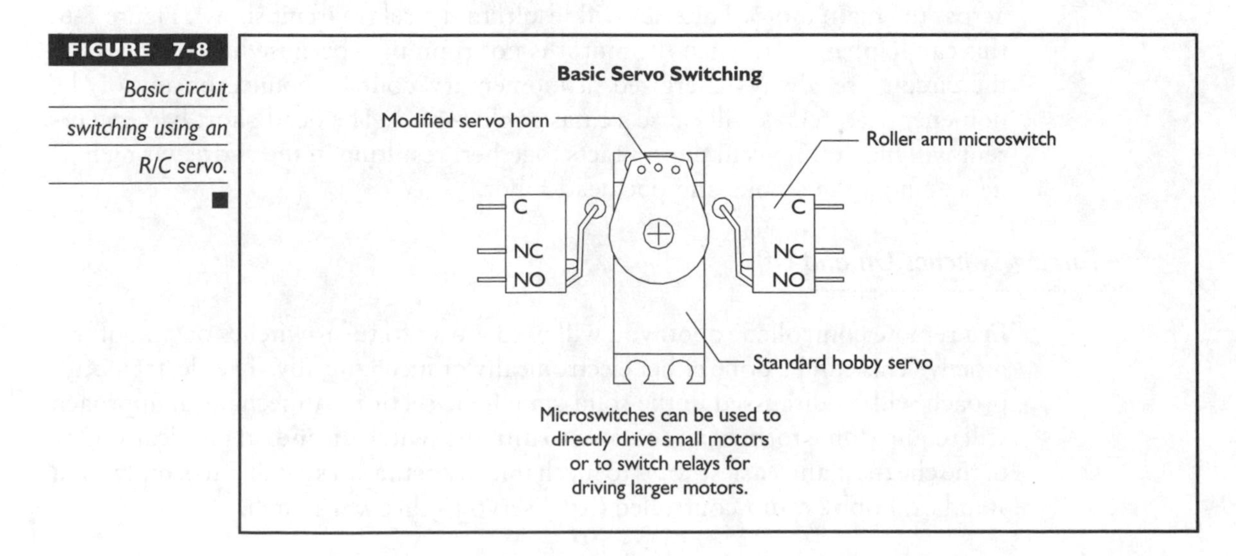

FIGURE 7-8 *Basic circuit switching using an R/C servo.*

Servo switching was quite common in the early days of robotic combat, but using it has many drawbacks and is not recommended.

- The response time of a servo is fast, but the time it takes the servo to rotate and trigger the lever switch will add a perceptible lag to the motor's activation. A half-second lag in your robot's response can make a big difference in the arena.
- Servo switching introduces extra moving parts into your control system that can break or jam and cause the motor to stop working or, even worse, turn on and refuse to turn off.
- A servo switching system will have trouble meeting fail-safe requirements present in most competition rules. Depending on your radio type, loss of signal may result in all servos connected to the radio simply locking in place. If the motor was on when contact was lost, it'll stay on until you can switch the bot off manually. Even if your radio has the feature of returning all the servos to the neutral position if radio signal is lost, loss of power in your radio receiver or a severed connection between the receiver and the servo can still result in a motor stuck running.

caution *For safety reasons, servo switching should not be used for controlling drive motors or weapons that can injure someone if the servos or relays should fail.*

Remember that you must have absolute control of your robot at all times and you must be able to shut it off remotely even if internal control parts break inside. Servo switching can be used for applications in which failures are not safety issues, such as for an arm that turns your robot right side up or an electrically driven lifting arm.

Solid-State Logic A better method to control the relays is to use *solid-state logic* to interpret the control signal from the radio and trigger the relays when the appropriate signal is received. You can use a programmable microcontroller, such as the Basic Stamp from Parallax, Inc., and program it to receive the command signal from the R/C receiver and convert that signal into an output signal. The output signal is then used to turn a transistor on or off, and the transistor is used to supply power to the relay coils.

Figure 7-9 shows a simple schematic that illustrates transistor-relay control. In the figure, a low-voltage signal is used to turn a transistor on and off. The schematic drawing shown on the left is an NPN transistor. A positive voltage to the transistor base (shown as a B on the transistor) will turn it on and the relay will be energized. The schematic to the right uses a PNP transistor. In this schematic, the relay coil is energized when there is no voltage signal to the base. An NPN transistor is analogous to a NO-SPST switch, and a PNP transistor is analogous to a NC-SPST switch. A "*flyback*" diode is required to protect the transistor when the relay is de-energized. At the instant a relay coil is de-energized, the magnetic field in the coil collapses. A collapsing magnetic field will create a momentary current spike, which will induce a voltage spike that will exceed the original voltage that was in the coil. This spike can damage the transistor. By adding a diode in parallel with the coil, the diode will allow a path for the current flow back to the original source, thus protecting transistor. When a diode is used in this application, it is called a *flyback diode*.

Another solution is to use solid-state relays instead of using the transistor approach. Solid-state relays come in small plastic enclosures that are about 2 inches square in size. A low-current, 5-volt signal will open or close the circuit. Depending on the model, it can handle currents up to 40 amps. For low-powered applications, a solid-state relay can be used instead of electromechanical relays such as solenoids.

Fortunately for the less electronically astute, off-the-shelf solutions are available. For example, Team Delta (*www.teamdelta.com*) sells four types of simple remote controlled switching boards that are used in many combat robots. The R/CE200 is a single-output control board that uses a transistor driver to run a load of up to 9 amps—enough to run most relays. The R/CE210 is a relay module that can switch a load of up to 24 amps, enough to run smaller motors. The R/CE220 and R/CE225 interface boards are dual-relay controllers with ratings of 12 and 24 amps, respectively. These controllers can switch two independent motors or can be wired in an H-bridge configuration to run one motor in forward and reverse. The R/CE220 and RDE225 boards can also be used as a switch to control the coils on larger solenoids to control a higher-powered motor, or they can be configured as an H-bridge for low-powered motors. Figure 7-10 illustrates this type of a setup.

When using relays to drive motors, it is recommended that you use fuses between the relays and the batteries for all non-drive motors. Due to the harsh environment combat robots operate in, shock impacts of weapons damage may cause a relay to momentarily short out. If this happens, the batteries will be destroyed.

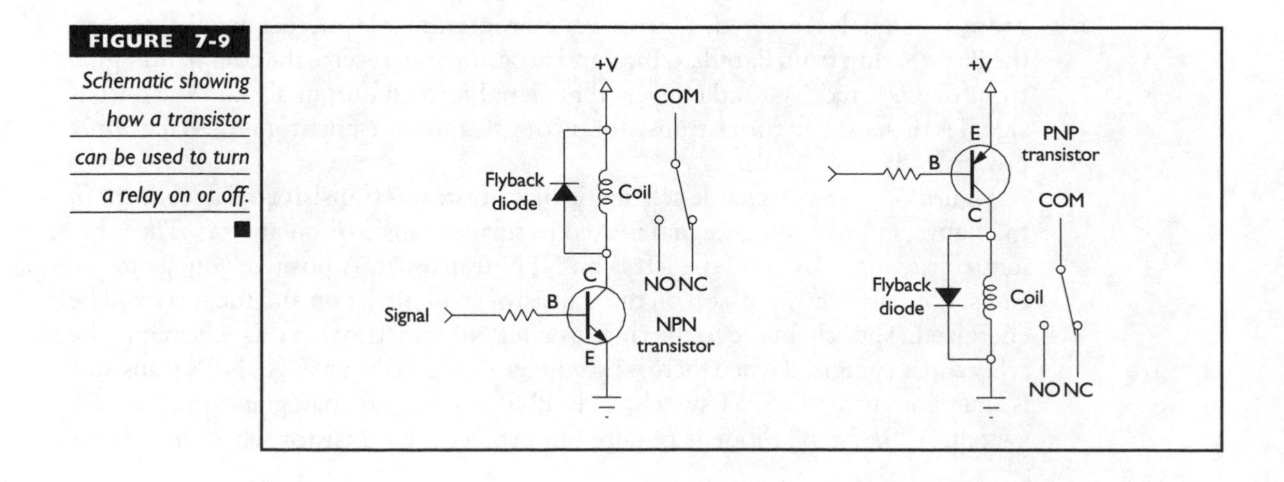

FIGURE 7-9 Schematic showing how a transistor can be used to turn a relay on or off.

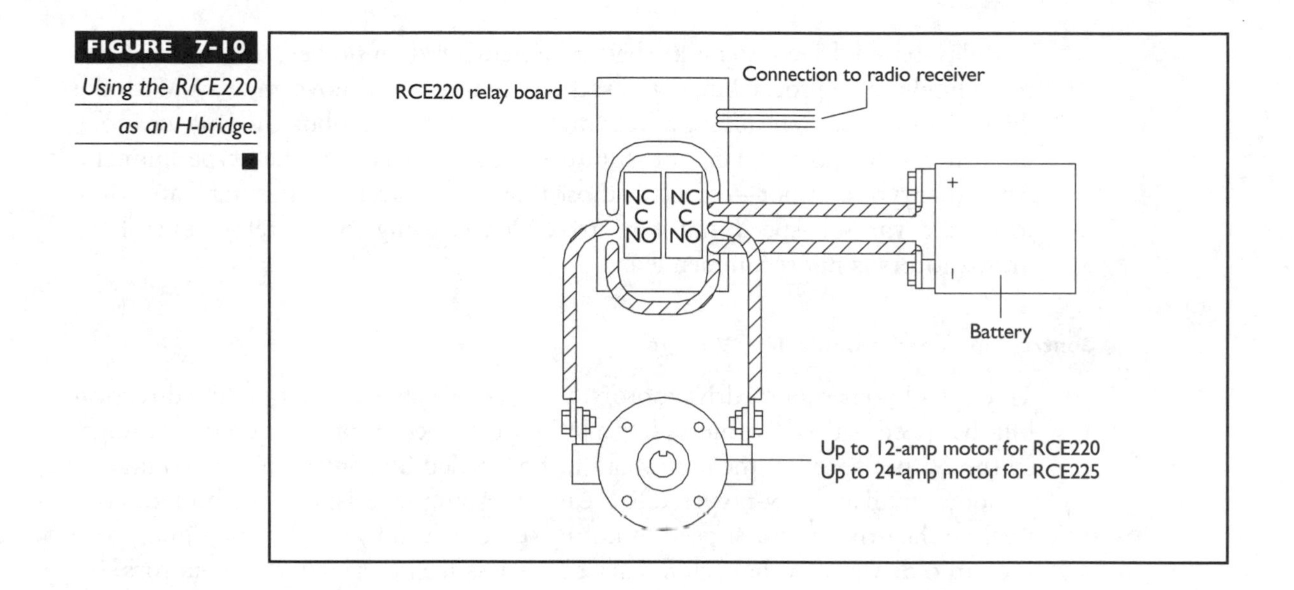

FIGURE 7-10 *Using the R/CE220 as an H-bridge.*

The fuses should have a higher amp rating than the maximum amp rating on the motors. The fuse(s) should be placed where it will cut power only to the single relay-motor set—in other words, use one fuse per non-drive relay controlled motor. You can place fuses on the drive motors; but most experienced robot builders do not do this because, if a drive motor fuse blew, the motor will stop and the robot will immediately lose the match. Many combat robot builders would rather lose a match due to a burned-out motor or battery than a blown fuse.

When testing the robot, you should use fuses with the drive motors. You do not want to take the chance of damaging drive motors and batteries during testing runs. Weapon systems are a different matter, however. A burned-out weapon system doesn't mean the robot loses the match. It can still continue to fight on. So, using fuses with weapons systems to protect the rest of the robot is highly recommended.

Variable Speed Control Basics

If you use relay control with your drive motors, your robot will need to drive at full speed whenever it's moving. This might not seem like a great disadvantage, but turning your robot around when going full-blast and accurately lining up on your opponent is a difficult task. Relay-only drives should never be considered for a two-wheeled robot because turning accurately would be extremely difficult. Four-wheeled robots are more amenable to relay-controlled drives, since their steering usually has a higher amount of friction when turning because all wheels are slip-steering. This higher amount of friction helps reduce the overshoot from relay-controlled drives.

Relay-based drive systems are better implemented on slower robots, which are more likely to be proceeding at full speed whenever they move anyway. With the difficulty of accurately aiming a weapon on a relay-based robot, the only weapons used should be those that do not require aiming, such as large shell-type spinners. Any other type of robot—especially those that require accurate steering—are going to need a variable-speed motor control. Hence, using simple relay control for drive motors is not recommended.

Controlling Speed = Controlling Voltage

To control your robot's drive motors, you need to change not only the direction but the speed of the drive motors. In a DC motor, speed is proportional to voltage, so the output speed of the motor can be controlled by controlling the voltage.

Some small and low-powered R/C cars use a simple resistance method for controlling the drive motor's speed. A hobby servo driven by the throttle signal from the radio drives a mechanism to vary the resistance in series with the motors. Either a sliding wiper arm on a variable resistance strip or a set of contacts to switch the motor power through fixed resistors is used to give a varying speed. This method works for small motors with large amounts of airflow available for cooling, but it should never be considered for combat robot systems. A motor used in a combat robot could draw continuous currents in the tens to hundreds of amps—a variable resistor or bank of fixed resistors large enough to handle the required power levels would be impractically large and fragile.

One method for changing the voltage to a motor is to use a bank of batteries tapped at multiple locations within the battery bank to obtain multiple voltage levels, and to use relays to switch by which voltage point the motor is driven. For example, if your robot is powered by a 24-volt motor that is broken down into two 12-volt packs, you could use a single Type C relay to switch your motor between running off a single 12-volt battery or both in series. This would give you high- and low-speed settings. If you break your battery pack into more segments and add additional relays for each voltage tap, you can approximate the effect of continuous control over your robot's speed. This method has been used by several teams, with usually only two or three different speeds.

It does have the advantage of reliability if done correctly. The downside is that each relay must be rated for the full stall current of the robot's drive motors, and the large number of relays needed for good multistep control can make this an expensive approach. The wiring and control logic involved can also get pretty complex when combined with an H-bridge setup for direction control. In addition, unless the robot is operating at full speed most of the time, the extra batteries are just dead weight that could otherwise be better put to use in weapons or armor.

Pulse-Width Modulation

Most combat robots use a method known as *pulse-width modulation (PWM)* for controlling motor speed. A PWM control fools the motor into thinking it's being fed a variable voltage by switching the motor power on and off many times per second. The frequency of the switching is usually held constant while the percentage of time the switch is on or off is used to vary the desired output voltage. Figure 7-11 shows a typical PWM signal.

The percent of the time the switch is on is known as the *duty cycle*. The duty cycle is defined as the on time, t_{on}, divided by the sum of the on time and the off time, t_{off}. See Equation 1. The PWM frequency is the inverse of the time for one complete on-off cycle.

$$\text{Duty cycle} = \frac{t_{on}}{t_{on} + t_{off}}$$

The duty cycle is generally expressed as a percentage. For 10-percent duty cycle, the switch will be on 10 percent of the time and off the other 90 percent of the time. Fifty percent duty cycle will have the switch on half the time and off half the time, and with 100-percent duty cycle, the switch will be on all the time.

Because the windings inside the motor act like an inductor, when the power is cut off to the motor, the magnetic fields inside the windings collapse. The changing magnetic field induces a current through the windings for a short period of time. When a source voltage (the battery voltage, for example) is pulsed to the motor, the motor will, in effect, time average that voltage. When the frequency of the pulsed voltage to the motor is high enough, the voltage time average will be proportional to the duty cycle. Thus, the average voltage is equivalent to the source voltage multiplied by the duty cycle.

To produce the effect of a smooth output voltage, the PWM switch must be switching thousands of times per second. This is much too fast for any mechanical relay to function. PWM applications with relays have been attempted, with a switching speed of about 10 times per second, but this gives poor control and quickly destroys the relay contacts. Power switching at the speed required for good PWM control requires a high-speed, high-power transistor.

Transistors act like switches or simple relays. They are reliable and can switch thousands to millions of times per second. Most transistors cannot handle the high currents that relays and solenoids can handle without burning up. The two most popular types of transistors that are designed for high-powered applications

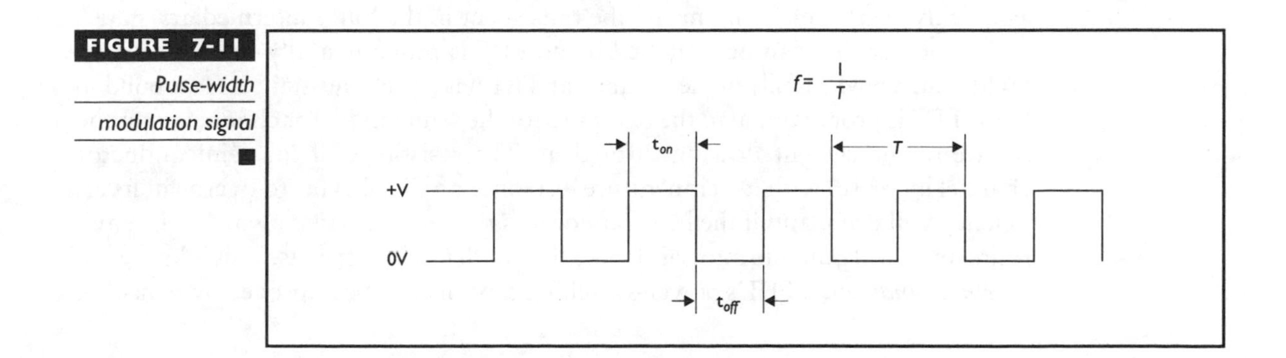

FIGURE 7-11 Pulse-width modulation signal

are called the Field Effect Transistor (FET) and the Metal Oxide Semiconductor Field Effect Transistor (MOSFET). For the following discussions, FET will be used as a generic term to represent both MOSFETs and FETs.

Field Effect Transistor

An FET works something like a semiconductor implementation of a relay. An FET has two leads, known as the *source* and the *drain*, connected to a channel of semiconductor material. The composition of the material is such that current cannot normally flow through it. A third lead, called the *gate*, is connected to a conductive electrode that lies on top of the semiconductor junction but is insulated from it by a thin non-conducting layer. When voltage is applied to the third electrode, it creates an electric field that rearranges the electrons in the semiconductor junction. With the field present, current is able to flow between the source and drain pins. When the gate is driven to a low voltage, the electric field reverses and current is unable to flow. The FET acts as a voltage-controlled switch, where an applied voltage to the gate will control the current flow between the drain and source.

The layer of insulation between the gate and the source/drain channel must be very thin for sufficient field strength to reach from the gate into the semiconductor channel. This thinness makes the FET vulnerable to being damaged by too high a voltage. If the voltage between either the drain or source and the gate exceeds the breakdown voltage of the insulation layer, it will punch a hole through the layer and short the gate to the motor or battery circuit. This can be caused by connecting the FET up to too high a voltage, or simply by zapping the FET circuit with static electricity. You should be careful when handling FETs and attached electronics to avoid accidentally discharging static electricity into them. It is also good practice to use FETs with a voltage rating of twice the battery voltage you wish to run your motors on to avoid the possibility of inductive spikes momentarily exceeding the FET breakdown rating.

When using an FET as a high-current PWM switch, it is important that you switch the gate from the off voltage to the on voltage as quickly as possible. When at an intermediary state, the FET will act as a resistor, conducting current inefficiently and generating heat. Commercial PWM FET-based controllers use specialized high-current driver chips to slam the FET gates from low to high voltage and back as quickly as possible, minimizing the time spent in the lousy intermediary state.

The power that can be switched by an FET is fundamentally limited by heat buildup. Even when fully in the on state, an FET has a slight resistance. Heat buildup in the FET is proportional to the resistance of the semiconductor channel times the square of the current flowing through it. The resistance of the semiconductor channel increases with its temperature—so once an FET begins to overheat, its efficiency will drop; and if the heat cannot be sufficiently carried away by the environment, it will generate more and more heat until it self-destructs. This is known as *thermal runaway*. A FET's power-switching capacity can be improved by removing the heat from it more quickly, either by providing airflow with cooling fans or by attaching the FET to a large heat sink, or both.

The current capacity of an FET switching system can also be increased by wiring multiple FETs together in parallel. Unlike relays, FETs can be switched on and off in microseconds, so there is little possibility of one FET switching on before the others and having to carry the entire current load by itself. FETs also automatically load-share—because the resistance of an FET increases with temperature, any FET that is carrying more current than the others will heat up and increase its resistance, which will decrease its current share. Most high-powered commercial electronic speed controllers use banks of multiple FETs wired in parallel to handle high currents.

Bi-directional and variable-speed control of a motor can be accomplished with a single bank of PWM-control FETs and a relay H-bridge for direction switching, or with four banks of FETs arranged in an H-bridge. A purely solid-state control with no relays is preferable but electronically more difficult to implement. Building a reliable electronic controller is a surprisingly difficult task that often takes longer to get to work than it did to put the rest of the robot together. The design and construction of a radio controlled electronic speed controller is an involved project that could warrant an entire book of its own.

Commercial Electronic Speed Controllers

Fortunately, several commercial off-the-shelf speed controller solutions are readily available for the combat robot builder. Several companies make FET-based motor controllers designed to interface directly to hobby R/C gear; and many brands of commercial motor drivers and servo amps, with some engineering work, can be adapted to run in combat robots. Building a motor controller from scratch will usually end up costing you more money and more time than buying an off-the-shelf model, so there is little reason for a robot builder to use anything other than a pre-made motor control system.

Hobby Electronic Speed Controllers

Hobby ESCs were originally designed to control model race cars and boats. Early R/C cars often had gas-powered engines, but refinements in electric motors and the use of nickel-cadmium rechargeable batteries saw a switchover to electric drive cars. The first systems used a standard R/C servo to turn a *rheostat* (a high-power version of a potentiometer) in series with the drive motor to control the speed of a race car. This system had a bad feature, in that the rheostat literally "burned away" excess power in all settings except for full speed. Needless to say, this did not help the racing life of the batteries.

There had to be a better way to conserve battery life and allow better control of the motors. The result was the hobby electronic speed controller. All of the major R/C system manufacturers are now producing various styles and capacities of

ESCs. These controllers typically have only one or two FETs per leg of the H-bridge, and most use a small extruded aluminum heat sink to dissipate the heat from the FETs.

These controllers are intended for use in single-motor models. The initial units had only forward speed as model boats and cars rarely ever had to reverse. Their technical specifications were geared for the model racing hobby using NiCad batteries and were written accordingly for non-technical people. To this day, most of the manufacturers still specify the "number of cells," rather than the minimum and maximum voltage requirements of a particular ESC, and use the term "number of windings" (on the motor's armature) as a measurement of current capacity. This can be confusing to those who feel comfortable with the terms "volts" and "amps."

Figure 7-12 shows a block diagram of a hobby electronic speed controller.

The number of cells designation literally means you can multiply that number by 1.2 volts to get the actual minimum and maximum voltage requirements of the particular ESC. You must remember that many of the cars used stacks of AA or sub-C cells packaged in a shrink-wrapped plastic cover and were rated at about 9.6 volts (eight cells) maximum. Few cars used 10 cells to arrive at 12 volts, the basic starting point for robot systems.

Many model boats use motors that draw relatively high currents, as do most competition race cars. Most of the specifications for standard ESC's speak of "16-turn" windings for the DC permanent magnet motors as being the norm. This

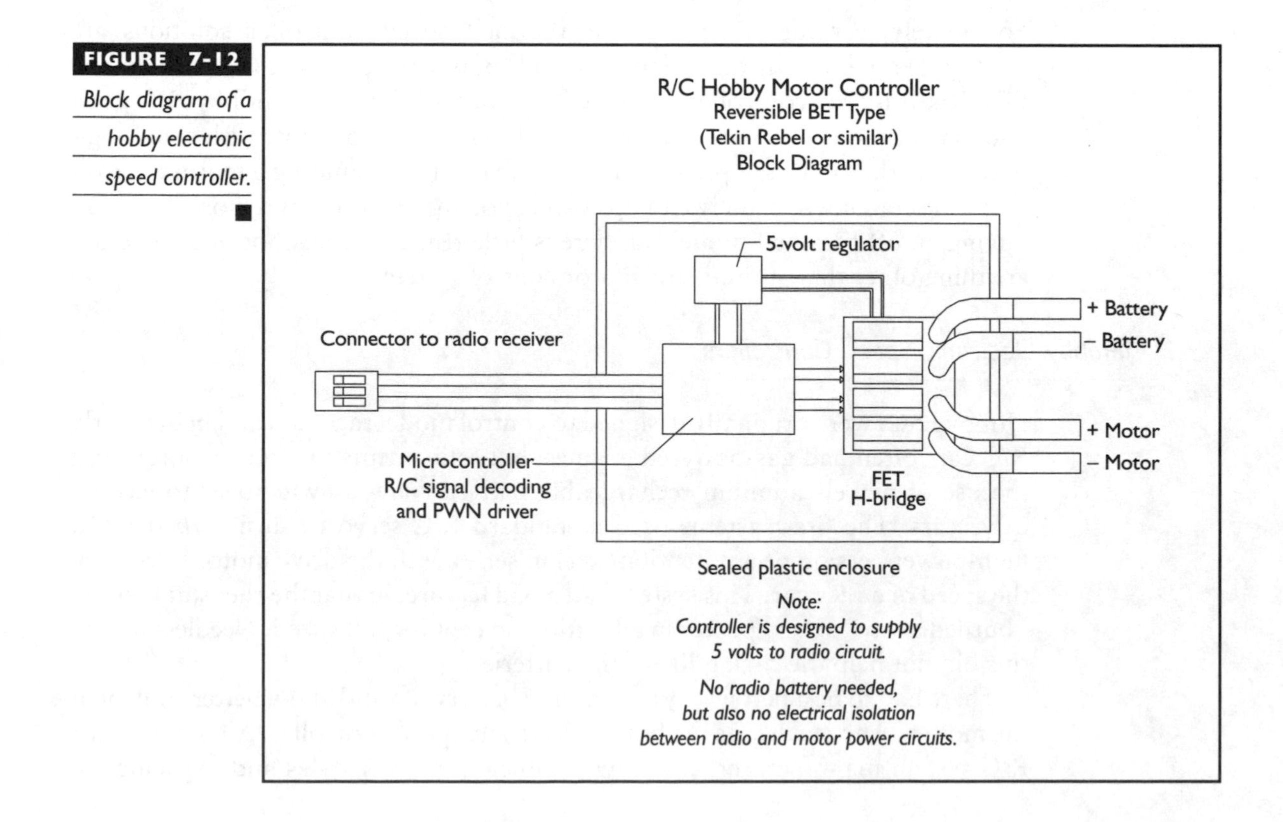

FIGURE 7-12 *Block diagram of a hobby electronic speed controller.*

means that each of the poles of the motor's armature has 16 turns of wire wrapped around the pole. As the number of turns decreases, the diameter of the wires increases, which results in a higher torque motor that has a higher current draw.

Current Capacity in Hobby ESCs True current capacity of a hobby ESC can be difficult to determine; and the ratings given by the manufacturer are generally misrepresented, since they reflect the instantaneous peak current capacity of the semiconductor material in the FETs rather than a realistic measure of the current the controller can handle. Real current capacity of a hobby motor controller will be determined largely by the builder's ability to ensure that the little heat sink on the speed controller stays cool enough to keep the electronics inside from cooking. Since most hobby controllers are designed for low-average currents and with a high airflow in mind, continuous high-current operation will likely cook a hobby controller even with cooling fans installed.

Many of the cheaper hobby controllers are non-reversible, which means that they're designed for running the motor in one direction only. These controllers should not be used in a combat robot. Hobby controllers that are reversible usually have a lower current rating in reverse than in forward—the FETs used in the reverse-going side of the H-bridge have a lower current capacity than the forward-going FETs. Many hobby controllers designed for R/C car or truck use have a built-in reverse delay, so that, when the throttle goes from forward to reverse quickly, the controller will brake the motor for a preset interval before starting to reverse. In an R/C car, this helps controllability and lengthens the life of the motor and geartrain; but in a combat robot, it can make smoothly controlled driving difficult—if not impossible.

Many hobby-type controllers have what is known as a *battery eliminator circuit (BEC)*. The speed controller contains an internal 5-volt *regulator* that generates the power for the electronics inside the speed controller. This power is then fed out through the ESC with the intention being the ability to power the R/C receiver from the main drive batteries. While this is a great help in an R/C car, where the extra weight of a radio battery can make a real performance difference, the more powerful drive motors of a competition robot create a lot more electrical noise that can cause radio interference in the receiver. A robot builder can defeat the BEC by popping the power pin out of the ESC's servo connector and then use a separate battery pack to supply power to the receiver.

Hobby ESCs in Combat Robotics Hobby ESCs have been proven to be usable in small combat robots. These are usually seen in weight classes of 30 pounds and under, but rarely in larger robots. Determining the appropriate hobby controller can be a challenge. If you enter a larger hobby shop that specializes in model boat and car racing, or check out catalogs or Web pages of some of the main suppliers, you will find literally hundreds of models to choose from. Your first instinct may be to talk with an employee for advice, but keep in mind this person might know a lot about cars and/or boats but absolutely nothing about the use of ESCs in robots.

You may hear about number of cells, maybe number of windings on your motor, and raves about how tiny the ESC is to fit in a small model. But, as a robot builder, you don't really care about these specs—you need an ESC that can handle extreme current loads without frying.

The hobby ESCs that have been proven to be usable in small combat robots are the Tekin Titan and Rebel models and the larger Novak speed controllers. Larger robots need more current than hobby grade controllers can deliver.

When selecting a hobby ESC, you need to select one with a voltage rating that is higher than the voltage your robot's motors need. Since these speed controllers are rated in terms of cells, you can divide your actual motor voltage by 1.2 to give you an equivalent cell rating. Choose a controller that has a higher cell rating.

Next, find a controller that has a current rating that is higher than what your robot's normal current draw will be. This is the hard part of the selection process. You will have to obtain detailed specifications of the ESC—most likely, direct from the manufacturer, since their current ratings are usually theoretical instantaneous ratings. Most hobby ESC's reverse current rating is lower than the forward current rating, so the selection process should be based on the reverse current rating. Although this may be a challenge, the hobby ESCs work well when used within their designed operating ranges.

Table 7-1 shows a short list of several electronic speed controllers. The maximum current rating is generally the advertised current rating. In practice, the continuous current rating for these types of controllers is approximately one-fourth the maximum current rating.

Manufacturer	Model Number	Voltage	Max Current
Associated	FI Reverse	4.8–8.4	100
Associated	FI Power	4.8–8.4	170
Associated	FI Pro	4.8–8.4	270
Duratrax	Blast	6.0–8.4	140
Futaba	MC230CR	7.2–8.4	90
Futaba	MC330CR	7.2–8.4	200
HiTec RCD	SP 520+	6.0–8.4	560
Novak	Reactor	7.2–8.4	160
Novak	Rooster	7.2–8.4	100
Novak	Super Rooster	7.2–12.0	320
Tekin	Rebel	4.8–12.0	160
Traxxas	XL-1	4.8–8.4	100

TABLE 7-1 *Hobby Electronic Speed Controllers* ■

Victor 883 Speed Controller

A more serious option is the Innovation First (IFI) Robotics Victor 883 speed controller (*www.ifirobotics.com*). The Victor 883 is an offshoot of technology developed for the FIRST robotics competition. The competition needed a heavy-duty speed controller, usable for drive motor or actuator duty, that would fit in a small space and lend itself to high design flexibility. Built like a hobby-grade controller "on steroids," the IFI Robotics Victor has a built-in cooling fan and uses three FETs in parallel for each leg of its motor control H-bridge, for a total of 12 FETs. Figure 7-13 shows the Victor 883 alongside a hobby ESC.

The IFI Robotics Victor controller can handle 60 amps of continuous current and up to 200 amps for short duration, and it is designed for up to 24-volt motors. Because the Victor 883 was designed specifically for competition robot use, it gives consistent and matched performance in forward and reverse.

The Victor was originally designed to be used exclusively with the IFI Robotics Isaac radio control gear. Following marked demand, IFI Robotics released a new version of the controller that is compatible with hobby-grade radio gear. Some R/C receivers, such as the Futaba receivers, do not deliver enough current to drive the opto-couplers in the Victor 883. Because of this, IFI Robotics sells an adapter that boosts the signal. Knowing whether your radio will need the signal booster or not

FIGURE 7-13 *Associated Runner Plus hobby ESC and Innovation First Victor 883 speed controller.*

is difficult without testing it—simply buying the booster cable and using it is probably the best idea.

Like a hobby-speed controller with a battery eliminator circuit, the Victor 883 controller uses a voltage regulator to produce a 5-volt power source for its control logic. But, unlike the hobby-grade controllers, the Victor 883 does not feed power back to the radio receiver, and uses an *opto-isolator* for full electrical isolation between the controller and the radio to prevent electrical noise generated by the motors from getting into the receiver power circuit. Figure 7-14 shows a block diagram of the Victor 883 electronic speed controller.

The electronics on the Victor 883 are contained on a single small circuit board, which is encapsulated inside a sealed plastic housing. The controller is highly impact resistant and does not need special mounting to be safe from impact shocks, although it's still a good idea to protect all onboard electronics from large shocks. Take care to ensure that the cooling fan has access to ambient air; the 60 amps continuos rating assumes that the fan has a constant source of external room-temperature air to blow over the FETs. Sealing a Victor 883 inside a box will have it circulating the same air over the cooling surfaces again and again, which will reduce the effective current capacity.

As a final safety measure, Victor 883 controllers ship with auto-resetting 30-amp thermal breakers. Intended to be wired in series with the motor, these heat up and disconnect the power at a current rating well under what the controller itself can handle. After a few seconds, the breaker will cool off and reconnect the motor. While these will ensure that the controller will not be damaged by over currents or shorts, they effectively cut in half the maximum current that the controller can source. While most motors used by robots in weight classes under 60 pounds usually don't draw more that 30 amps continuous, many motors in the larger

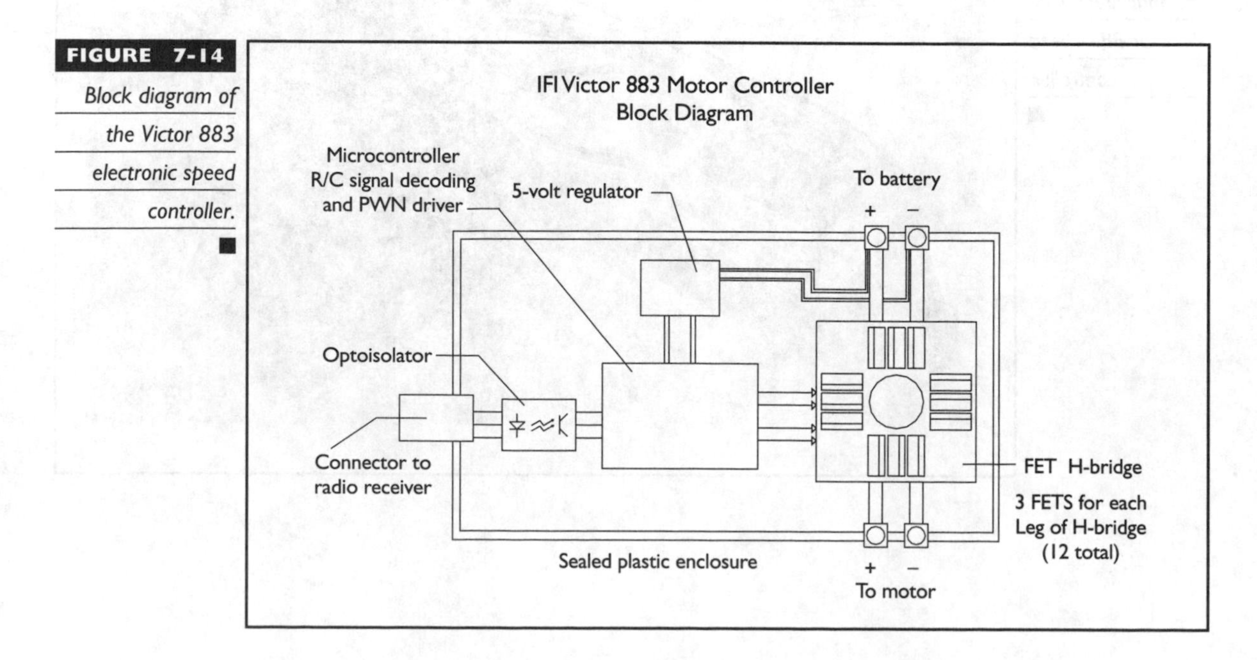

FIGURE 7-14 *Block diagram of the Victor 883 electronic speed controller.*

weight classes will exceed this limit regularly. Because of this, many robot builders do not use the thermal breakers.

The Vantec Speed Controller

Some of the most-popular electronic speed controllers used in combat robots are the Vantec RDFR and RET series controllers (*www.vantec.com*). The Vantec RDFR series controller has two speed controllers in one package that are designed to control a robot with separate left- and right-side drive motors. The Vantec includes a microcontroller signal mixer that automatically generates left and right motor signals from steering and throttle input from the radio gear. This allows the Vantec unit to be used for tank-steered robots without an external mixer or a radio transmitter with a built-in mixing function.

The RET series controllers are used to control single motors. They are ideal for applications in which a single DC motor is required to actuate a weapon system, a flipper arm, an end-effector, or a similar motor. The Vantec controller was originally developed for industrial application, such as bomb disposal robots. Table 7-2 shows a list of Vantec ESCs and their specifications.

Part Number	Voltage Range	Continuous Amps	Starting Amps
For four-cell to 24-volt DC systems:			
RDFR21	4.5–30	14	45
RDFR22	4.5–30	20	60
RDFR23	4.5–30	30	60
For 12–36-volt DC systems:			
RDFR32	9–43	24	65
RDFR33	9–43	35	95
RDFR36E	9–43	60	160
RDFR38E	9–32	80	220
For 42–48-volt DC systems:			
RDFR42	32–60	20	54
RDFR43E	32–60	35	95
RDFR47E	9–43	75	220
For single-motor systems:			
RET 411	4.8–26	12	30
RET 512	4.8–26	18	50
RET713	4.8–26	33	85

TABLE 7-2 *Vantec Electronic Speed Controllers* ■

All Vantec speed controllers are built in a similar manner. Two circuit boards are separated by standoffs—the upper board contains the radio interface, control logic, and 5-volt power supply, and the lower board contains masses of FETs wired in parallel and arranged in two separate H-bridges. The FETs are all mounted flat to the bottom of the Vantec's aluminum case, which acts as a heat sink for the controller. The physical nature of the controller—two separate boards and many discrete components—makes the Vantec controllers particularly susceptible to impact shock. It is best not to mount the Vantec unit directly to your robot's frame. Instead, use rubber insulation bumpers or padding to protect the Vantec ESC from impact shock. Figure 7-15 shows a Vantec electronic speed controller.

The Vantec controller does not have a sealed case but is mounted in an open aluminum frame. Before mounting it in your robot, you must make a cover to seal over the open boards and keep foreign matter off the exposed printed circuit boards. Combat arenas are full of metal chips just waiting to get inside your robot and short exposed electrical connections. The larger Vantec controllers are C-shaped extruded aluminum cradles with the circuit boards mounted inside. A piece of thin aluminum or Lexan (a polycarbonate plastic) bent into a C shape will cover over the open frame of the controller. Use tape to seal the seam between the edges of the shield and the frame and the hole for the radio signal wires.

The smaller series controllers are mounted in an aluminum box with only one side open. While this might make them seem more protected, in practice, the box tends to act as a trap for any bits of metal that do find their way in—letting them rattle around until they cause a fatal short. These can be sealed with a bit of tape, although a nice Lexan plate cut to fit the box opening looks nicer. With either Vantec, you should line the inside of the box and cover with double-sided tape to catch any bits of metal that do make it inside. Don't be concerned about the shielding's effect on the Vantec's heat dissipation. The power-switching transistors inside are mounted

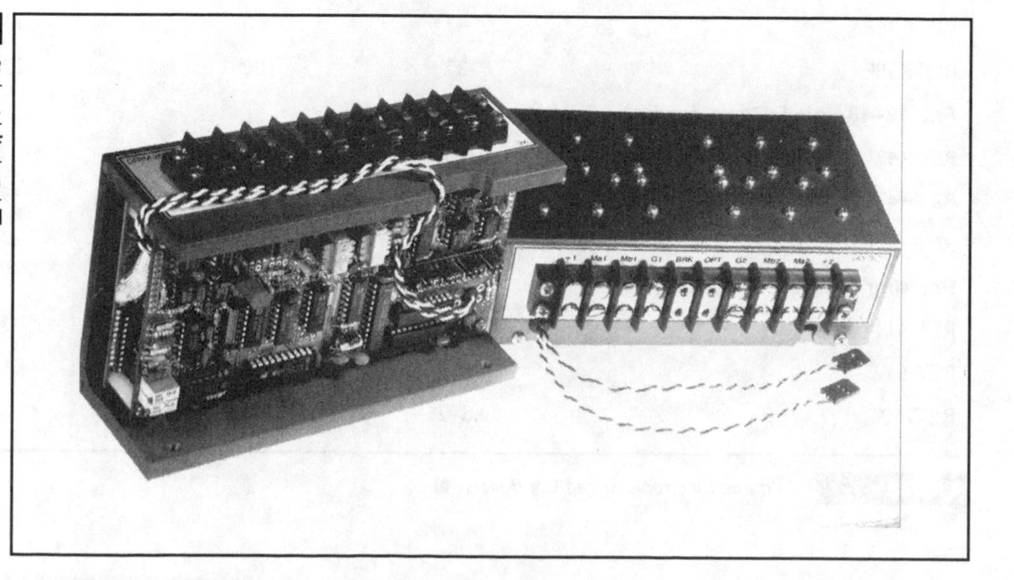

FIGURE 7-15 Vantec RDFR-23 motor controller. (courtesy of Vantec)

to the aluminum case, so enclosing the drive logic boards will not make the unit overheat.

A Vantec RDFR series controller has separate power connections for the left- and right-side motors and batteries. The high-current terminals—eight in all—are arranged on a single terminal strip on one end of the controller. This terminal strip, and the wiring connections to it, can be the weak point in your power train if not properly connected. The larger Vantec controllers (RDFR32 and above) have standard barrier blocks with eight screws to fasten down wires. Use ring-type crimp connectors on your wires to prevent accidental shorts or connectors pulling free of the terminal blocks. It is also a good idea to replace the soft screws used in the Vantec terminal strips with alloy-steel, cap-head machine screws to prevent accidentally twisting a screw head off by over tightening, and apply Loctite to keep the screws from vibrating loose during combat.

Figure 7-16 shows a block diagram of a Vantec RDFR series motor controller.

The smaller Vantec RDFR21-23 speed controllers have terminal blocks that use screw-down captive blocks to clamp the wires in place. The per-contact current rating of these terminal blocks is only 15 amps, not sufficient to handle the 30-amp current rating of the controller, so the Vantec ESC uses two adjacent contacts for each terminal. The lazy builder may think he can get away with using only one of these terminal points for each connection, thus running the risk of overheating and melting the terminal block by running over 15 amps continuous—a current level that the electronics of the Vantec unit can handle without difficulty.

To get the full capacity out of a small series Vantec controller, you *must* use both terminal block contacts for each connection. The easiest and most secure way to do this is to use a fork-type crimp connector fitting into two adjacent slots on the Vantec terminal. The exact side of the prongs on crimp connectors varies from manufacturer to manufacturer, so you may have to bend or file down the fork to fit snugly into the terminal block.

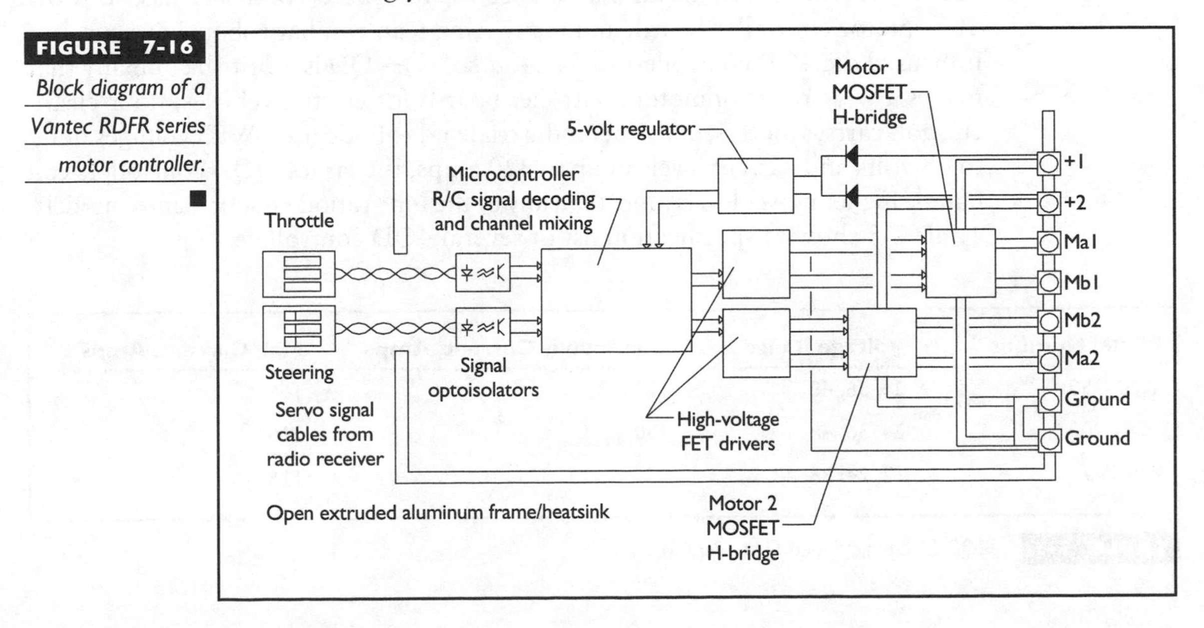

FIGURE 7-16 *Block diagram of a Vantec RDFR series motor controller.*

Like the IFI Robotics Victor, the Vantec draws its 5-volt logic power supply from the motor drive power and uses opto-isolators to prevent electrical noise from feeding back into the radio receiver. The low-voltage regulator circuit automatically draws power from whichever battery input is at the highest voltage. The negative sides of both batteries are connected together internally, but the positive sides are not, and the Vantec can be used to independently control two motors of different voltages if desired.

Vantec also makes a product known as the "Bully" power servo amplifier that accepts a standard from an R/C receiver to control a large motor just as if it were a very large servo. The signal is fed into the "Bully," along with a potentiometer input. The potentiometer is used to monitor the actual rotational position of a geartrain's output shaft or an actuator arm's position. The Bully can be used to control an arm where the actual position control is required, such as leg positions in walking robots.

The biggest challenge with the Vantec speed controller might be dealing with the company. Lead times on a Vantec controller can be weeks or months in times of high demand, and repair times on a damaged controller sent back to the company are similar, so you might want to keep these lead times in mind when testing and competing. You may find that most of their models are a bit expensive, but this company is one good example of "getting what you paid for"—its products are well built. Vantec stands by its products and has a reasonable "repair deposit" policy that allows users who have "fried" the Vantec products for whatever reason to have them repaired at a significant cost savings over purchasing a completely new product.

The 4QD Speed Controller

For British robots, the traditional choice for the speed controller has been the 4QD motor controller board, and many American combat robots have successfully used the 4QD controllers (*www.4qd.co.uk*). 4QD is a British company that makes a wide range of motor controller boards for electric vehicles, floor-cleaners, golf carts, scooters, and other industrial and robotic uses. With voltages of up to 48 volts and current levels of up to 320 amps, the largest 4QD controllers can handle higher power levels more than any of the Innovation First or Vantec models. Table 7-3 shows a specification list of several 4QD controllers.

Model Number	Voltage Range	Continuous Current, Amps	Max Current, Amps
4QD-150	24, 36, 48	120	160
4QD-200	24, 36, 48	150	210
Pro-120	12, 24, 36, 48	30	115

TABLE 7-3 *4QD Electronic Speed Controllers* ■

4QD controllers ship as open printed circuit board assemblies, so the end user will have to make his own housing and mounting arrangement to keep the 4QD board isolated from impact shocks and protected from debris. The 4QD controller is physically much larger than the Victor and the smaller Vantec controllers, and is generally used in weight classes of 100 pounds and greater. It does offer great reliability, built-in automatic current limiting, and a better power-to-cost ratio than other variable speed controllers.

The downside of the 4QD boards is that they are not compatible with hobby radio gear. The 4QD board has a purely analog input logic, and it is designed to directly connect to analog throttle and direction control signals. Getting a 4QD board to talk to traditional R/C units is the biggest challenge in successfully implementing this design. One method to generate an analog signal is to connect a potentiometer to the output shaft of an R/C servo. Feed 5 volts through the potentiometer to the 4QD controller, and then drive the servo with the regular R/C transmitter set. Although this works, it is not recommended because it adds more parts that can become damaged during a combat match.

The ideal way to generate the analog voltage is to use a microcontroller to read in the transmitter's signals and convert them into an analog signal to drive the 4QD controllers. Getting the signal conversion just right is a challenging task if the builder wants consistent and reliable control out of his 4QD board. The 4QD boards offer a lot of power for the price, but the difference between smooth control and spastic twitching can take a lot of control-system troubleshooting.

The OSMC Motor Controller

The Open Source Motor Controller (OSMC) was developed by robot builders for robot builders (*www.robot-power.com*). The OSMC is a modular control system that offers the high current capacity of the 4QD with the plug-and-play interface of the Vantec and Victor controllers. The OSMC was developed as a collaborative effort between robot builders to develop a high-powered, low-cost speed controller alternative to the then-limited supply of commercial controllers.

The OSMC is a modular system, available fully assembled in kit form or as bare boards. The controllers can be assembled with several different FET configurations to give current capacity of up to 160 amps continuous and voltage capacity to 50 volts. The controller is made up of two separate circuit boards the logic board and the power board. The logic board is the interface to the radio receiver and handles channel mixing. The power board contains the FETS and associated driver circuitry. One logic board can drive two separate power boards, allowing for complete drive-train control over a tank-steered robot.

The open source nature of this controller means that the full development details—schematics, parts list, and control code—are freely available to developers. The hobbyist nature of the controller means that a lot of rapid changes have occurred in the development of the software and documentation of the controller logic, and different versions of the control board with different features are available.

At the moment, using the OSMC controller successfully means committing to learning the ins and outs of the system in some detail and being prepared to do your own programming and modification.

The OSMC shows great potential as a high-powered motor controller; but at the time of writing this book, the OSMC lacks significant combat testing. If the current momentum on the project is maintained, the OSMC could become the choice for high-power motor control. Keep your eye on this one in the coming years.

---

caution *When using any ESC, you must carefully inspect and test all the wiring before powering up your robot for the first time. It takes only a momentary short circuit, reversed polarity, or over voltage to destroy the controller, batteries, and in some cases even the motor—which can cost hundreds of dollars and weeks of precious time to replace.*

Most combat robots will use a traditional radio control system that was originally designed for R/C airplanes, cars, and boats for controlling the robot's motion and actuators. Because they are so widely available, combat robot components are being designed to accept the standard R/C servo command signal, such as the Vantec and Victor speed controllers. Some robot builders prefer to build their own remote control units but use regular R/C servos and speed controllers that accept the standard R/C servo control signal.

Some robot builders even build servo-mixing circuits to help improve the driving control of the robot. Servo mixing is common with robots that use tank-type steering. Instead of having one stick on the radio transmitter controlling the speed and direction of one motor and the other stick on the transmitter controlling the other motor, by combining both of the signals together, one stick on the transmitter can be used to control the velocity of the robot and the other stick can be used to control the direction of the robot. In fact, one joystick on a transmitter can be used to control both direction and speed. This frees up the robot driver's other hand to control weapons on the robot. Servo mixers are commonly called elevon mixers, veetail mixers, or v-tail mixers.

To develop custom controls for driving R/C servos or speed controllers, you must understand how the R/C command signal works. Many people call the R/C command signal a pulse-width modulated signal. Though technically correct, it is nothing like the true variable-duty-cycle–controlled PWM signal that is used to vary the speed to a motor. A true PWM signal is a square wave signal that has a duty cycle that can range from 0 to 100 percent. The R/C control signal is a variable 1 to 2 millisecond pulse that must be repeated every 15 to 20 milliseconds. The internal circuitry of a R/C servo is designed to interpret the 1- to 2-millisecond pulse and convert it into a position command. A pulse width of 1.5 milliseconds represents the neutral position of the servo, or zero degrees. R/C servos rotate approximately +/– 60 degrees from the neutral position. A 1.0-millisecond pulse width represents an approximate –60 degree position, and a 2.0-millisecond pulse width represents an approximate +60 degree position. Figure 7-17 shows a graphical representation of the R/C pulse control signal. The servos are also designed to shut off if they do not receive a signal every 15 to 20 milliseconds.

The repetitive nature of the signal can be advantageous to the robot builder. If the repeated signal stops, this is an indication of a power loss, a broken signal line, a failed receive, or a failed or turned off transmitter. If any of these events were to happen, you will want your robot to immediately shut down. The Victor and Vantec speed controllers will automatically shut down if they stop receiving the repeated signal. This shutdown feature is known as a *failsafe* in the combat robotics community. Most competitions require robots to demonstrate the fail-safe feature.

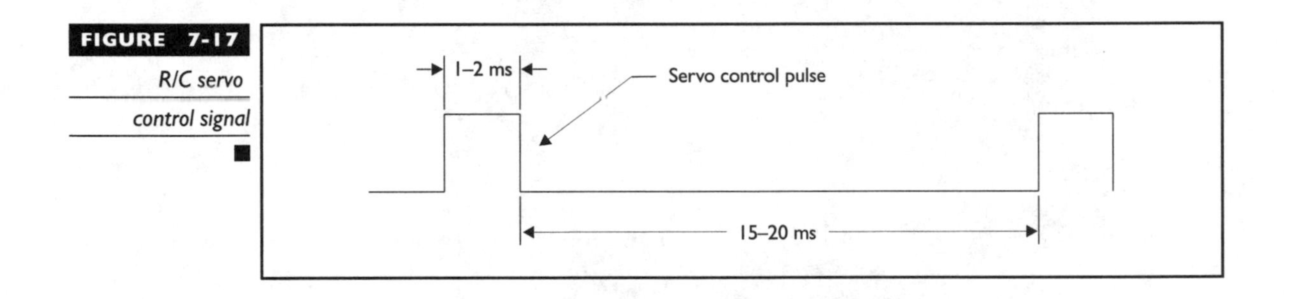

FIGURE 7-17 R/C servo control signal

chapter 8

Remotely Controlling Your Robot

THE control system you use for your robot must fulfill several requirements. It should be reliable and reasonably immune to interference. It should have at least enough range to communicate to your robot in the far corner of the arena—and preferably much more to be safe. The receiving system should be small and able to withstand a lot of vibration and shock. It should be able to command multiple systems on your robot simultaneously. It should be capable of varied degrees of control so that your robot does not have to drive at full speed all the time. And, finally, it should be available as a reasonably inexpensive off-the-shelf, solution so that you do not have to spend more time engineering the radio control (R/C) gear than the rest of the robot.

In the early days of robotic competition, robot builders attempted to use everything from garage-door–opener radios modified for multiple command channels to radio gear sending commands encoded in audio tones, infrared remote controls, tether-line controls, and networked computers running over wireless modem links. The most effective technology turned out to be hobby radio control (R/C) gear, the relatively low cost, off-the-shelf R/Cs intended for use in model cars and planes. Today, nearly every robot in major competitions uses some form of commercial hobby R/C, and competitions have based their R/C rules around this standard control system.

Traditional R/C Controls

All R/C systems, whether AM or FM radio systems or high-end computerized transmitter and receiver sets (which are all discussed later in this chapter), use essentially the same electrical signals to transmit control information from the radio receiver to the various remotely controlled servos and electronic motor controllers. See Figure 8-1. A three-wire cable runs from the radio receiver to each speed controller and servo in the robot. One wire provides about 5 volts of power to run the servos. A second wire is a ground reference and power return line. The third line carries the encoded 1- to 2-millisecond pulse train signal that commands the motion.

Movement commands are encoded with a pulse position modulation system (some people call this "pulse-width modulation"; Chapter 7 explains the difference

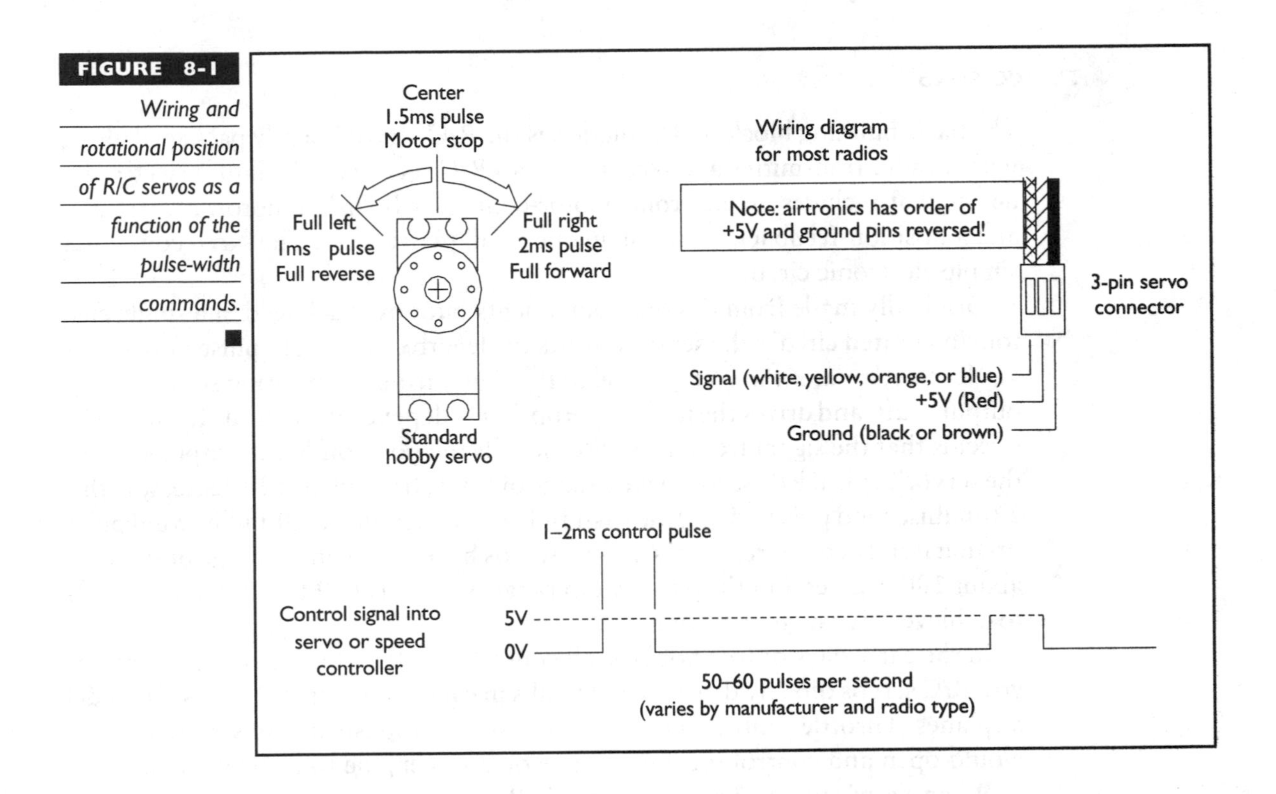

FIGURE 8-1 *Wiring and rotational position of R/C servos as a function of the pulse-width commands.*

between the two). A signal pulse is sent from the radio receiver to each servo approximately 50 times per second. The exact pulse frequency can vary from 50 to 60 times a second, depending on the manufacturer and model of the radio. The length of the pulse encodes the movement data in the range of 1.0 to 2.0 milliseconds, with a pulse of 1.5 milliseconds being a neutral or center position command.

The R/C Controller's Interface

Although the electrical interface has been standardized, manufacturers use their own color codes and connectors to attach the radio receiver to the servos. The color-coding of the wires always follows a similar motif: the ground wire is black or brown; the power line is almost always red; and the signal line will be white, yellow, orange, or occasionally black. The order of the control pins is the same in nearly all manufacturers' units—the wire closest to the notched edge of the radio connector is the signal wire, the center wire is the 5-volt power, and the last wire is the ground wire. (Airtronics brand connectors use a unique wire arrangement that's worth mentioning here. The wire next to the notched side of the connector is the signal wire (blue), the ground wire is in the middle (black), and next is the 5-volt power (red) wire.) Electrically speaking, most manufacturers' systems are compatible, so the connectors can be easily cut off and swapped with another style of connector to convert servos or speed controllers from one system to another.

The R/C Servo

The basic building block of R/C models is the R/C servo. Usually packaged along with a radio transmitter and receiver set, an R/C servo is a miniature electronics device that includes an electronic motor-controller board, a motor, a geartrain, and a position-feedback sensor all in one small plastic case. The servo contains a simple electronic circuit.

Originally made from discrete components but now packaged in a single custom integrated circuit, the servo converts the length of the input pulse into a voltage level, compares the voltage level to the signal from the position sensor on the output shaft, and drives the motor appropriately depending on the difference. The effect is that the signal from the radio controls the position of the output shaft of the servo. Typical R/C servos have a range of travel from 90 to 120 degrees, with a 2.0-millisecond pulse driving the shaft fully clockwise and a 1.0-millisecond pulse driving it fully counterclockwise. Most servos have a maximum range of travel of about 180 degrees, but the pulse-width range will be from 0.8 to 2.2 milliseconds to achieve this range of motion.

In the early days of R/C hobbies, all controls worked through mechanical servos. R/C servos directly drove steering links in cars and control surfaces on model airplanes. Throttle control of motors was also accomplished with servos. A servo would open and control the intake valve on a gas engine to control its power.

When electric motors became popular in R/C cars, the same hobby control servos were used to control them; but instead of opening and closing a throttle valve, the servo arm would slide along a set of contacts to make or break the power circuit to the motor. When Field Effect Transistor (FET)–type electronic speed controllers entered the market, they duplicated the interface of the earlier mechanical speed controllers, with what had been a position control signal to control a servo's output shaft now being a speed and direction control for an electric motor.

Control Channels

Traditional R/C systems are rated by the number of *channels* they can control. Channels refer to the number of independent servo signals the system can send simultaneously to the receiver. Most of the low-cost radio sets meant for R/C cars are two-channel radios. The radio transmitter can send command information for two separate servo positions at once to the receiver to control both steering and motor speed (or throttle) simultaneously. The next level for R/C cars is three-channel radios; the third channel is intended to control a gearshift, air horn, lights, or other on-board accessories. Most of these radio transmitters use a pistol-grip configuration, in which a gun-style finger trigger controls the throttle channel and a miniature wheel on the side of the transmitter controls the steering channel. A pistol-grip transmitter is shown in Figure 8-2.

FIGURE 8-2 *Futaba's top of-the-line 3-channel, pistol-grip–style computer transmitter.* (courtesy of Futaba)

The next step up is the model aircraft radios that typically have four channels. The transmitters have a two-stick type configuration. The primary control is conducted through the two sticks, called *joysticks*. More advanced transmitters include additional channels that consist of switches and knobs for extra R/C capabilities. Figure 8-3 shows a stick-style transmitter.

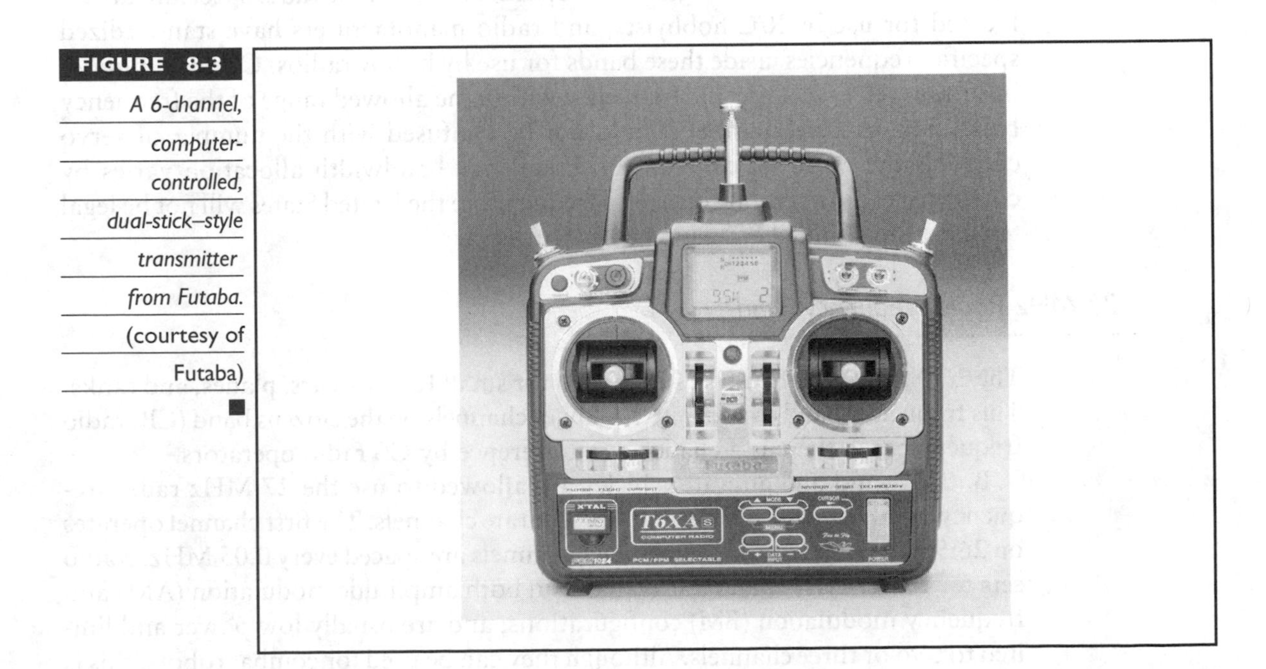

FIGURE 8-3 *A 6-channel, computer-controlled, dual-stick–style transmitter from Futaba.* (courtesy of Futaba)

Each of the two joysticks controls two channels—one channel with the horizontal direction, and one with the vertical. The top-of-the-line R/C sets, usually intended for the R/C helicopter market, can have up to nine channels of servo control. Most of the high-end radio sets also have computerized control interfaces that allow the driver to configure the channel allocation, and change mixing settings, and the R/C system can be programmed for custom control sequences.

Whether you are independently controlling each of the channels that control the left and right motors, or you are controlling the robot speed with one stick and steering with the other stick, two channels are the minimum needed to drive a robot in a controlled fashion. Some more-complex robots that involve omni-directional wheels or multi-legged walking mechanisms need more than two channels for drive control.

Most competitions require that weapons are controllable via remote control, so you will need to include at least one channel for each weapon. Complex weapons—such as saws on moveable arms or spring-loaded rams with separately controlled release mechanisms—will need more than one channel. Gasoline engines may require several control channels—one for the throttle, a second to start the engine remotely, and a third to shut down the engine remotely. A general rule to remember is that you will need a separate servo command channel for each action that you want to control separately.

Radio Control Frequencies

The frequency bands for R/C systems are established by Federal Communications Commission (FCC) regulations. Specific bands of the radio spectrum are allocated for use by R/C hobbyists, and radio manufacturers have standardized specific frequencies inside these bands for use by hobby radios. *Channel number* in a radio refers to a specific frequency within the allowed range of the frequency band. The channel number should not be confused with the number of servo channels the radio set can control. Frequency bandwidth allocation varies by country; a radio operating on a legal frequency in the United States will not be legal for use in the United Kingdom, and vice versa.

27-MHz Radio Frequency Band

The 27-MHz radio band is usually used for small R/C toy cars, planes, and tanks. This frequency band crosses into the lower channels on the citizens band (CB) radio frequencies, so there is a chance of interference by CB radio operators.

Both ground and aircraft vehicles are allowed to use the 27-MHz radio frequency band, which is divided into six separate channels. The first channel operates on 26.995 MHz, and each of the other channels are spaced every 0.05 MHz. Radio sets for the 27-MHz band are available in both amplitude modulation (AM) and frequency modulation (FM) configurations, and are usually low power and limited to two or three channels. Although they can be used for combat robots, this is not recommended. The antenna on the radio transmitter must have an attached flag indicating the frequency with which they are transmitting. The flag colors for channels 1 through 6 are brown, red, orange, yellow, green, and blue, respectively.

50-MHz Radio Frequency Band

This channel band is licensed for use by air or surface models, although it is usually used for R/C airplanes and helicopters. The 50-MHz band is divided into 10 frequency channels starting from 50.800 MHz and spaced every .020 MHz. Although several high-quality radios are available for this band, use of them requires a ham radio amateur license from the FCC. Although this band is rarely used for competition, the individual lucky enough to use this channel will be virtually assured of a clear channel, with no other robot builders using the same frequency.

Two flags must be flying on a 50-MHz radio transmitter antenna: a flag with a number between 00 and 09 to identify the frequency number, along with a black streamer to identify the 50-MHz radio frequency band.

72-MHz Radio Frequency Band

The 72-MHz radio band is reserved by the FCC for aircraft use only. In other words, ground vehicles, including combat robots, are not allowed to use this frequency band. A total of 50 different channels are available in the 72-MHz radio band with frequencies ranging from 72.010 MHz to 72.990 MHz, and with each channel number spaced every 0.020 MHz. The channel numbers range from 11 to 60. The channel identification flags include one with the channel number and a white streamer, attached to the transmitter's antenna.

For all modern 72-MHz radios, changing the frequency requires changing the frequency crystals. The transmitter uses a crystal marked with "TX" and the receiver's crystal is marked with "RX." When changing the crystals, they must both have the same radio frequency. (More on crystals in the upcoming section "Radio Frequency Crystals."

75-MHz Radio Frequency Band

The 75-MHz radio band is reserved by the FCC for ground use only. Thirty different channels are available in the 75-MHz radio band with frequencies ranging from 75.410 MHz to 75.990 MHz, and with each channel number spaced every 0.020 MHz. The channel numbers range from 61 to 90. The channel identification flags are the ones with the channel number and a red streamer.

Changing the channel frequency or channel number within the 75-MHz frequency band also requires changing the frequency crystals, as with the 72-MHz radios. However, you cannot change a 72-MHz band radio into a 75-MHz band radio by swapping frequency crystals. Although the crystals look identical in size and shape, swapping the crystals between the two radio frequency bands will not work. Switching from one band to another requires retuning the radio, which should be done only by an FCC licensed technician.

If your robot is going to use a traditional R/C system, the frequency bands that you are allowed to use by law are 75 MHz, 27 MHz, and 50 MHz. The dilemma in this scenario is the fact that R/C systems that are meant for ground applications usually have only a few channels available for driving two or three servos. The high-quality, multi-channel radios are almost exclusively made for aircraft use. In the early days of robot competition, many robot builders used aircraft frequency (72 MHz) radios exclusively, because good-quality ground frequency (75 MHz) radios were not available. In recent years, however, competition organizers have begun enforcing FCC regulations about channel number and frequency band use, forcing robot builders to switch to non-aircraft frequencies.

Most 72-MHz R/C systems can be converted to operate on 75 MHz, but only after an extensive retuning process. Legally, retuning for 75 MHz has to be done by an FCC licensed technician. In most cases, this is most easily done by the radio's original manufacturer—although some third-party shops, such as Vantec, can do the conversion process. For a nominal fee, some radio manufacturers will retune a radio for the 75-MHz ground frequency band when the radio is sold.

United Kingdom Radio Frequency Bands

Radio control systems in the United Kingdom are similar to those in the United States, but the particular radio frequencies used are different. The UK hobby radio control system runs on the 35-MHz and 40-MHz bands. The 35-MHz frequency band is reserved for aircraft use, and the 40-MHz band is reserved for ground applications such as combat robots. The 40-MHz band is separated into radio control channels every .010 MHz, from 40.665 to 40.995 MHz. As with those in the United States, robot builders in the U.K. must either purchase a 40-MHz ground radio or have a 35-MHz aircraft radio set converted into a 40-MHz system for ground channel use.

Radio Frequency Crystals

Within the frequency bands is a set of individual channel numbers that can be used for R/C applications. For example, 30 different radio channel numbers can be used in the 75-MHz frequency band. The specific channel number frequency is controlled by an oscillator called a *frequency crystal*, which is shown in Figure 8-4. The frequency crystals come in pairs: one for the transmitter and one for the receiver. To change the channel number on your radio, you simply replace the frequency crystals. Both the transmitter and receiver must use the same channel number, or the system will not work. The 72-MHz and 75-MHz crystals look identical, but the crystals are not interchangeable between frequency bands. In other words, putting a 75-MHz crystal in a 72-MHz radio will not work.

FIGURE 8-4 *Typical radio frequency crystal pair.* (courtesy of Futaba)

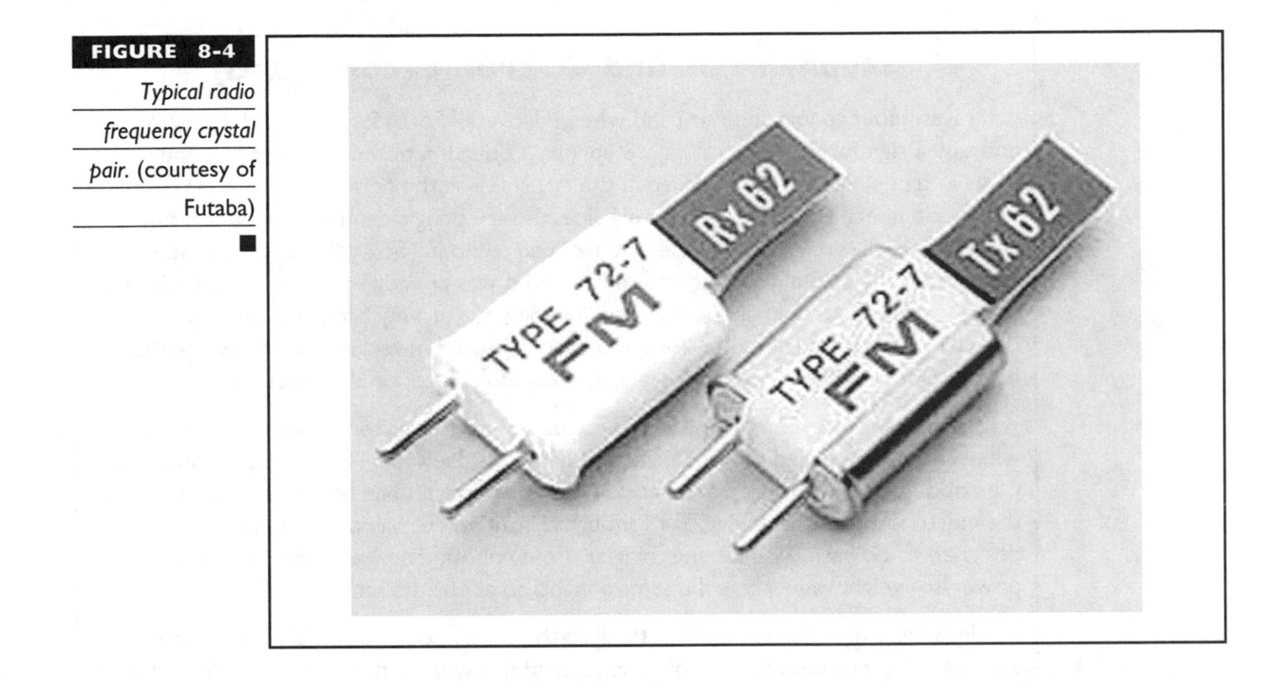

When selecting a radio system, make sure it will allow you to change the transmitting frequencies. Because it's likely that at least one other person at a competition will be using the same frequency that you want to use, you will want to be able to change the frequency of your R/C equipment to avoid frequency conflicts. When this happens at a match, everyone loses control of their robots. This is also why you display the frequency number flags on your transmitter's antenna so that everyone else will know what frequency you are currently using.

At some matches, organizers control the frequencies that can be used and will issue the appropriate frequency crystals prior to each match. Other organizations, such as *BattleBots*, will impound your transmitter when you show up. Your transmitter will be returned to you prior to a match, during the 15-minute testing session and safety inspections, and after the event is over. Impounding transmitters is an extreme, but effective, method for preventing radio frequency interference.

Prior to competing, you should have at least two different sets of crystals so that you can change them to avoid frequency conflicts during the competition—especially if you are competing in multiple-robot rumbles.

True Story: Stephen Felk and Voltronic

Stephen Felk had a wild ride on the way to his most memorable fight—his very first.

Although Stephen started out as an engineering student at Northwestern University way back in 1970, engineering studies didn't keep his attention. He soon switched to the arts, and found himself in San Francisco dabbling with a variety of artistic endeavors: sculpture, music, even acting. But then a chance event changed his life.

Stephen Felk and Voltronic (continued)

"I was about to join another band when I drove by Fort Mason one Friday night and saw a sign for *Robot Wars,"* says Stephen. "I bought a ticket and went in. I knew I was in trouble as soon as I walked in the door. It was the perfect combination of wrong elements. It had the engineering side, the sculpting side, the competition side, and some really great camaraderie. I knew I had to do it." So addicted was he, Stephen tried returning on Sunday. "I got there late and it was already sold out. I kept badgering the guy at the door, and finally, he says, 'I'm going to turn my back. Whatever you do is up to you, but just leave me alone.' So I snuck in—I'd never done anything like that before. But I watched the whole event, and couldn't sleep for like a week afterwards."

Stephen started working on his first robot shortly thereafter, beginning with a wheelchair he managed to pick up second-hand for just $100. "In this sport, sometimes the robot guys shine on you and I got off to a great start. I had no experience at all with the electrical/mechanical thing. But I thought about wheelchairs, and realized they're designed to do basically the same thing as these robots. They're designed for the same power-to-weight ratio, carry the same weight, go about the same speed."

Unfortunately, Stephen underestimated the time needed to build his creation; and while he worked obsessively right up until the weekend before *Robot Wars '97,* he simply couldn't get his creation completed in time. "I got in completely over my head. It was way too complicated, I had to learn too much, and a few days before the competition I thought, 'My god, I'm not going to make it.' Nothing could ever be as terrible as that." The following year wasn't to be either; but by 1999, he and *Voltronic* were ready to rumble.

"My very first match was against *Razer,* a really famous English robot, and it was far and away the best match I've ever been in. It was a really, really great battle. There were four or five major turning points, points where we switched superiority, and it was incredibly exciting."

Unfortunately, at its debut, *Voltronic* had a sheet metal skirt, a design element that Stephen describes now as "a really stupid idea. *Razer* comes slamming into me and rips the sheet metal right off. I'm driving around with these three pieces of sheet metal skirt just flapping in the wind." The fight turned around, though, and Stephen says, "It ends up with *Voltronic* picking up *Razer* and slamming him into the wall. And that's how the match ended: I had him two feet up in the air, pinned against the wall."

Despite the triumphant ending, the winner was declared by audience vote—and *Voltronic* officially lost to *Razer*. "But it was so exhilarating," says Stephen, "going through this three-year ordeal, all that frustration, maxing out all my credit cards, and the battle was so incredible and so addicting, it was such a great reward and a vindication that this whole thing was really worth it."

Stephen adds that he understood—even at that moment—why he lost. "He had a great-looking robot, and I just had a simple wedge. Worse, the entire time we were fighting, he was tearing off great sheets of sheet metal. It looked like I was torn up even though he didn't really hurt me. But I was so proud to have this great fight against these great guys. They were great competitors, great sportsmen . . . and that first match instantly justified all the work that I'd put into it. It erased any doubts I ever had."

AM, FM, PCM, and Radio Interference

While all the R/C sets use the same electrical signals for communicating with the servos and motor speed controllers, they differ in how they deliver that information from the radio transmitter to the radio receiver. Most R/C sets use a single radio frequency to transmit the control information from the transmitter to the receiver. To deliver information to drive multiple servo channels, the servo pulse information is transmitted serially, one pulse following another on the radio signal.

The transmission of control information between the transmitter and the receiver is usually sent as radio waves in one of two different ways: AM or FM.

Amplitude Modulation

In an AM radio system, the strength of the transmitted radio signal is varied to encode the control information. This means that the radio signal is being switched between high and low power output levels to encode the pulse data stream. AM radio transmission is inexpensive and easy to implement electrically, but it is highly susceptible to radio interference.

The AM transmitter sends each channel's servo position as an analog pulse with a width that varies from 1 to 2 milliseconds. All the pulses are transmitted as a continuously "on" radio frequency (RF) carrier, with each channel's beginning and ending marked by an "off" for 0.35 millisecond. All the channels are sent sequentially with the .35-millisecond end mark between each channel serving as the beginning mark of the next channel. A special framing pulse designates the beginning of the channel series by resetting the receiver. The receiver uses the marks to determine which servo to control based on the proper 1- to 2-millisecond command pulse. Any radio interference could be interpreted as a marker and cause the servos to go to a wrong position or to sit and "jitter" erratically.

Using AM, any electrical noise from electric motors, fluorescent lights, or gasoline engines, for example, can cause unwanted movement of the robot because the electrical noise can be added to the original AM transmitting signal. Because AM receivers interpret the intensity of the incoming radio signal as specific information, they have trouble distinguishing electrical noise from the actual transmitted signals. This results in the receiver sending false signals to the motor controllers and servos. Because AM radios may cause uncontrolled movement in combat robots, most competitions prohibit the use of AM radios entirely.

Frequency Modulation

A more robust and reliable method for transmitting control signals is to use frequency modulation (FM). In an FM radio system, the amplitude of the signal is held constant, and the transmitted information is encoded by varying the frequency of the transmitted carrier signal. The FM receiver locks onto the constant transmitted signal and is much less likely than AM to be distracted by random electrical noise from the environment. This does not say that FM systems are immune to radio interference, though, because all radios are subject to radio interference. However, FM radio signals are far less susceptible to radio interference than AM radio signals.

Pulse Code Modulation

To further improve the reliability of FM radios, a more advanced system of signal transmission known as Pulse Code Modulation, or PCM, can be used. A PCM radio signal uses an FM radio transmission similar to an ordinary FM radio set, but the servo commands are transmitted as a digital data stream rather than time-coded pulses.

A PCM receiver contains a microcontroller to develop and interpret the pulse code for servo control. PCM systems form the servo commands using a set of algorithms and precise code timing. PCM allows accurate signal reception, even when severe radio frequency interference (RFI) or other noise is present.

The process begins in the transmitter by converting each joystick, switch, trim knob, and button position into a 10-bit digital word, plus the extra bits to enable the receiver to verify the word. The PCM radio system compacts this data representing 1,024 servo positions per channel into the FCC-specified radio bandwidth, while maintaining responsive real-time control. The PCM data is transmitted synchronously; each bit has a particular position in time, within a frame. The frame continuously repeats. A crystal-controlled clock in the receiver locks onto the transmitted signal to maintain synchronization with the data, bit by bit. Thus, the receiver can process data immediately after interference instead of waiting for a framing pulse.

Received data is evaluated channel by channel. When the microcontroller detects an error, previously stored valid channel data is used. If an error persists, failsafe servo operations previously specified by the operator are initiated until accurate commands are again received. The microcontroller converts the proper data into pulse widths to command the servos, and you no longer have servo "jitters." Some receivers can be programmed to shut down if they receive bad data, or they can be programmed to output specific commands so that the robot enters a controlled and safe state. Because the actual data signal and a data *checksum* signal are sent at the same time and compared together at the receiver, it is nearly impossible for a robot to move out of control accidentally because of radio interference.

The other advantage of PCM radios is that they grant you the ability to customize the control interface. Because the signals are being digitized and encoded, it is easy for the internal computer to perform custom mixing and scaling operations on the data before transmitting it. Known as *computer radios,* these units have a liquid crystal display (LCD) screen and a miniature keypad that can be used to write custom programs for the controller interface. Typical settings include custom gain, and center and end points on individual controls, as well as custom mixing of two channels to generate left and right motor drive signals from a single joystick for driving skid-steer robots.

When choosing a radio system, you may want to consider more than just the robot you are currently using. While the rest of a robot may be scrapped, recycled, or even completely destroyed in combat, your R/C system can be reused on robot after robot. If you intend to participate in robotic combat competition year after year, it makes sense to spend a little more on your R/C system at the start, rather than buying a low-end radio and then having to pay more on a better radio down the road. If you buy a PCM radio with at least seven channels, you will probably never have to buy another radio for as long as you are competing. Most veteran combat robot builders will recommend that if you use a traditional R/C system, you should use a PCM radio with your robot. It will save you a lot of headaches when testing and competing with your robot, since you will know that erratic motion is not due to radio interference.

Tables 8-1 and 8-2 contain short lists of the available R/C systems. The column under "Band, MHz" lists the frequency bands these systems can use. If two different

Manufacturer	Model	Channels	Band, MHz	PCM Available
Futaba	3PDF	3	27 and 75	No
	3PJS	3	27 and 75	Yes
Airtronics	CX2P	2	27 and 75	No
	M8	3	27 and 75	No
Hitec	Lynx 2	2	27 and 75	No
	Lynx 3	3	27 and 75	No

TABLE 8-1 *Pistol-Grip–Style Radio Control Systems* ■

Manufacturer	Model	Channels	Band, MHz	PCM Available
Futaba	4VF	4	72 and 75	No
	6VH	6	72	No
	6XAS	6	50 and 72	No
	6XAPS	6	72	Yes
	8UAPS	8	50 and 72	Yes
	9ZAS	9	50 and 72	Yes
Airtronics	VG400	4	72	No
	VG600	6	72 and 75	No
	RD6000	6	72	Yes
Hitec	Ranger 3	3	27 and 75	No
	Laser 4	4	72	No
	Laser 6	6	72	No
	Eclipse 7	7	72	Yes

TABLE 8-2 *Stick-Style Radio Control Systems*

frequencies are listed, a system can be obtained to operate under either frequency, not both frequencies. The "Channels" column shows the number of servo channels the R/C system can control at once, and the "PCM Available" column lists whether the system uses PCM error-correction controls.

Radio Interference and Reliable Control

Model aircraft radios are designed to control airplanes at ranges over thousands of feet; yet in the arena, robots less than 50 feet away from their controllers can go wildly out of control or fail to move at all. The difference between the two environments is in the ambient radio interference and the antenna placement. Installing a radio that was designed to be run inside a balsa wood or plastic airplane with only small servos and a single glow-plug engine, and making it run inside a metal-cased combat robot with large noisy electric or gasoline motors, is more difficult than you might think.

The first challenge to overcome is radio interference, most of which will come from inside the robot itself. As a brush-type DC motor turns, the sliding contact of the brushes over the commutator segments is constantly making and breaking circuits and reversing the flow of current in the motor's armature winding segments. This constant arcing creates high-frequency electrical noise whenever the motor is running. This noise can be picked up by the radio system and can jam or interfere with the normal control signal. If your robot's weapons unexpectedly actuate by themselves when you drive it, or if your robot twitches back and forth by itself when you trigger the weapon, you may be experiencing radio interference from your motors that is altering your radio control.

To combat this interference, start by neutralizing it at the source. You cannot do anything about the arcing at the terminals, but you can divert most of the noise before it leaves the motor. Small ceramic capacitors can be attached to filter the noise from the brushes (see Figure 8-5). Capacitors have a low impedance to high frequencies and can short-circuit the noise before it even leaves a motor's case. You should use non-polarized ceramic capacitors in the range of .01 to .1 μF, with a voltage rating of at least twice your motor's running voltage. If possible, use three capacitors—one from each brush terminal to the motor case, and one across each of the two motor terminals. The capacitors should be connected as close to the actual brushes as possible, ideally inside the motor case itself, and they should be mounted carefully and secure to avoid the chance of shorting out the motor if one comes loose.

What noise that does manage to escape from the motor will radiate from the motor power wires like a broadcast signal from an antenna. You can minimize this by twisting the motor wires together (leave the insulation on the wires); the noise emitted by the motor leads will be significantly reduced. Placing these twisted wires within a braided shield grounded to the robot's structure also helps. You can also reduce the transference of noise from the power system to the radio by placing your receiver as far as possible from the motors and their wires. Placing the receiver in a shielded metal container will also help reduce the noise interference.

note *Do not run the lines from your radio receiver to the servos and speed controllers near or parallel to the motor power lines, if you can help it. As current goes through a wire, a circular magnetic field is generated. If a wire is running parallel to this wire, and it is inside the magnetic field, the field can induce a current flow in the adjacent wire. The physics behind this is why motors and transformers work in the first place. Twisting the servo leads and power leads also helps minimize their tendency to pick up electrical noise from the motor system.*

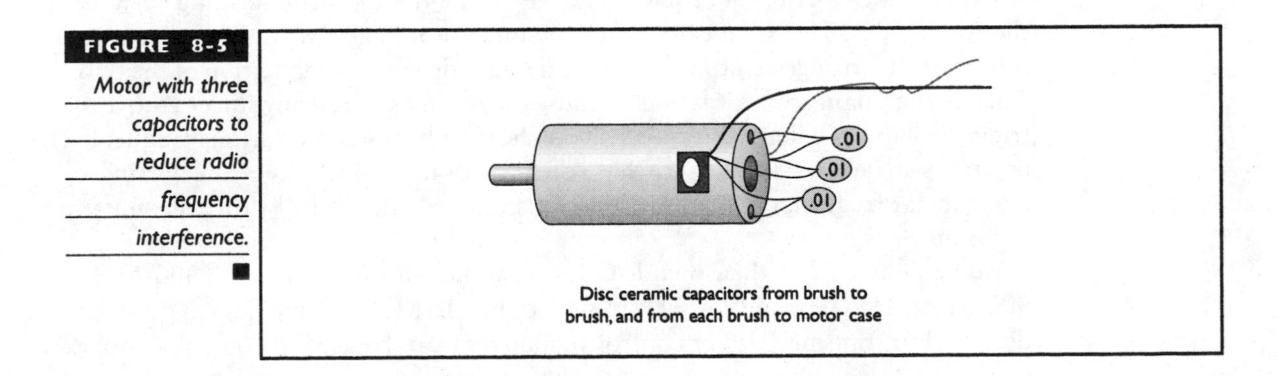

FIGURE 8-5 *Motor with three capacitors to reduce radio frequency interference.*

Of course, minimizing the transmission of noise from one system to another does no good if your radio control and power circuits are not electrically isolated. No common ground or shared power source should exist between your radio and your drive motor power. Electronic speed controllers (ESCs) that make a direct electrical connection between the servo signal line and the motor battery, or those that tap power off the drive batteries to feed to the radio (known as a *battery eliminator circuit*, or *BEC*), should not be used. Electrical isolation through opto-isolators or relays should be mandatory. A separate battery should be used to power the radio. If a power converter is used to provide power to the radio from the motor batteries, it should be a type with full electrical isolation, such as the Team Delta's R/CE85-24.

note *If speed controllers with BEC must be used, the power pin connecting the ESC to the receiver can be removed from the connector and insulated to prevent an electrical connection. A separate battery should then be used to power the receiver.*

Gasoline engines can be a huge source of electrical noise—particularly the small, high-RPM, two-stroke motors used in chainsaws and lawn trimmers. The high-voltage pulses generated by the ignition system can play massive havoc with a nearby R/C system. To prevent noise from the engine from getting into the radio circuitry, place the radio control system in a metal box, test the servo leads for interference, and keep the distance between the radio receiver and the engine's electrical system as far as possible in the robot. The electrical noise that is radiated from the motor can be minimized by using resistor-type spark plugs and replacing the ignition wire with a shielded line. Resisting this sort of electrical noise is where PCM radios really prove themselves to be worth the extra money. The error-checked digital transmission system is much better at rejecting extraneous noise than simpler non-PCM setups.

Radio to Radio Interference

Radio interference commonly occurs when two radios transmit on the same frequency. In such a case, your robot will have a difficult time distinguishing between the two signals. The robot can stop responding, or it might respond to whichever radio has the strongest output power, or it might do some combination of the two. This can be a dangerous situation, because the robot can suddenly start to move or trigger weapons when it shouldn't. You should always carry various frequency crystals with you, and make sure that you are the only robot driver transmitting at a particular frequency. As noted, this is ensured at some events by the transmitter impound.

Some people build their own R/C systems that transmit under the 300-MHz, 900-MHz, 1.2-GHz, and 2.4-GHz frequency bands. Many companies sell products designed to transmit data or control signals that can be used to control a robot.

Some of these systems offer more control flexibility than traditional R/C systems. The drawback of using these frequencies is that other ground-use systems also transmit at the same frequencies. For example, cordless phones transmit at the 900-MHz and 2.4-GHz frequencies. A cordless phone near your robot could cause radio interference with your robot. Because of this, it is recommended that you use only radio transmitting equipment that has built-in error correction methods that can filter out unwanted information, such as the IFI Robotics system.

Antennas and Shielding

Antennas are used in combat robots to transmit data from the hand-held transmitter to the receiver on the robot. Without the antenna, you cannot communicate with your robot. One of the biggest problems most robots have with reliable control is not electrical noise but improper antenna setup.

The ideal antenna configuration would be a vertical wire of a length equal to one wavelength of the radio wave used for communication. This works out to nearly 14 feet, which is not practical for most combat robots—or most model aircraft or cars, for that matter. Most 72- and 75-MHz radios come with a 1/4-wave antenna attached, with a length in the range of 37 to 42 inches. Most robots do not have the length or convenient mounting room to carry an external antenna of this size, so the usual antenna length and placement are far from optimal.

A 1/4-wave antenna means 1/4 of the wavelength of the transmitter/receiver system's operating frequency. It's a unique characteristic of the physics of antenna design. The higher the frequency, the shorter the wavelength and, of course, the shorter length a 1/4-wave antenna will be. Light and radio waves travel at 300 million meters per second, so a 75-MHz signal will have a wavelength of 300,000,000 meters per second divided by 75 million cycles per second, resulting in a 4-meter-long wavelength—or about 157 inches. A 1/4-wave antenna should be 1/4 this wavelength, or about 39 inches.

A very important fact about antennas is that *they should be mounted vertically*. This not only applies to the receiver's antenna on the robot but also to the hand-held transmitter's antenna. These types of antennas emit their energy in a pattern much like a flattened doughnut, with the antenna passing through the doughnut hole. The greatest thickness of the doughnut, as well as the most significant signal from the antenna, is at the sides. Conversely, the "thinnest" part of the doughnut is the hole, which is what you see when you look straight down on it. And the thinnest signal comes straight out the end of the antenna.

If the transmitter's and the receiver's antennas were placed in space where there are no reflections, no signal would be created if they were pointed at each other. The greatest signal would be created when they were parallel to each other. In situations on Earth, especially in a room with a metal floor, the signals bounce around and reception can be accomplished with almost any orientation. You should always keep in mind that these reflections are far weaker than a direct signal, though, and you should never "point" the transmitter's antenna directly at your robot. The antenna on your robot and your hand-held transmitter should always point straight up for optimum signal transmission and receiving.

tip *You should develop a habit of holding the transmitter vertically in tests and trial runs so the strain of a hot battle won't have you accidentally pointing the transmitter at your machine or, worse yet, shorting out the antenna on the metal rail or supports of the arena.*

Antenna Placement

You may have seen some combat robots zipping about the floor in competition with what appears to be an antenna protruding out the top. It probably *was* an antenna—perhaps with a little flag attached so the operator can see the orientation of his machine for control purposes. This certainly is the ideal placement electrically, but it's a pretty bad thing when a flailing robot severs the antenna with a weapon. Sometimes you cannot find an adequate vertical location for the antenna, especially in a small, flat machine, so you are forced to place the antenna in a horizontal position. Fret not, though, because most model airplanes also have to place the antenna in this orientation. If this is the case with your machine, you should mount a nonconductive (nonmetal) strip of material on the robot's shell, under which you can place the antenna. Do not attempt to cut the antenna wire a bit (or add more wire) to make it fit in an area or try to improve the signal; the wire is cut at the factory to accommodate the appropriate frequency.

A rookie bot builder might simply pile the antenna wire loose inside the robot, or cut it short and tape part of it to the outside of the robot's shell. While the radio reception will be far from ideal, at a typical combat range of less than 50 feet you might get away with it. A better setup, though, is to have a flag or post extending out the top of the robot, and run the antenna up it to get it away from the main body of the robot and get better exposure to the radio signals. Even this is not an ideal antenna setup, but it will work for most bots.

The ideal antenna setup for a combat robot is to use a *base-loaded* antenna. Base-loaded antennas get away with having a short length of actual antenna by embedding a tuned resonance circuit in the base of the antenna module. Base-loaded antennas have to be purchased for a specific frequency band, but they save a lot of room over standard antennas: a base-loaded, 72- or 75-MHz band antenna can be as short as 6.5 inches. In some cases, the base-loaded antenna can be mounted inside the robot's body next to the radio, although this is not recommended. As mentioned, the antenna should be mounted vertically on the top of the robot. The base of the antenna should be at least 1 inch away from any metal parts on the robot's frame, and the wire from the antenna to the radio should be as short as possible and not run near any motor power lines. W.S. Deans sells a base-loaded antenna that is popular with veteran robot builders and can be obtained at most hobby stores.

The “pit” for between-battle repairs.

Jamie Hyneman, a.k.a. “Adam Savage,” in the pit sharpening his robot “Blendo.”

Razer, built by Ian Lewis and Simon Scott, was designed to be "a machine of ultimate offensive capability."

Jon Washburn and his robot "Scorpion" with a heavy spiked metal tail.

Carlo Bertocchini's "Biohazard," an
all-star champion of robot combat.

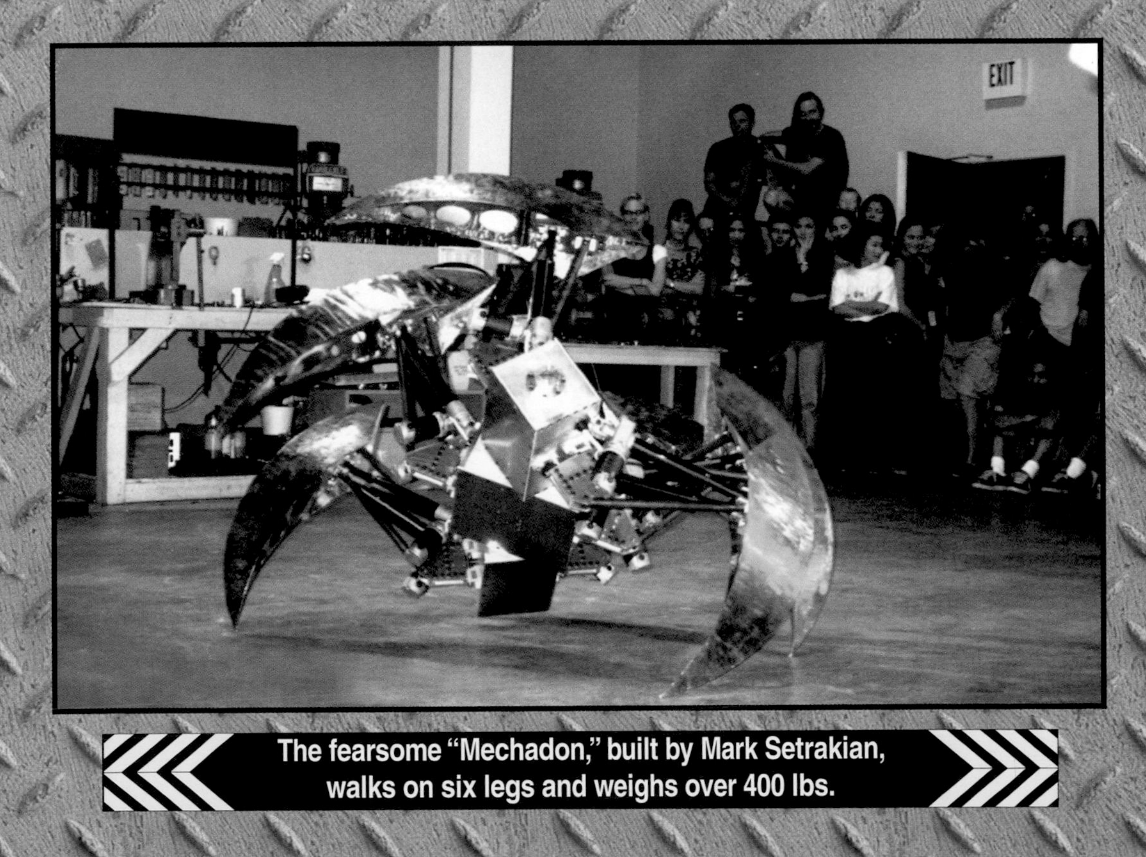

The fearsome "Mechadon," built by Mark Setrakian,
walks on six legs and weighs over 400 lbs.

Mark Setrakian's "Snake" and other robots in the melee round of competition.

"The Grinch" fighting "Dough Boy," a robot made with a steel lawnmower blade.

Cassidy Wright, 11, controls "Fuzzy" in the competition area.

"Tazbot" immobilizes an opponent.

"Vicious I," built by Mike Regan and the rest of Team Vicious, swings its sharp pointed rotors against its opponent.

Robots running amok during a melee round of competition.

Autonomous class competition: "Thumper" vs. "Gladiator II."

Innovation First Isaac Robot Controller and Other Radio Modems

The IFI Robotics Isaac R/C system was originally developed for the FIRST (For Inspiration and Recognition of Science and Technology) robotics competition. FIRST robots are designed to participate in a competition requiring rather complex mechanisms with jointed arms, telescoping grabbers, and complex omni-directional movement, which makes their control needs a lot more involved than that of a typical combat robot.

The FIRST system is built around a 900-MHz, bi-directional radio modem, which transmits high-rate serial data between the control gear and the robot. The transmitter gear is a modular design that is capable of using standard PC-compatible game-type peripherals such as joysticks, steering wheels, foot pedals, or custom user-built control gears. The receiver contains a user-programmable radio, which can control complex functions on the robot in response to commands from the transmitter. Digital and analog inputs to the receiver board can be used as feedback to the control system, or they can gather telemetry data to send back to the transmitter for driver displays or recording on a laptop computer.

The IFI Robotics system uses the 900-MHz band to transmit its control signals. The data packets traveling between the transmitter and receiver are coded with a team number to ensure that one IFI Robotics radio set does not interfere with another IFI Robotics radio set, which is a tremendous advantage over hobby R/C gear that has no way of distinguishing between one radio and another on the same frequency. The coded team number is custom settable by the users and the event's organizers. The bi-directional data transmission also gives the operator a clear indication of radio signal integrity, diagnostic lights on the operator interface tell the operator the status of the receiver, and a button on the transmitter control board can be used to reset forcibly the receiver's user-programmed computer system.

IFI Robotics sells two types of robot controllers—the Isaac16 and the Isaac32—that are similar except the Isaac32 has twice the number of output channels and onboard sensor inputs, and the radio modem is a separate item and not built into the system as is the Isaac16. Table 8-3 shows a list of the number of inputs and outputs in the two robot controllers.

Feature	Isaac16	Isaac32
Digital sensor inputs	8	16
Analog sensor inputs (0–5 volt, 8-bit A/D)	4	8
PWM outputs	8	16
Solid-state relay outputs	8	16

TABLE 8-3 *IFI Robotics Isaac Robot Controller Input/Output Specifications* ■

The "PWM" outputs are the same type of 1- to 2-millisecond signals that R/C servos and electronic speed controllers such as the Victor 883, Vantec, and traditional R/C car ESCs, understand. With this system, you could control 16 different high-powered motors—double the number of motors you could control with top-of-the-line traditional R/C systems. Then you can add up to 16 additional relay controls for weapons, actuators, lights, or just about anything else you would like to control.

What makes this system different from traditional R/C systems is its ability to analyze digital and analog inputs. In your robot, you could include tachometers on the motors to monitor actual rotational speed to implement closed-loop speed control. You could add thermocouples or resistive temperature sensors to the motor housing to monitor the temperature of the motors and help prevent them from overheating. In the robot controller is a Basic Stamp that can be programmed to read in the input values and send signals out to control the corresponding actions of the robot. Not only can the robot perform some semiautonomous actions, but the robot controller can send back information to the main operator interface so that the operator can be notified what the robot is doing internally. One set of data could be a self-diagnosis to monitor the health of the robot during a combat match, and you could even monitor the charge on the batteries in real time.

Table 8-4 shows a list of input and output features of the operator interface. The operator interface for the Isaac system is different from traditional R/C transmitters. With the traditional R/C transmitter, the radio frequency (RF) transmitter, joysticks, knobs, switches, and all the electronics are enclosed in one single hand-held package. The Isaac operator interface consists of a general electronics module and a separate RF transmitter/receiver module. All the joysticks, switches, knobs, and displays have to be added. The drawback to this system is that the entire operator interface has to be built. The advantage to this type of setup is that you could build an interface that has all the control features you want in the robot, and the features can be located where you want them. So, for example, the same joystick used with computer games can be used, or a simple potentiometer joystick found in traditional R/C transmitters can be used. The light emitting diode (LED) indicators

Input/Output Device	Quantity
Joystick ports	4
Digital inputs	16
Analog Inputs (0–5 volt, 8-bit A/D)	16
LED indicators, user defined	8
LED output drivers	8

TABLE 8-4 *IFI Robotics Isaac Operator Interface* ■

are user programmable so that they can provide feedback from the robot. The operator interface has a port called the dashboard port that can be connected to a PC so that the operator can get total feedback from the entire robot.

An interesting feature about this control system is that multiple operators can use the same controller to control the same robot. For example, one operator could be using one joystick to drive the robot around the ring, a second operator could be using a switch panel to control weapons on the robot, and a third operator could be monitoring system readouts and controlling a third panel for defensive weapons. Or the entire system could be set up so that one person drives the robot and sensors on the robot automatically control the weapons.

Figure 8-6 shows a block diagram of the Isaac operator interface and the robot controller showing component functionality.

As you can see, the IFI Robotics control systems are more powerful and flexible than the top-of-the-line PCM computer radios. The added abilities make the Isaac systems more expensive than the PCM computer radios, and many single-robot competitors in the smaller weight classes will find the price prohibitive. However, because the same Isaac system can be easily used on multiple robots, it's a good investment for a team with many entries. The Isaac radio receiver is physically larger than a typical PCM receiver. The smaller system—the Issac16—will fit in most robots

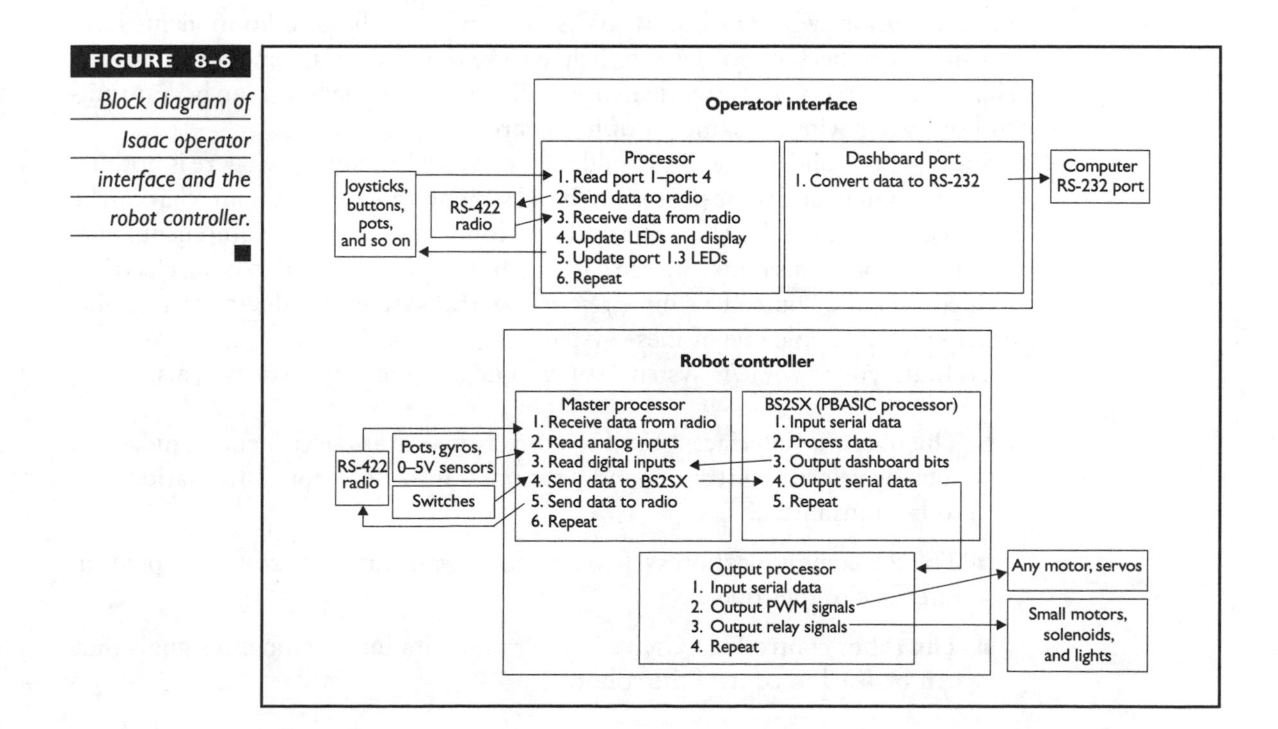

FIGURE 8-6 Block diagram of Isaac operator interface and the robot controller.

of 60 pounds and larger, but is generally too big for smaller robots. The Isaac system also requires that robot builders have more skills in electronics and software programming than those who use off-the-shelf R/C systems.

The reliability, amazing flexibility, and competition-friendliness of this radio system has made the Isaac system a hit among many top *BattleBots* teams. Because the Isaac system has proven to be reliable and resistant to radio interference from other radios, *BattleBots* is heavily encouraging the use of this controller in its events.

note *The* BattleBots *organization has reserved the 902- to 905-MHz and the 925- to 928-MHz frequencies for the IFI Robotics Isaac robot controllers. For most people, this doesn't mean much; but to those robot teams that want to build their own R/C systems using 900-MHz radios, this means that they will be prohibited from using these frequencies at a* BattleBots *tournament.*

Radio Modems

The actual RF transmission for the Isaac robot controllers uses a pair of RS-422 radio modems made by Ewave, Inc. (*www.electrowave.com*). RS-422 is a serial communication protocol that is more reliable than the standard RS-232 serial communication with which most of us are familiar. These radio modems have built-in error-checking software to help ensure that the data being transmitted is reliable and correct. These modems are bi-directional so that data can be transmitted both ways with the same set of hardware.

Some robot builders prefer to build their own R/C equipment. There is nothing wrong with that, and some people can build systems much better than what can be purchased off the shelf. The subject of developing reliable R/C equipment is beyond the scope of this book. Suffice it to say that you do not need to be an electrical engineer to build yourself a simple remote control system. Products are available to help you assemble one of these systems.

To build your own R/C system, you will need three major subsystems:

- **The operator interface** Used to convert operator control commands—such as velocity, direction, and weapons—into electronic information to be transmitted.
- **The RF communication system** Used to transmit data from the operator interface to the robot.
- **The robot controller** Converts the radio data into command signals that can be used to control the robot.

One of the easiest ways to establish RF communication with your robot is to use a radio modem. A radio modem sends serial data from a host device to a remote device—from an operator interface to the robot. All computers and virtually every microcontroller can receive and transmit serial communications data. Because of this, operator interfaces and robot controllers can be designed to transmit and receive serial communications, and the radio modems can be used to transmit the data between them.

A simple operator interface can be a microcontroller, such as the Basic Stamp or the Motorola 68HC11, to read in analog data from a joystick and digital data from a weapons switch, and to convert that data into serial communications data that can be transmitted. The robot controller can also use the same type of microcontrollers to convert incoming serial data to output digital signals for turning on and off solid-state relays for weapons and generate the 1- to 2-millisecond pulse modulation that motor controllers use to drive the robot's motors. The details of how to create the specific subsystems is outside the scope of this book, but in Appendix B you'll find several references to books that will explain how to build the various components that can be used in your own custom combat robot R/C system.

It is recommended that beginning robot builders use either a traditional R/C system or the IFI Robotics Isaac system. If you try to build your own R/C system, you will eventually end up with something that is functionally similar to the Isaac system, and you might end up spending most of your time building the remote control system.

For those of you who really want to build you own custom remote-control systems, research FCC rules on radio communications, seriously consider using radio modems, and remember safety is the number-one consideration that must be built into controllers. You must have failsafe and interference-handling features built into the control system, or you will not pass safety inspections. In addition, some competitions require noncommercial custom radio systems be separately pre-approved, far in advance of the actual event.

Failsafe Compliance

Whichever radio setup you use, most competitions have strict rules on failsafe compliance that must be met for your robot to pass safety inspection. Your robot must stop moving and deactivate all its weapons when it loses radio contact. This shutdown must occur even if the robot was in motion or had its weapon running at the time it lost radio contact.

Radio systems respond differently when a loss-of-signal condition occurs. AM and low-cost FM receivers simply stop transmitting servo pulses when they stop receiving a valid radio signal. Most electronic speed controllers shut down when they stop receiving a valid servo pulse, and R/C servos will simply freeze in place. The ESCs that shut off when a loss-of-signal condition occurs will fulfill the failsafe requirement with nearly any non-PCM radio. Mechanical speed controllers that use a servo to trigger relays to run motors will not pass a failsafe requirement test, as the servo will remain in its last commanded position when the radio shuts down.

Some AM and FM radios have the unfortunate habit of transmitting a few garbled servo pulses when they are switched on or shut off. Known as *chirp*, this behavior can cause the robot to twitch or fire its weapon when the radio is switched on or off. When using this kind of radio, the operator should adopt a policy of never switching the radio on or off when the robot is powered up; instead, the radio transmitter should be switched on before the robot is turned on, and it should stay on until after the robot is powered down.

Many of these problems can be solved with a failsafe board (Figure 8-7). Several manufacturers of radio control equipment sell modules, such as Futaba's FP-FSU1 Fail Safe Unit, which is connected between the radio receiver and the R/C servo or electronic speed controller. The failsafe board monitors a signal from the radio receiver; and in the event of a lost or badly garbled signal, the board generates a servo signal output that commands the servo to move to a preset level, or it shuts off the attached electronic speed controller. Some failsafe boards will even store enough power to center a servo in the event of battery failure.

Radio systems with computerized receivers, such as the PCM-type receivers, are smart enough to recognize when the radio signal has been lost and take appropriate action. Depending on the controller type and parameter settings, the shutdown behavior might be to return all outputs to a preset level or to keep all outputs at whatever level they were in when radio contact was lost. The latter is the default behavior on many model airplane and helicopter radios because it will keep the plane or helicopter in stable flight until radio control is regained. But this is not the behavior you want in a combat robot radio; it will cause your robot to keep moving on radio contact loss. This behavior is usually programmable. For a combat robot, the failsafe units should be programmed to shut down all motors, apply brakes to spinning weapons, and move servos to a safe position.

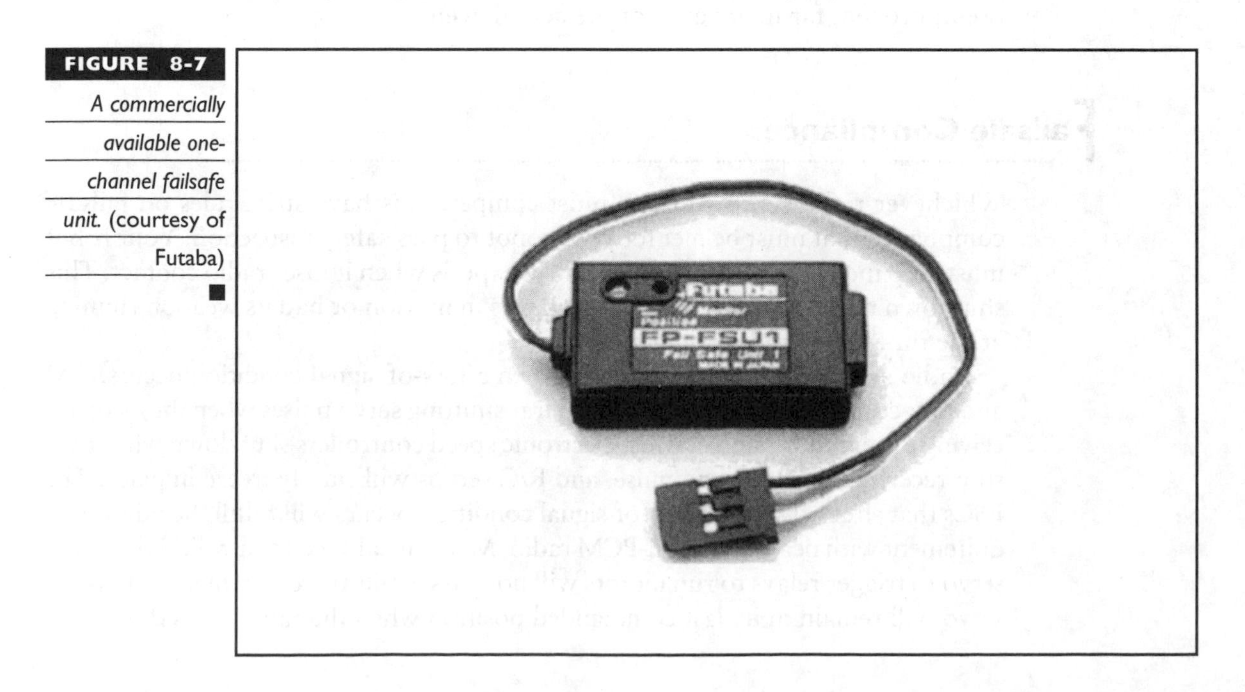

FIGURE 8-7 A commercially available one-channel failsafe unit. (courtesy of Futaba)

chapter 9

Robot Material and Construction Techniques

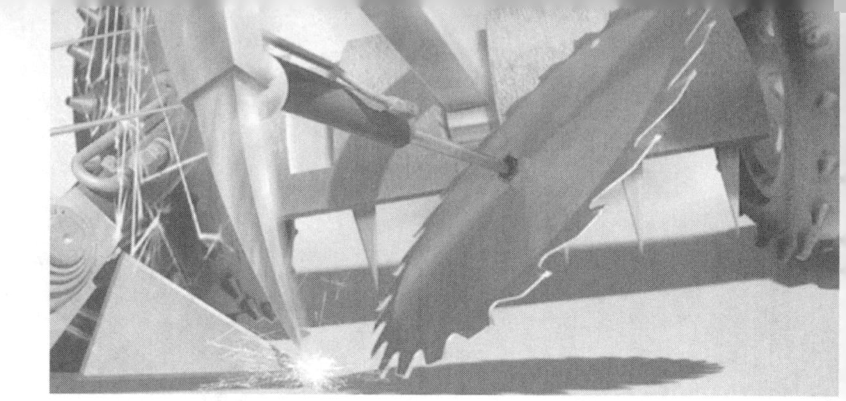

WHEN we human beings experience an injury or sickness, it's frequently our skin and bones that really keep us together. Carefully applied skin grafts after a serious burn or injury can mean the difference between life and death. Likewise, if you've watched a robot combat match, you know that a robot is doomed if its skin is ripped off by an opponent. The same follows for the failure of fasteners for a wheel assembly, a weapon, or a strategic internal system. If any of these are torn off in the arena, that robot is most likely going to lose the match.

The information in this chapter will help you make your own decisions about what materials and construction techniques you will use after thoughtful consideration of the many types of elements and fasteners available. Each material has a best application. Before you begin building, you should look up specifications in suppliers' catalogs and use logical design practices in the layout and construction of your combat robot. Use common sense. Talk with friends who have done mechanical design. Look at successful designs and determine just what made the design work so well, or what caused others to fail. Don't be afraid to ask others for advice. Get on the Internet and converse with those who have built a robot similar to what you have in mind.

Metals and Materials

When you think of durability, you probably think of metals first. However, some of the newer plastics offer many advantages over metals when it comes to building robots for competition.

High-Strength Plastics

With virtually unmatched impact resistance, outstanding dimensional stability, and crystal clarity, Lexan polycarbonate resin continues to be one of the popular types of materials for use in combat robots. The product is a unique thermoplastic that combines high levels of mechanical, optical, electrical, and thermal properties. GE Structured Products is one of the leading suppliers of Lexan sheet material.

At a recent *BattleBot* competition, GE handed out hundreds of hand-sized samples of Lexan 9034 to robot designers, some of whom immediately put it to use on their creations as protective armor or spacing material. Technical demonstration videos were on display and product specification sheets were made available.

Even the *BattleBox* was designed with four "layers" of protection using Lexan material to keep the deadly robots and flying parts from injuring spectators. Even this material is not impervious to all types of damage, as a large chunk of one of the Lexan panels had a large chunk torn out of it by a wayward robot in a recent match. Your local plastics supplier may have the material on hand, can order it, or can direct you to the GE Structured Products division (*www.gestructuredproducts.com*) nearest you.

Metals

Despite Lexan and other materials, metals are the material of choice for most robot structures and armor, and numerous types of metals are available for robot construction. While newer experimenters are often confined to using only those materials they can find at the local hardware store, surplus store, or junkyard, we recommend using the highest grades of materials you can get your hands on to construct your combat bots. (Appendix B at the end of this book will point out vendors that can help you get the best materials.)

Metal supply companies are available in larger cities, but many potential robot builders are not familiar with the best metal and materials to use for a particular type of project. Although we don't cover modern ceramics, plastics, and composites in this chapter, a plethora of alternative options such as these are available out there.

The word *strong* as applied to the various durability characteristics of metals and materials is often misused. For example, rather than look for a *strong* metal, you might want a metal for a particular weapon design that can take a lot of bending after being struck and not break, and you'll find that a piece of spring steel works well for that. Another part of your robot might call for a stiff rod, and you select an alloy of stainless steel. Your wheel hubs must be light, tough, and easily machined on your small lathe, so you select aluminum alloy 7075. Two nice pieces of brass seem to work fine as heat sinks for your drive motors. A thick piece of Kevlar you find in a surplus yard is destined to be your robot's sub-skin, to be covered by a sheet of 304 stainless steel bonded to it. All of these materials have their *strengths* and *weaknesses*.

Aluminum

Aluminum is probably the most popular structural material used in experimental robot construction. It offers good strength, though it's certainly not as tough as steel. Its best characteristics are its ability to be machined, its availability, and its light weight. You might be able to go to a junkyard and ask for aluminum, and the sales person will lead you over to a pile of twisted metal. Enter a metal supply house, and you'll be asked "what alloy, what temper, and do you want sheet stock or extruded?"—and a host of other questions. Extruded geometries include angle-shaped bars, tee-shaped bars, I-beams, C-channels, and square and rectangular tubing.

You can choose from among at least nine common aluminum alloys: 1100, 2011, 2017, 2024, 3003, 5052, 6061, 6063, and 7075. If that list makes your head spin, add to that numerous *tempers* for each of the alloys. Don't despair, for even though each of these alloys has an application where it fits best, we'll discuss only the few that seem to be best for robots—considering just how well you can machine it, its cost, and its availability.

Alloy 6061 at a temper of T6 seems to be one of the most versatile and readily available aluminum types for sheet stock. This popular aluminum alloy comes in sheets from 1/32 inch (0.032 inch) to several inches in thickness (the thicker version is called *plate* rather than *sheet*) and can be up to 48-by-144 inches in size. This alloy is available at aerospace surplus yards, metal supply houses, and the better specialty hardware stores, and it is fairly good for robot skin covering and excellent for internal structures. It welds, drills, and taps well. Alloy 6061 also comes in extruded angle stock, which is useful for fastening two pieces of sheet stock together at right angles for structures. Alloy 6063 is similar to 6061, yet it offers better corrosion resistance for wet applications.

Alloy 7075 is one of the hardest aluminum alloys and is an ideal material for machining high-stress parts. It is popular in aircraft and aerospace production. It also comes in sheet stock tempered at T6 and makes good robot skin. 7075 can be found at most metal houses and aerospace surplus yards.

Alloy 2024 is another "aircraft-grade alloy that offers high strength and is fairly machinable. 2024-T3 (T3 is a temper number) comes in extruded stock such as rounds and squares. Alloy 2011 is also easy to machine and comes in rounds and hexagonal stock. It is probably the best for threading and machining on a lathe and milling machine. Robot hubs, shafts, and similar items can be easily made from this alloy.

Aluminum alloys are easy to mill, cut, and drill, but the careful application of cutting fluid to these operations will greatly assist your machining operations. This is especially important in tapping aluminum. Tapping fluids used for drilling and tapping of steels should not be used. AlumiTap and special compounds designed for aluminum should be the only types used. This also applies to cutting large holes with a fly cutter or in sawing with a band saw. As always, use a good pair of goggles or a face-mask when machining any material.

Aluminum, as well as stainless steel, requires special talents and equipment to weld properly. Both require what are commonly referred to as *wirefeed welders*, also called MIG (metal inert gas) welders, or TIG (tungsten inert gas) welders. You might have seen cheaper varieties of these types of welders in cut-rate tool catalogs or stores. This is an area where more money means a better job, and cutting corners just to own a MIG welder will cost you in the end with poor and weak welds. If you want to save money, go to a welding shop that specializes in aluminum and stainless steel welding and have a professional do it right the first time.

What you'll pay for the job will cost you far less than what you might pay for a cheap TIG or MIG welder, and you won't have to go through a learning curve and deal with joints that may fail. Welding is covered more extensively later in this chapter in the section "Welding, Joining, and Fastening."

Stainless Steel

Aluminum is certainly not the only material available for robot construction, and nobody can say it is the *best* structural material for all applications. Stainless steel is popular for many applications with robot construction, especially for tough robot skin uses. Alloy 304 is one of the most popular forms of these alloys and is used in many applications where formed sheet steel is best, such as for sinks (and robot shells). It typically comes in 36-by-36-inch sheets from 0.024 inch to several inches in thickness. It welds well, providing you have a good TIG welding system. Again, we recommend that you have your welding done by an expert who deals with stainless steel, such as professional welders who make food-processing equipment.

Stainless steel sheet metal is usually recognized by someone who does not know metals as a "steel-like" metal that is weakly magnetic or totally non-magnetic, though some high nickel steel alloys are magnetic. Stainless steel alloys contain iron as the basic element plus a small amount of carbon. They also contain the element chromium and are sometimes called *chrome steel*. At least a dozen alloys can also contain various amounts of nickel, cobalt, titanium, tantalum, manganese, molybdenum, silicon, and even sulfur that give the different alloys specific properties for particular uses. The most desired property of stainless steel is its resistance to corrosion and rust.

Stainless steels are usually categorized in three groups: austenitic, martensitic, and precipitating-hardening alloys. Austenitic stainless steel alloys are low-carbon based with nickel added to enhance workability. They are hardened by cold working and are slightly magnetic. They have excellent corrosion resistance and are easily welded. Alloy types 304/304L are some of the most popular alloys and are easily welded, and these are used extensively in food processing equipment. This alloy can be purchased as round stock from 1/8 inch to several inches in diameter in 3- to 6-foot lengths. Sheets are available from 0.024 inch to several inches thick, and in sizes from 12-by-12 inches to 36-by-96 inches. It welds well using a good TIG welding system and a good welding professional. Another useful alloy in this series, type 347, has tantalum and cobalt added for greater hardness and is used as machinable rounds and in pressure vessels.

Martensitic stainless steels are not popular in most robot applications because of their lower corrosion resistance and poor weldability. Type 440C is a high-carbon alloy that is used in gears, bearings, and shafting. It is available as round stock and can be heat treated. (Heat treating is done to change the mechanical properties of the metal.) It is hard, giving good wear and abrasion resistance.

Precipitating-hardening stainless steels are particularly useful for high-strength applications after heat treating. Alloy types 17-4, also known as type 630, and 15-5 are the most popular alloys in this group. One of its greatest uses is for springs, but it also finds uses in gears and shafting. It is available in round stock from 3/16 to 4 inches in diameter.

Cold-Rolled and Mild Steel

Standard cold-rolled steel is frequently used in robot construction, especially in combat robot–style machines. This can be as extruded galvanized 1011 angle used for base or weapon construction, or it can be used as sheet stock for various applications. Alloy 1018 is probably the best steel for welding and machining. *Plain* steel, if unprotected, has the bad habit of rusting, even in air. It is harder to machine and saw than aluminum, but it is stronger for most applications.

Most of the stock cold-rolled steel is not galvanized and is ideal for welding. These alloys are also prone to rust, which can cause you a lot of grief after the robot is completed. After your robot structure is completed, whether by welding or by nut and bolt fasteners, it is a good idea to sandblast the structure and immediately coat it with a preservative such as anodizing or a thin plastic conformal film. This will protect the surfaces and allow quick and secure electrical ground connections on parts of the structure, providing the coating is removed at the electrical point of contact. Sandblasting is particularly important before welding, and further hand filing may be necessary to prepare the surfaces to be welded.

Most of the softer steel alloys such as cold-rolled steel are easy to machine, though not quite as easy as aluminum or brass. Slower drill speeds are recommended, which can be found in most shop handbooks, such as the *Machinery's Handbook,* or in the lids of many drill indexes. Keep the operation well lubricated with a good-quality cutting fluid. You should take care to feed drills, mill cutters, and saw blades slowly to the metal. As mentioned earlier, always use a good pair of goggles or a face-mask when machining any metal.

Brass

Brass is another alloy that has useful applications in robotics, particularly in smaller machines. Most brass alloys are easy to machine. Alloy 260 sheet stock is readily available in sizes up to 24-by-96 inches, and in thicknesses from 0.10 to 0.250 inch. Alloy 360 is another brass alloy that many metal supply houses carry. It is also called *free-machining brass*, and, as the name implies, it is best for machining of small parts, fixtures, hubs, and similar items.

Brass also has an excellent property of being able to be brazed or soldered by simple, easily obtainable home shop tools. The low-cost, Bernz-o-matic–style hand torch can be used to braze brass (and bronze fittings) to similar alloys. The use of a larger Presto-lite torch might be needed to braze larger sheets of stock that carry the heat away too fast. A large soldering iron or soldering gun can be used to solder small brass pieces together, but these should not be used in high-strength areas or where shock may be present.

Many hobby shops carry miniature brass extruded sections in 12-inch and 36-inch lengths that are great for small robot construction. They come in square, rectangular, hexagonal, and round tubes that fit closely within each other for telescoping applications, as well as channels, solid sections, and sheet stock. Sizes vary from 1/32 to about 1/2 inch. Note, however, that brass has a poor strength-to-weight ratio, and is therefore not a good choice for most combat applications.

Titanium

Titanium is finding more use in combat robots. Though "heavier" than aluminum at a ratio of 1.7:1, it does not really compare with aluminum—or any other metal, for that matter. Long used by the military for lightweight armor and jet engine parts, it is finding uses for consumer applications such as combat robots. It melts at a temperature of almost 1000 degrees Celsius higher than aluminum, and can withstand deformation and bending much better than that alloy or most steels. Its main drawback is its extremely high cost and difficulty to machine and form, but it is becoming more popular for so many uses that the cost is dropping rapidly.

Titanium alloy 6AL-4V is a general-purpose, high-strength metal that is available in round bars and flat sheets. As with all titanium alloys, it requires patience in machining. Ample lubricant and slow feed speeds are necessary. The 40,000 psi yield strength alloy is an easier-to-machine alloy. Each can be found in lengths of 3 and 6 feet, and diameters from 1/8 to 2-1/2 inches.

Using Extruded Metal Stock for Robot Structure

In discussing the many types and alloys of metals available for robot construction, we mentioned the many forms in which the metal is available. Careful thought in design can make use of these forms not only to add to the structural integrity of the robot, but to simplify the construction. Co-author Pete Miles made use of a wide piece of aluminum C-channel stock to form the sides of his robot *Live Wires*. This heavier piece of preformed metal not only offered much greater side strength from possible puncture by an opponents weapon, but it offered him a simple and secure way to fasten the upper and lower plates to form the overall structure. Figure 9-1 shows how *C*-channel extrusions can be used as external robot structures.

The most common form of extruded structural shape is the angle, or *L*-shaped, piece of metal. These shapes can be used in two different ways to achieve a stout and robust structure for your robot. Each of the sides of the robot's frame can be constructed of pieces cut to form the edges. If either of the metals is to be welded, individual end welds will not have sufficient strength without the help of a "gusset" welded into the corners. These triangular pieces of metal add tremendous strength to the overall structure. Figure 9-2 illustrates a simple gusset arrangement.

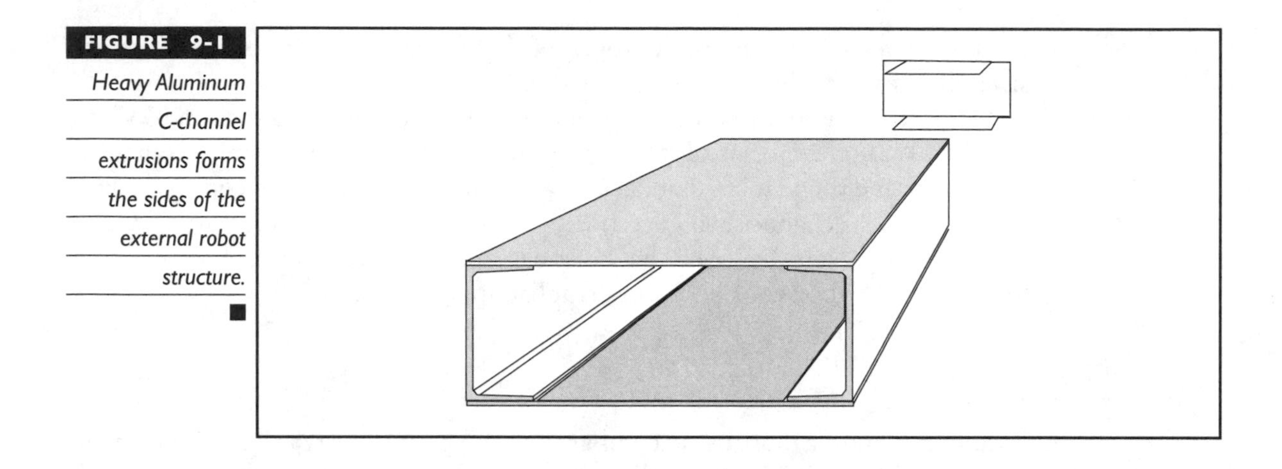

FIGURE 9-1 *Heavy Aluminum C-channel extrusions forms the sides of the external robot structure.*

Angle extrusions are not the only method used for attaching pieces of sheet stock to each other. Extruded square and rectangular tubing and even various sizes of C-channel offer the same edges to which you can attach sheet stock. C-channel is available in thicknesses of 1 inch to 15 inches. In selecting the extrusions to be used, you must remember that the stock must have walls of the appropriate thickness for the robot you're creating—that is, as thick as possible. You gain little weight to obtain the greatest bending resistance.

As mentioned, most robot designers have relied upon the common steel angle iron pieces to form a robot structure. This is an excellent approach, as long as you take care to examine the load paths encountered in the robot as it operates in the battle environment. You do not need to go into a complex stress and structural analysis program to determine potential load paths within the overall robot structure. For example, if you expect to encounter an extreme load from a type of weapon striking downward upon the center of your robot, you might consider placing a central tubular column within the robot to help transfer loads into the

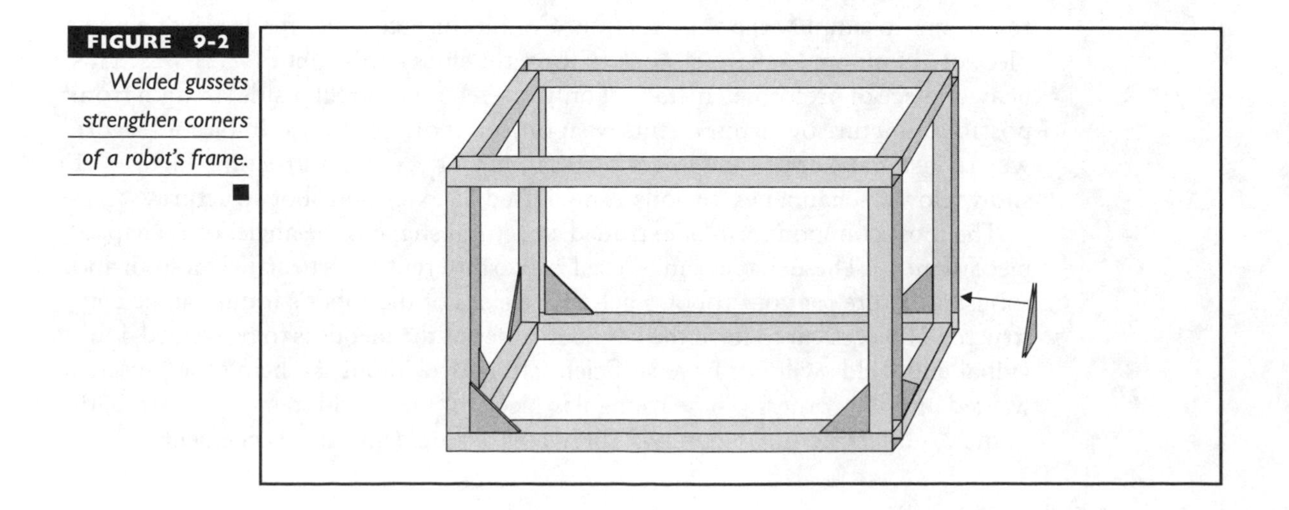

FIGURE 9-2 *Welded gussets strengthen corners of a robot's frame.*

base. An excellent book on structures and how they bend when loaded is *Design of Weldments*, by Omer Blodgett.

How to Know When You Need a Sponsor

Building and maintaining a robot for competition is expensive. Many builders admit to spending tens of thousands of dollars in pursuit of their robot dreams, and that's in addition to the hundreds or even thousands of hours of personal time they invest as well. Indeed, Team Coolrobots' Christian Carlberg finds that each robot requires him to learn a new skill. "One robot was parts intensive, so I learned the value of using a CNC milling machine to spit out parts. Another robot had a lot of steel, so I learned to weld."

Robots are so time and money intensive that you might want—or need—a little help. Following in the footsteps of sports like auto racing that meld technology, sheet metal, raw human skill, and intense competition, many robot builders have embraced sponsorships to help defray expenses. Sponsors come in two flavors: part sponsors contribute free or highly discounted gear to builders, while financial sponsors deliver direct financial support that allows builders to buy parts and equipment, as well as travel and pay for other incidental expenses. In return, sponsors get their name associated with the robot, which can be a valuable asset when it, or you, appears on television.

If you're interested in getting your own sponsor, many veteran builders caution that it takes effort; a professional, business-like approach; and, in many cases, an established track record with a completed robot. *Diesector* builder Donald Hudson acknowledges that sponsorships are more difficult to land in today's competitive environment. "It's certainly tougher to get sponsors nowadays. A few years ago maybe 40 percent of the robots would be shown on TV. Today, if you have a brand-new robot, the chances of getting on TV are kind of rare. Sponsors want their name to be seen, so it's like other racing—it's a tough sell if you don't have any rankings yet."

Christian Carlberg says, "Team Coolrobots is one of the best-funded teams in the competition, but it didn't happen overnight. I first developed a reliable track record. Then I put together a package of our accomplishments and made a strong argument why 'Company Blank' should fund us in exchange for advertising space. Then I searched out possible sponsors. It takes a lot of time to find someone interested, and then it takes a lot of time to convince the company that it would get a lot of exposure on TV."

To begin with, you'll need to make contact with a company representative. When dealing with a smaller or local business, you may find yourself talking directly to the owner or CEO. At larger businesses, you'll probably talk to a marketing manager. In general, larger companies will be more receptive. Says Team Blendo's Jamie Hyneman, "The larger the business the more likely they'll feel enticed by national TV coverage, and the more money they'll have."

How to Know When You Need a Sponsor (continued)

Team Nightmare's Jim Smentowski doesn't think impersonal correspondence is effective. He always recommends meeting in person. "Show your robot to your potential sponsors in person. Don't just e-mail or call them; you need to meet with them in person. Hype your bot and explain how much publicity the show gets, and the potential for your robot to be on TV and toys."

Sponsorship meetings aren't the time for humility or modesty. Be proud of your robot; be up-front about your talents and combat record; and back up your sales pitch with visuals, such as videotape from a televised event. Donald Hutson, of *Diesector*, says he went equipped with pictures of his robot and video clips of his appearance on the Tonight Show. "That was all they needed to see; they said 'that's cool' and became a sponsor." You may also want to emphasize that you already use the company's product in your robot. This demonstrates that you understand the company's product, that you're not just looking for random acts of generosity, and that the company's widget has a track record in combat.

If you dislike "selling" yourself and prefer to be relatively self- reliant, sponsorships can also be somewhat uncomfortable business propositions that take some adjusting to. Says *Deadblow*'s Grant Imahara: "The best part about having sponsors was e-mailing a list of parts and getting them in the mail in a few days. The worst part about it is actually mailing the list, trying not to feel guilty for asking for too much."

Most builders agree that part sponsors should be your first goal; don't bother trying to get direct financial sponsorships until you have established yourself and your robot. Financial support is essential to your plans to reach the next level. Not only is it often easier for a vendor to divert a few products off of its production line than to write a check outright, it can cost them less as well, since they're donating only the presales cost of the product, which is a lot less than retail.

Carlo Bertocchini, *Biohazard*'s papa, says to build your robot first. "Then enter it into a competition and get a national ranking number. Getting a company to consider a sponsorship proposal will be a lot easier with a proven robot. Even if it ranks low, it is a lot better than going to a sponsor with nothing to prove you are serious and capable of building a robot. Trying to get sponsorship without a robot is like trying to get a job without a resume."

Christian Carlberg agrees. "Gaining sponsorship is difficult. The best way to get a sponsorship is to first build a successful robot, then go after sponsorship money. It is much easier to find a company that manufactures the parts you need and then ask them if they are willing to donate parts in exchange for sponsorship. Over time your minor sponsors might grow into major sponsors."

A financial sponsorship has an extra layer of complication: what is the sponsorship worth to both you and to the company giving you the money? Jamie Hyneman says to avoid exclusive sponsorships unless you're getting a fortune, and not to tie sponsorship payments to specific competition results, since winning is far from predictable. He also says to tailor the amount you ask for to the size of the sponsor. "Bob's Auto Parts isn't going to give you $10,000 unless Bob happens to be your uncle; Microsoft might."

We've lightly touched on some of the more popular metals in common use for robot experimenters. The actual machining and use of these materials is covered in many textbooks and shop manuals. *The Home Machinist's Handbook*, by Doug Briney, and other books offer valuable hints and instruction for home machinists and mechanical experimenters. This particular book is geared around small table-top lathes and hand tools available to the hobbyist. A few words should be mentioned about the machining of metals with hand power tools and drill presses, tools often found in the shops of robot builders.

General Machining Operations

When it comes to constructing your robot, keep a few "golden rules" in mind: Keep your tools sharp, lubricate cutting operations, clamp your work piece and tool if possible, always use safety goggles, and use common sense for shop safety. Drilling larger holes in harder metals, such as steel, requires slower speeds and continual lubrication using Tap Magic, Rapid Tap, or similar products. Aluminum cutting and tapping requires different lubricants, such as Tap Magic for aluminum. Remember that sanding, grinding, and filing of softer metals such as aluminum can "load up" your sandpaper or wheel, so plan accordingly. You will be amazed what you can machine and construct in a home shop with simple home tools and a bit of ingenuity.

Tools You Might Need to Construct Robots

You certainly do not need a machine shop outfitted with a top-of-the-line milling machine (upward of $5000), a heli-arc welder, a 16-inch metal band saw with blade welder, and a floor model 12-by-36-inch machine lathe to build a competitive combat robot. Hiring out the complex machining can save you a lot of money over the purchase of these machine tools. You do need a certain amount of basic tools to be able to build the robot's structure, drill holes, and apply fasteners, however. After some experience, you may want to buy more specialized power and hand tools.

Obviously, a set of basic hand tools such as screwdrivers, open-end wrenches, socket wrenches, and various pliers is a must. Most home car mechanics already have a great start on many of the required hand tools. The extra tools that might be considered as musts are the metal handling tools such as files and deburring tools for smoothing rough edges, rasps for roughing out holes and slots, pin punches for inserting and removing pins, and a good drill set.

Drill indexes come in various sizes and qualities. A first set might be a fractional set of high-speed steel drills. A better set is a larger numbered set with extra lettered drill bits included. Most of the sizes you will use fall within the 1–60 number sizes. A 60–80 set is used only for drilling tiny holes. The lettered sizes are used for sizes larger than a quarter inch. You might want to spring for a few extra bucks to buy a titanium-nitride set of drills that last a lot longer. As you find your most used drills beginning to dull, you can also buy a drill-bit sharpener.

Of course, to use the drills you need a drill motor. If you're on a budget, you might consider buying a good cordless drill such as ones made by Makita, Bosch, or DeWalt. These tools can serve you well during construction and then later in the back areas of the various competition sites where electricity may not be available. For small work only, you might consider a Dremel high-speed drill set.

The next power tool should be a small bench-top drill press used to drill multiple layers and keep all holes perpendicular to the surface you're drilling. These can be found in some of the import tool shops for low prices—$40 or less. A drill press offers a lot of advantages over a hand-held drill. It can be used with a fly cutter to cut large holes in sheet metal, and it can handle larger drill bits that cannot be accommodated in a smaller hand-held drill. Other attachments can be used for polishing, sanding, deburring, and grinding. A helpful tip when drilling multiple parts that have to be fastened together is to drill one set of holes and attach the fasteners before drilling the next hole. This will ensure that all sets of holes are kept in alignment should something slip a bit during construction.

Cutting metal can always be accomplished with a hacksaw, but larger cuts can be tiring if done by hand. Some builders have used a hand-held saber saw fitted with a fine-toothed metal cutting blade to cut large pieces of thick sheet metal. A better way to go is to use a reciprocating saw such as the Sawzall, which can rip through sheet metal, bar stock, tubular extrusions, and pipes quite easily. Metal band saws can be quite expensive, but you can buy a metal band saw made for small stock materials for under $200. These saws can cut in the horizontal or vertical positions and can be fitted with a small table to guide small pieces of metal to be cut.

Bench sanders help make metal edges even and smooth, and a bench grinder is useful for working with metal forming. Pneumatic hand tools such as drills, impact wrenches, and sanders are inexpensive and offer a different approach to power tools. Woodworking tools such as routers, planers, and wood saws help form nonmetallic workpieces. A good bench vise is useful to hold any type of work piece.

As you become more proficient at working with metal, you will probably want to buy more tools. Rather than invest in larger power tools, you might consider buying tools to help you in the construction process and wait on larger machine tool purchases. It has been said that "you can never have too many clamps," and this certainly applies to building metal structures. Clamps come in handy to hold pieces together while you drill and screw them together, or even for welding. The standard 3-, 4-, and 6-inch C clamps can serve a lot of purposes. Several large bar clamps or furniture-style clamps can help hold together large structural pieces while fastening.

Yes, you can end up spending a lot on tools; but after the battle is over and you are ready to build that new machine, your tools will be waiting for you. Take care of your tools and they will take care of you. Always remember, safety for yourself and those nearby is very important when using any tools.

Welding, Joining, and Fastening

We're not about to tell you all there is to know about fasteners in these few pages or give you a course in Fasteners 101. The McMaster Carr industrial supply catalog has more than 250 pages of fasteners for sale. We cannot even tell you which particular fastener is best for your particular robot project because so many varieties of robot designs are built for so many purposes. We will attempt to list and describe those fasteners that have proven useful in robot projects we've been involved with or that have had positive feedback.

Structural Design for Fastener Placement

Before even laying out the design and figuring out where you need fasteners, you need to have an idea of the load paths that are present in the robot's normal operations, as we discussed earlier for structural members. You determine a load path by examining every possible location where a load may be placed, and then determine just what pieces of structure might transfer that load.

As your robot sits on a workbench or shop floor, it must bear very little weight; but once a robot begins to operate in and out of the arena, stresses build up, especially in a combat robot. You don't need complex finite element analysis or failure-mode analysis software to determine load paths and stress analysis. You can imagine that the robot was made of sticks and cardboard and held together with thumb tacks and consider this: "What would happen if I pressed here or struck it here?" You might want to construct a model made of balsa wood and cardboard to determine where you might want to place welded fillets or support brackets.

Some of the failures of a combat robot occur as a result of a failed structural design. The robot's skin is peeled off because the designer did not contemplate all of the potential stress areas. A weld breaks, a screw is sheared in half, or a weapon comes loose and flies across the arena only to have the robot disabled due to an unbalanced condition. A designer sees his robot flattened by a weapon because an internal member was fastened with cheap pop rivets, and $2000 worth of electronics is fried in the resulting short.

Once you've got your robot's design all worked out, you can start to think about the best ways to assemble it. If you're building a combat robot, words like *strong*, *tough*, *resilient*, and similar phrases come to mind. Your creation will leave your workshop and enter an unfriendly battlefield where every opponent is trying to smash it to bits, not to mention the actual arena itself with its many hazards. Your machine has to stand up to a lot of abuse.

If you look at heavy off-road equipment, you see that its sturdiness comes not from fasteners, but from heavy steel construction. Large machines weigh many tons, far above even the heaviest robot. Heavy steel forgings and castings are welded together or connected by huge bolts and pins. Battle robots contain heavy batteries, weapons, and motors and have a minimal amount of mass left to apply to structural needs. Careful design using strong but light fastening methods is important.

Arc, MIG, and TIG Welding

Welds seem to be the first thing that comes to mind when considering a sturdy robot's construction. You might successfully build a neatly welded robot and try it out in your driveway, deftly spearing your trash can filled with a hundred pounds of trash and tossing the whole can into the neighbor's yard. You spin the robot in a series of victory circles and yell, "Yeah! I'm ready!"

At your first bout, though, you're up against a machine made of unforgiving steel and it pounds your robot silly. Several welds split and your bot limps into a corner, smoking. "What happened," you ask? You think back to the test run. The thin aluminum or plastic test trash can gave easily when you slammed into it—and it didn't fight back. A better test would have been to have your neighbor, who's still a bit ticked at you for all the mess in his yard, take a sledgehammer to your robot.

Some home robot builders might have a cheap MIG welder available to weld aluminum, and possibly a gas or arc welder for steel work. The oxyacetylene and standard arc welder that you bought at the large warehouse hardware store are keepers, but the MIG/TIG welder you choose should not be a cheapie, as mentioned earlier.

MIG and TIG welders do not use a welding rod with a coating that burns off to protect the joint like in an arc welder; instead, they use an inert gas flowing from a nozzle to bathe the hot joint and protect it from atmospheric oxygen contamination. This gas, which is usually argon, helium, or sometimes dry nitrogen, comes through a regulator and hose connected to the welding nozzle or gun. In the MIG, a welding wire from a reel in the welder is fed through the center of the gun. The wire is selected for the particular type of metal being welded. A trigger in the gun feeds the wire to the joint being welded at a speed controlled by the person welding. The rest of the system is similar to a standard arc welder, a transformer feeding a high current and lower voltage to the wire that arcs to the metal being welded.

In TIG welding, a small tungsten rod is mounted inside the welding gun. Wires of various composition and thickness are hand fed and mixed into the pool of metal created by the heat, or arc, of the hot tungsten rod.

Other wire-feed welding units actually melt the wire to form a fillet of metal from the wire. Some types of welding systems, such as plasma arcs and heli-arc systems, are used for special, high-strength joints but are generally inaccessible to most robot builders.

Welds look great and hold tight when the welder is a pro and can make a smooth, seamless weld along the joint of two pieces of metal. A properly welded robot structure is usually far more stout than a similarly screwed one. Amateurs who build robots generally have talents that run more to the mechanical or electronic areas, and they can make pretty amateur welders. Welds in the lighter sheet metal used in robots are not always as strong as they look and can break under shock loads.

Welds also have another bad feature in that they are difficult to repair, especially in the field. You might think that simply rewelding the same broken weld will repair it as the metal melts in the seam. But Unseen oxidation may have taken place, or some liquid may have entered the crack in the weld, and the resulting repair will be poor, at best. Unless you have a large mobile van filled with welders and tools on site, manned by a team of mechanics, your better bet is to use some type of removable fasteners to attach your bot together. Welds, when properly made, are quite often the best, and sometimes the only way to attach two pieces of metal; but home experimenters should concentrate on nuts, bolts, and screws.

Screws, Bolts, and Other Fasteners

Fasteners such as screws, bolts, and rivets have the ability to give a bit when stressed and still retain their fastening strength. This may seem like a weakness, when, in fact, it is a strength. Of course, the ability to easily remove a fastener to disassemble a part of your robot for repairs or replacement is priceless in the field of battle.

A rule of thumb for bolts and machine screws is that the thickness of the material that has the threads tapped into it must be at least four times the thickness of the thread pitch (or the length of four threads). All the loads in a machine screw or bolt are supported by the first four threads. The rest of the threads do not support the loads until the fastener starts to stretch. When using screws in thin materials, the machine screw or bolt diameter should be selected based on the thickness of the material they are being screwed into—not just the diameter of the fastener.

Most fasteners that we commonly think of in robot construction are screws, bolts, and rivets, with the needed nuts and washers. Many other types of fasteners and many varieties of the above-mentioned fasteners, such as cotter pins, blind or "pop" rivets, nails, threaded rod stock, set screws, retaining rings, and so on, are also important. These are all important mechanical construction fasteners, but we'll focus on bolts and machine and self-tapping screws for our robot building.

If you look in industrial supply catalogs, you'll see items sometimes listed as bolts, and other times called screws. For argument's sake, we'll called the threaded items that usually require a screwdriver or an Allen wrench to install *screws* and the other items that generally require a wrench to install a *bolts*. Generally, screws are of the smaller variety from 4 to 40 and even smaller, to about 1/4 to 20 in size. Bolts are larger. (More about these sizes a little later.)Two types of screws are used in robot construction that involves fastening to metal: the sheet metal or self-tapping screw that looks something like a wood screw, and the machine screw that normally uses a nut to complete the fastening. Of course, you can drill and tap a hole in a piece of metal and insert the type of screw that normally uses a nut to fasten pieces of metal together.

The machine screw is available in numerous configurations; some are so similar that most people can't tell them apart. The round-head machine screw is probably the most common and has a partially spherical head that fits entirely on top of the piece of metal it's fastened to. The pan-head machine screw is a common variation that is similar to the round head but slightly flattened. The flat-head screw requires a counter-sunk hole and the round head screw head is sunk into the metal with the top flush to the metal.

The oval-head screw is a combination of the flat head, in that it is counter-sunk, and a pan head that is not flush. These screws usually are of the most common slotted-head or Phillips variety, with many available with hexagonal sockets for Allen wrenches. Many other types of screws can be used for security and other purposes, which we won't cover here.

Unless you have access to aerospace-quality fasteners, when you need to select machine screws for robot construction purposes, your best sources are your local larger hardware store or maybe a surplus store. Quite often, you will find that round-head screws are not of the highest quality. Their steel may be of lower quality and the screws tend to break easily. They are also not the best fasteners for attaching the robots "skin" to the internal structure, as they protrude outside the skin and can be struck by a swinging weapon.

Flat-head machine screws that can be countersunk into a robot's protective skin usually prove to be the best. They are made of a higher quality steel, usually 18-8 stainless steel or other steel alloys, and the better varieties are of the Phillips type. Drill the center hole and then counter-bore the hole to accept the recessed head of the screw. Drilling to the correct depth takes a bit of practice, and the use of a drill press is recommended because most have adjustable stops to keep the operator from making the hole too deep.

The countersink usually used for flat-head screws is 82 degrees, and you can buy drill/countersink combinations at larger tool supply places and from mail-order catalogs. Most experimenters find that a three- or four-flute countersink with a half-inch diameter works well with aluminum. One bad feature with using flat-head screws with countersunk holes is the chance of going a bit too deep and ruining that location for fastening. Another bad feature is that countersunk flat-head machine screws provide the least "holding power" due to the weak rim of the countersunk hole. Nevertheless, when properly machined, these screws seem to be the best for external robot skin applications.

Most cap screws are also one of the strongest types of screw. They are about the same strength as "grade 8" hardware. Flat-head cap screws rather than flat-head machine screws may be used when the protruding screw head is not an issue. The hexagonal drive type for cap screws is the most common variety because an Allen wrench can use a lot of torque for tightening. You won't find a wide variety of cap screws in a small hardware store, but larger suppliers will have a good selection for your project.

The pan-head machine screw seems to be the best for internal structural assembly. Most of the better varieties are made of 18-8 stainless steel and are of the Phillips type. This screw has excellent holding power due to the large head and larger flat area touching the metal. The pan-head machine screw, as well as the round-head, can use a washer to increase the holding area and, therefore, the *tensile strength* (the ability of the screw to prevent itself from being stretched apart or being pulled out of the hole).

All of the screw types mentioned here have either threads that are along the whole length of the shank or partially near the end. Either type will normally work fine for most robot applications.

Generally, most of the screws used in experimental robot construction are 6-32, 8-32, 10-32, and 1/4-20. Here's what these numbers mean: The 6-32 means screw size number 6, or 0.138-inch diameter with 32 threads per inch. This is a *coarse* thread for this size screw; likewise for a number 8 screw, but a *fine* thread is used on a number 10 screw. In the 1/4-inch sizes, 1/4-20 is coarse, and 1/4-28 is fine. Screws get much smaller, such as an 0-80, which is 0.060 inches in diameter with 80 threads per inch—or even as small as 000 size, or 0.034-inch in diameter.

If you're going through a surplus house and find a good buy on screws and bolts, make sure you locate the proper nuts for them because, for example, a 1/4-20 nut will not fit on a 1/4-28 bolt or screw. Bolts are generally larger and range from 1/4-20 or 28 to 1/2 inch or larger. Metric screws and bolts are becoming increasingly popular, especially on automobiles, and are designated in millimeters or fractions thereof; be careful not to mix the two types, though, as one will not fit on the other.

We mentioned tensile strength earlier as the ability of the screw to withstand stretching before breaking, but *shear strength* is probably the most important quality of a machine screw in most robot mechanical applications. High shear strength is the ability of the screw's shank to withstand shearing action—not the ability of the screw to be pinched in half or bent until it breaks. Hand-held crimpers for wire terminal lugs often contain screw cutters that allow a person to screw in a 4-40 to 10-32 screw and then shear it off to a desired length.

In a typical combat robot match, a robot can be struck repeatedly by an opponent's weapon(s) until its internal members literally start to shear the fastening screws in half. Many mild steel screws purchased in small plastic packages at hardware stores can easily fail the shear-strength test. You need to pay close attention to the type of steel used in the screws. You will certainly pay more for 18-8 stainless steel screws, or the even more expensive alloy steel screws; but large robot construction, especially combat robots, requires the extra strength.

Now that you've got a good idea of what fastener you're using on what parts of your robot, take care to install them correctly. If you're boring several holes in several pieces of metal that use multiple fasteners to hold them together, clamp the metal pieces together and bore the first hole through all the metal pieces. Insert your fastener through the hole and tighten a nut on it. Do this with each new hole. This way, the pieces of metal will have accurately matched sets of holes.

Don't hesitate to use washers on each side of the nut/bolt or nut/screw combination to spread the load, especially with softer metals such as aluminum and brass. Use a lock washer, where applicable, such as a typical split washer, rather than the lighter duty inside or outside washers. A fender washer that has a wider rim than a standard washer is useful to bind objects together, such as a pulley attached to the body of your bot.

In areas of your robot where vibration may be a severe problem, such as a *combat robot*, the use of a lock nut is preferred. These types of nuts offer resistance to screwing when tightening, but they also offer resistance to coming unscrewed during vibration. Some lock nuts derive their binding resistance from being slightly deformed (smashed), whereas others use a plastic insert that resists unscrewing. In addition, special liquids such as Loctite can be applied to nuts to prevent them from coming unscrewed at the wrong time.

The use of a torque wrench is common in automobile engine assembly and repair, but is rarely needed to determine bolt tightness in robot construction. The large, bending-bar type of torque wrench is generally in ranges too high for bolts used in even the largest robots, but the click type of torque wrenches can be useful in multibolt pattern tightening. A pattern of bolts with known tightness better distributes loads on the structure. In most cases, making a habit of tightening all bolts after assembly or repairs is more than sufficient for most designs. The use of a torque wrench set at a value you've determined from experimentation helps.

Self-Tapping or Sheet Metal Screws

As mentioned earlier, a self-tapping screw looks a lot like a wood screw, but the former is designed for metal and is the type of screw you see in common household electronic equipment. The threads are coarse like a wood screw, but generally the taper of the screw changes at the end, becoming narrow quickly. This allows the person assembling the item to start the screw easily in the pilot hole; then it becomes tighter as the screw cuts into the metal.

Many times, these screws have a hexagonal head for a nut driver and a slot for a screwdriver. Longer versions are also tapered but have two indentations at the bottom to aid in cutting into the metal like a drill (thus the self-tapping moniker). These types of screws are not recommended for any type of combat robot *BattleBot* that takes a lot of vibration, especially if you have to remove and insert them several times.

Blind and Pop Rivets

Rivets seem like a strong fastening method, and they really are. They look great on airplanes and tanks, and even on robots. When people finally decide to go the "rivet route," there are questions about just how to install rivets. Most builders finally decide to use the blind, or pop rivet. But using these rivets is a major mistake, especially in combat robots.

Rivets, just like welds, are pretty permanent, making it hard, if not impossible, to change them in the field. If you have to remove a pop rivet, it has to be drilled out—leaving bits of steel or aluminum shavings hiding in the corners of your robot's chassis, ready to sneak into your electronics at the wrong moment. Most pop rivets found in typical hardware stores are made of aluminum; and although basically "permanent," they are about the weakest way to attach two pieces of metal. They have poor shear strength, even the mild steel varieties.

When the rivet tool pulls on the pin to cause the rivet to deform and fill the hole, the pin breaks in half after the operation is over. Even though a rivet holds two pieces of metal together, the other piece of the metal pin can come loose during vibration and bounce around the inside the robot. The higher-strength aerospace"–quality blind rivets also have this extra piece of pin that can cause trouble. The best recommendation is to forget about pop and other types of blind rivets for robot construction.

Standard Impact Rivets

You've probably seen standard impact rivets on airplanes and tanks; these are even harder to install than pop or blind rivets. They require a heavy "bucking" piece of metal on one side of the rivet and a hammer to strike the other side. In WWII planes, construction crews sometimes used a small person to climb inside the wing to hold the piece of metal as the rivet was hammered flat. Bridge construction often used hot rivets that would swell inside of a hole and seize the rivet. Modern shops use a hydraulic press literally to squash the rivet. These things are hard to remove if you need repairs or make a mistake in construction. Forget about them.

When in Doubt, Build It Stout

An old engineering saying, "When in doubt, build it stout," reminds us that if you think some structure isn't going to be strong enough for combat, build it stronger with more material. If you have any doubt whatsoever if a particular technique or design might fail under extreme conditions, it probably will fail. You're building a machine for operation in an environment as harsh as deep space or the bottom of the sea.

Another thing that catches most robot builders by surprise is the final weight of their robot. When building your robot, keep in mind that your robot will always weigh more after you build it than you originally thought. Take this factor into consideration when you are in your preliminary design phase. Believe us, you'd rather add weight to a robot at the competition than have to drill holes in your precious fighting machine at a later date to reduce its weight.

BUILD YOUR OWN COMBAT ROBOT

chapter 10

Weapons Systems for Your Robot

BECAUSE robot combat has evolved from being a "backyard brawl" between a group of inventive engineering types into nationally televised sporting events, the rules governing the sport today are far more sophisticated than they used to be, and the types of weapons systems builders use have evolved over time. The majority of weapon regulations still focus on safety. However, a few of today's rules stem from instances in past matches in which a robot was judged as lacking in "fun"—an important factor for those who have plunked down their hard-earned money to come and see a robot rumble. For example, entangling devices such as netting, adhesive tape, fishing line, and chains are no-nos now, because they can slow down or even halt a battle.

Another disallowed item is noncombustible gases used to disable an opponent's fuel-burning engine. A heavyweight robot named *Rhino* once used Halon gas very effectively in its matches to starve its opponents' gasoline engine-powered weapons. As a result of that robot's inventive strategy, the preceding rule was added to the books the following year. The safety issues notwithstanding, seeing contestants lose because their engines got shut down as opposed to being immobilized due to getting their metallic guts ripped out and strewn all over the arena is not very fun to watch.

It is still possible to build a winning robot without having to resort to banned weapons like flame throwers, stun guns, and electromagnetic pulse emitters. In this chapter, we will discuss several types of weapons systems that are used in combat robots.

Weapon Strategy and Effectiveness

You have probably noticed that no single weapon is totally effective against all types of opponents. It is much the same as the old child's game "Rock-Paper-Scissors." The "rock" can smash the scissors but can be covered by the "paper." The "scissors" can cut the paper but can be smashed by the "rock." The "paper" can cover the "rock" but can be cut by the "scissors." Each has its advantage over one of the others but is at a disadvantage compared to another. The same goes for combat robotics. Some weapons seem to be able to demolish almost all other types of robots but fall short when paired with a particular type of machine. And the same applies to armor systems, as some protective measures are particularly effective against most machines but are shredded by others. Fourteen styles of weapons are listed here, and pros and cons are discussed.

Ram Bots

This type of weapon was first used in the *Julie-Bot* (*Robot Wars,* 1994). Other machines using a ram weapon include *Hammerhead, JuggerBot, Ogre,* and *Ram Force.*

The ramming robot features a powerful drive, big wheels with high traction, a strong frame, and good shock resistance. With no active weapons, this robot batters its opponent with brute ramming and shoving force.

Ram Design

A generic ramming robot design is shown in Figure 10-1.

FIGURE 10-1 *Ram robot design*

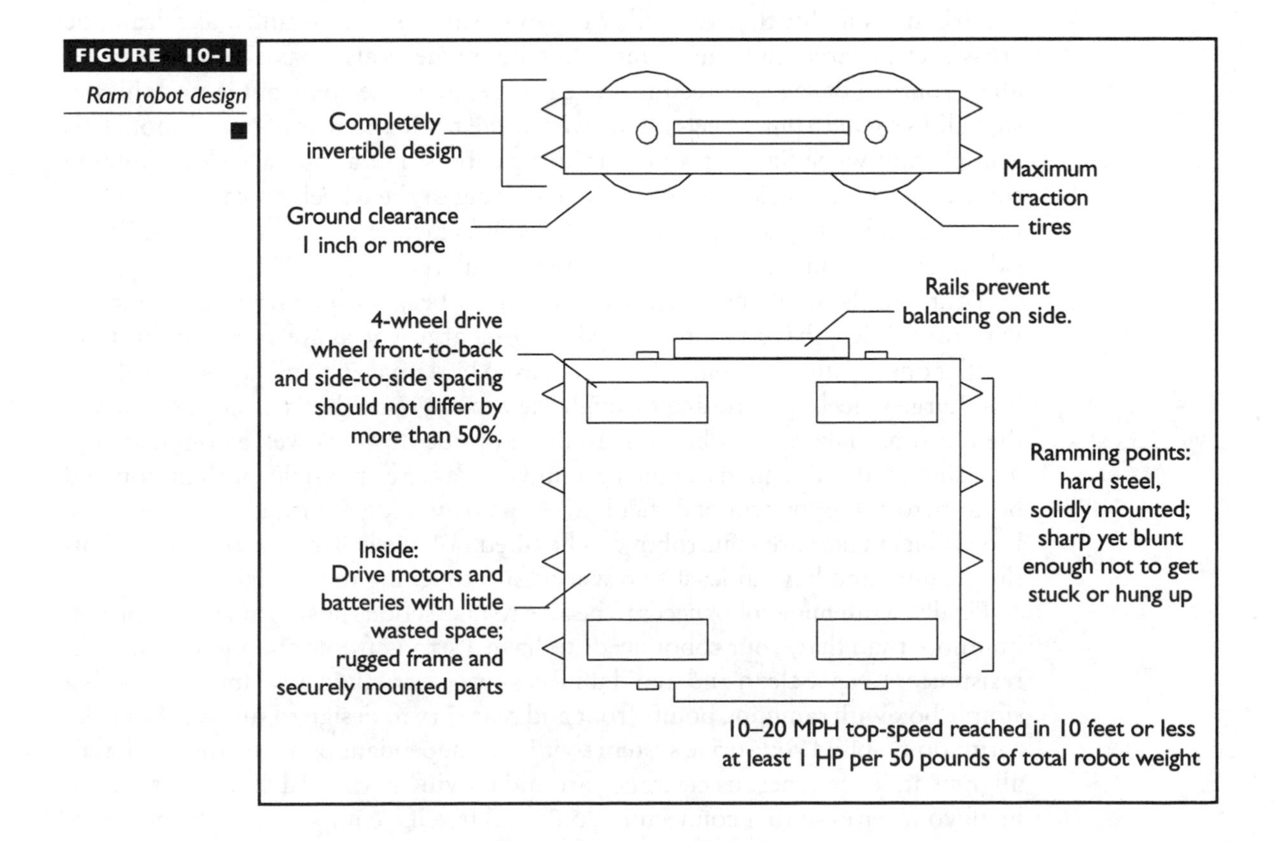

This type of robot lives or dies by its power, traction, and durability. Choose the largest drive motors and batteries and motor controllers to handle them, and base your frame around them. You should have as a minimum 1 HP of total drive power per each 50 pounds of your robot's weight. More is always better, as the strongest ramming robots have as much as 1 HP per 10 pounds of total weight.

Choose a gear ratio and wheel size that gives your robot a top speed of no more than 20 MPH—more than that will be uncontrollable. Low-end acceleration is very important, and you should aim to have your robot reach its top speed in a distance that's no more than three times its body length. Your robot's stall pushing force should be at least twice its own weight, as it not only has to accelerate but also overcome the opponent's mass and drive power.

To get as much of that power to the ground as possible, you need large, high-traction wheels. Soft rubber pneumatic go-kart or wheelbarrow wheels are best, but be sure to get them foam-filled if you want your robot to survive. Solid-foam power wheelchair wheels have slightly less traction but more durability. Avoid plastic wheels, solid-rubber castor wheels, or metal wheels with thin rubber treads—these wheels not only lack traction, but their lack of compliance will make your robot bounce and skip when it hits bumps or debris.

Four or six wheels are better than two for a ramming robot. Four wheels give much better stability than two, allowing you to line up a target and make dramatic cross-arena charges right into your target. Four wheels also make it possible to get all of your robot's weight resting on its tire tread, where you want it, and this design allows you to put wheels all the way at the front and rear of your robot. This is important when fighting wedge or lifting robots. For a four-wheeled ramming robot, you should make the side-to-side spacing of the wheels at least as much as the front-to-back spacing, as having the wheels farther apart front to back than side to side will make the robot turn awkwardly.

Your wheels should be large, with a diameter between a quarter and a third of your robot's length for a four-wheeled design. Large wheels are more durable than smaller ones, with more material that needs to be damaged to make the wheel useless. Large wheels, protruding through the top of your robot's armor as much as the bottom, make your robot able to drive upside down as well as rightside up. You should also design in as much ground clearance as possible, both on top and bottom, to make your robot difficult to hang up on wedges, lifting arms, or debris. If possible, make sure your robot can be tilted or have its front or back raised off the ground, and have at least two wheels still touching the ground.

Finally, a ramming robot needs to be able to take serious hits. Armor is important, but more than that, your robot needs to have a strong frame and internal impact resistance. Keep it clean and avoid unnecessary external details, and stick with a simple box with ramming points front and rear. Try to design to survive frame deformation—build your drive system so it is not dependant on your overall chassis alignment, leave generous clearance around moving parts, and leave a little slack in all your wires so that connectors don't pull free if a component shifts position. Heavy components like batteries and motors should be well secured.

Pushing Robot A similar form to the ramming robot is the pushing robot. The pushing robot concepts are similar to ramming robots, but they are focused more on traction and torque than speed. A pushing robot is usually designed to take advantage of traps and hazards built into the arena. Rather than try to damage the opponent through impacts, they simply use their pushing power to herd their opponent around the floor. While speed is not as important as in a pure ram design, a pusher should not be too slow, lest its opponent simply drive away. Many pushers use bulldozer blades or scoops, placing them more in the category of wedge robots.

Strategy

The ramming design is best against rotary weapons—spinners, saws, and drums. With no fragile external mechanisms, a strong frame, and the ability to take solid hits, a ramming robot can keep hitting a spinner until the spinner self-destructs. This design is weakest against an opponent that can lift its drive wheels off the ground. A wheel and chassis design that lets the ram still have two wheels on the ground even when one side is lifted will help this design get free from wedges and lifters, but being grabbed and lifted by a clamp bot will render a ram bot completely helpless. Use speed and lots of driving practice to keep that from happening to your robot.

The real weakness of a ram bot is its inability to knock out most opponents conclusively. With a ram bot, your victory can come by knocking out a weakly built robot in a collision, causing a spinner to knock itself out, or winning on judge points. When against an opponent built well enough not to be knocked out, and capable of damaging, flipping, or lifting the ram bot, the ram bot design will have a hard time winning.

Using (and Abusing) the Rules

If you look at the current crop of combat robots, you will see that conforming to the rules does not preclude creativity or battle-worthiness. Check out *Team CoolRobots* (*www.coolrobots.com*) headed by the highly imaginative Christian Carlberg. Carlberg's robot *Minion* won the Super Heavyweight title at *BattleBots* in 1999, and his other entries such as *Dreadnought* and *Toe-Crusher* are excellent examples of robots that not only conform to the rules but are also pretty darned effective as metallic harbingers of destruction.

The best way to show how to use the rules as the guideline that they were meant to be is to take you from inspiration to creation of a robot that won "most aggressive" at a recent *BattleBots* competition. One would assume a robot whose inspiration came from watching trench diggers and gigantic bucket-wheel excavators in action would have a hard time conforming to the specifications laid out in the manuals. However, Jim Smentowski's *Nightmare* passed muster with the rules lawyers. His 210-pound killing machine features a spinning blade that can deliver a 300-MPH uppercut to its opponents. Because the robot's main weapon is a spinning blade, it had to adhere to the rules regarding robots with spinning parts.

—Ronni Katz

Wedge Bots

The wedge was first used in the robot *Slow Moe* (*Robot Wars,* 1994. Examples of robots using a wedge include *La Machine, Punjar, Bad Attitude,* and *Subject to Change Without Reason.*

The wedge weapon features a thin, wide, ground-scraping scoop on the front, backed up by a strong frame and powerful drive system.

Wedge Design

Like the ram, the wedge's main weapon is its drive power and the ability to hit and push its opponent. Rather than simply impact, the wedge uses an inclined scoop front to lift the opponent on impact, breaking its contact with the ground and depriving it of traction. A well-made wedge bot can keep the opponent from escaping while being pushed by maintaining enough forward power to keep the scoop front under the opponent while shoving it across the arena.

The wedge design comes down to two things: enough power in the drive train and proper engineering of the scoop front. Power requirements of a wedge are similar to those of the ram—at least 1 HP per 50 pounds of robot weight (more, if possible), a drive gearing giving a top speed of 15 to 20 MPH, and pushing power of twice the robot's weight or more. Wedges should be two or four wheeled: a two-wheeled design will give faster turning rates; make the wedge more nimble; and allow a wide, short shape with maximum impact surface on the scoop. A four-wheeled wedge will be more stable and drive in accurate straight lines. Six-wheeled robots tend to be significantly longer than they are wide; this is not desirable for a wedge, which is better off wide and short. Figure 10-2 shows a classic wedge design.

The lower front edge of the wedge is the most critical part of the robot. This part should be thin and sharp to be able to get under other low-built opponents. Also, it should be as durable as possible because it will bear the brunt of full-speed impacts with the opponent; the arena walls; and any obstacles, arena hazards, or irregular spots in the floor that the wedge runs into. If possible, the lower edge of the front should be an integral part of the frame, rather than just an angled sheet of metal attached to the frame and not supported at the lower edge. Many wedges have been disabled when their own wedge front was bent downward, propping the front of the wedge off the ground and breaking the wheel's contact with the ground.

The surface of the wedge must also be strong enough to take the weight of the target robot held on top of it. If insufficiently supported, the wedge's front can be driven down into the ground, stopping it and preventing it from pushing the opponent further. Flexible materials should not be used for the leading edge of the wedge, as these will drag badly on the ground once an opponent is on top. If a hinged flap is used for the wedge, it must be rigid and supported from underneath with structural standoffs that limit its movement so that it won't drag on the ground.

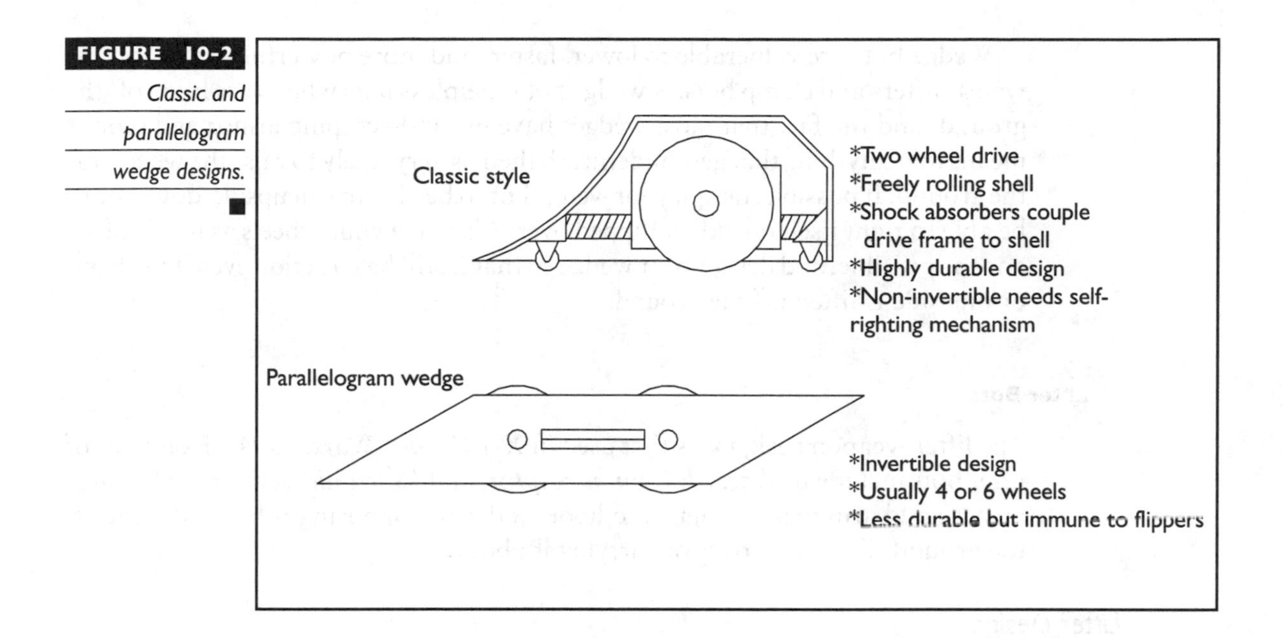

FIGURE 10-2 *Classic and parallelogram wedge designs.*

As with the ram bot, the wedge bot should be designed with maximum impact protection in mind because it relies on collisions to disable its opponent. One tactic used successfully with wedge bots is to make the entire outer shell, including the front scoop, a single shell of metal mounted through rubber bumpers to a separate inner frame holding the drive system, batteries, and electronics. The drive frame is isolated from the impacts, and damage and deformation to the shell should not affect the inner drive system. The only drawback to this kind of design is that it makes the wedge bot quite vulnerable to attacks from below, such as from another lower wedge, lifters, or floor-mounted arena hazards.

A variant of the wedge bot is the parallelogram-wedge bot. Typically four- wheel drive, these wedge bots have a normal angled wedge on the front, an inverted wedge on the back, and large wheels that protrude through the top of the body shell. This design can run as well upside down as right-side up, using the rear wedge, which becomes the front when the robot in inverted.

Wedge bots rarely inflict damage on an opponent. Their main goal is to control the match by getting the opponent's wheels off the ground and pushing the opponent into hazards and obstacles. A wedge bot may be able to flip over its opponent on a good hit.

Strategy

A well-armored wedge is a good tactic to use against spinners. If the front of the wedge is strong enough to survive hits from the spinner, it can be used to shove the spinner into a wall or hazard, or even—on a good hit—to flip the spinner completely over. A wedge also has an edge when fighting a ram, because with good power and good driving the wedge can get under the ram, denying the ram the traction it needs to push back or get away.

Wedge bots are vulnerable to lower, faster, and more powerful wedge bots, as well as lifters and clamp bots. A wedge bot is helpless if its wheels are lifted off the ground, and the fact that most wedges have ground-scraping armor and scoops means that anything that gets underneath them is very likely to raise the wheels off the ground. If possible, design your wedge bot to be able to run upside-down, or to be able to right itself quickly if flipped over. Also give your wheels as much clearance as possible, and design your wedge so that it still has traction even if the front or one side is lifted off the ground.

Lifter Bots

The lifter weapon design was first used in *X-1* (*Robot Wars,* 1994). Examples of lifter bots include *Biohazard, Gamma Raptor,* and *Voltronic.* A lifter bot features an actuated arm that's designed to hook under the opposing robot and lift it off the ground, flipping it over or carrying it about.

Lifter Design

Like the wedge, the lifter is designed to get underneath the opposing robot and lift its drive wheels off the ground. The lifter uses an active device to do so—an arm driven by hydraulics, pneumatics, a geared electric motor, a powerful spring cocked by a motor, or an electric linear actuator—with enough power and leverage to tilt or lift up the other robot. The end of the arm is often wedge shaped, or blended into a wedge-shaped front; and in many cases, it has grip-enhancing hooks or teeth. Figure 10-3 shows a lifter robot.

The advantage of a lifter over a wedge is the ability to lift the other robot's wheels off the ground independent of movement. While a wedge can only lift the opponent higher by shoving itself under its opponent, a lifter, once underneath the opponent, can lift it up as high as its arm can go while remaining stationary. A well-designed lifter can drag its opponent around the arena freely, while a wedge can only push its opponent forward.

Most combat bots are designed to be low and wide, and won't fall over until tilted 90 degrees or more. To flip opponents over with a lifter, you will need an arm with a maximum height comparable to the width of your targets. Usually, this means the pivot point of the arm is located nearly at the back of the robot, and the arm should extend on top of or down the middle of your bot to the front. Arms of this type can often double as self-righting mechanisms.

The most common drive systems for arms are linear drive actuators, either electric ball-screw types or pneumatic cylinders. Electric screw actuators, consisting of an electric motor driving a telescoping cylindrical assembly through a nut and screw mechanism, make for a slow but powerful lift. These devices have the advantage of being self-contained and functional in one unit, needing only an R/C relay or motor controller to extend and retract. Pneumatics is a faster option. A powerful pneumatic system can actually hurl the opponent into the air (see the description in the

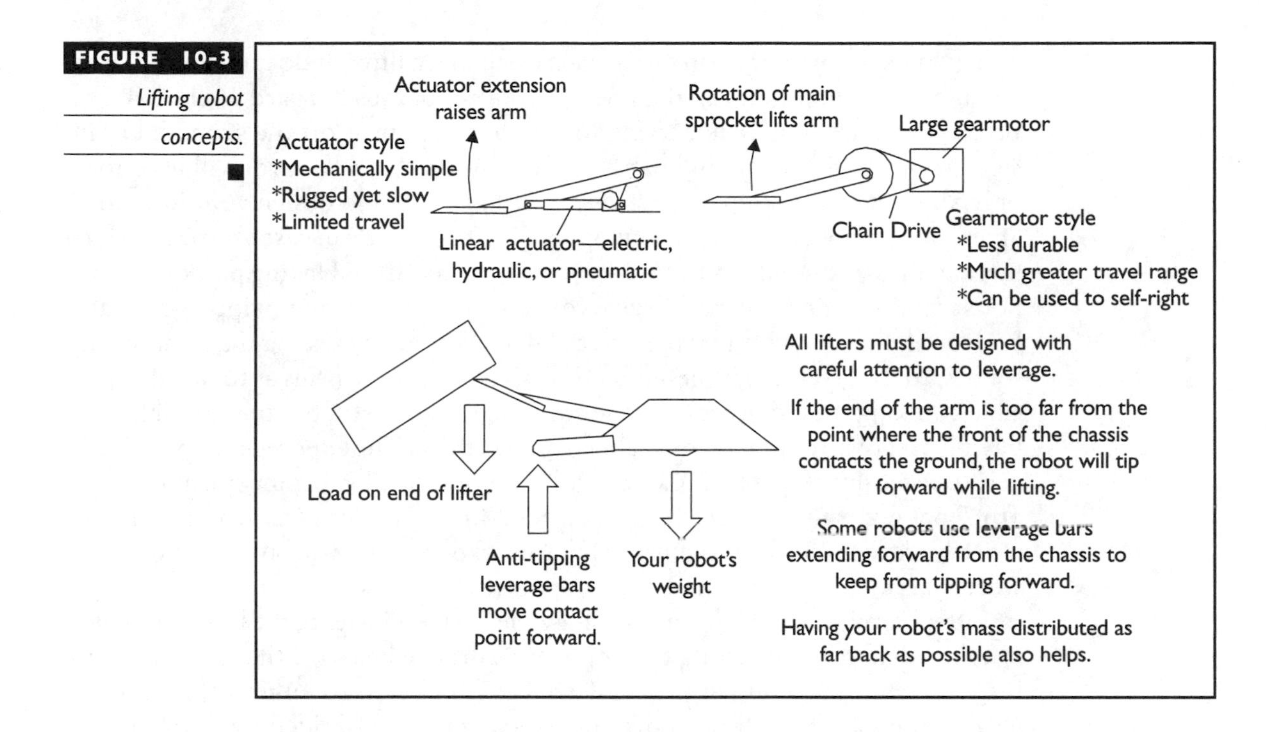

FIGURE 10-3 *Lifting robot concepts.*

upcoming section "Launchers"). Pneumatic systems are more complex than electric linear actuators, having the added bits of tanks, regulators, valves, tubing, and optional buffer chambers; but in the end, they can make for a weighty and more-flexible design. A pneumatic-powered lifting arm also has the disadvantage of being unable to stop in mid-stroke (barring a complex position-controlled feedback system), which makes it less useful if your tactic is to drag the opponent around the arena rather than flipping it.

Hydraulics have also been used for lifting arms, but the complexity and weight of the hydraulic system make this an unlikely option. Unless your robot already has a hydraulic system onboard for other reasons, an electric linear actuator will be a much cheaper and lighter weight solution than a hydraulic lifter.

Shaft-driven arms, with the output of a gear or chain reduction directly driving the arm's rotation, are a more challenging design for a lifter. Designing a motor drive capable of supplying, and surviving, the kind of torque needed to lift an opponent is difficult We're talking of 500 to 1000 foot-pounds of torque for heavyweights, here. Most designs of this type use a large-diameter gear or sprocket bolted straight to the arm as the final drive stage, rather than attempting to drive straight through a shaft. The advantage of this kind of arm is that the range of rotation possible is much greater than with a linear actuated arm—often enough to make the arm able to reach around behind the robot; reach down below it to push it off obstacles or lift while the robot is upside down; and even, in some cases, travel unrestrained 360 degrees.

You must keep leverage in mind when designing a lifter. It does no good to have enough power to pick up an opponent if your robot falls forward while doing so. Because you are usually not trying to get your opponent off the ground, but instead trying to get them off balance, the down force on your arm will need to be only about half your opponent's weight. Still, you should design your lifter to be able to lift the entire weight of your opponent, if not more, in case you need to lift a target with an unusually off-center center of gravity (CG). Having part of your robot's frame extend forward will give you more leverage to avoid tipping forward, but you should also consider the effect of that extra force pressing the front of your robot into the ground. The best lifter designs place drive wheels as forward as possible, flanking the lifting arm, to take advantage of the extra traction possible from having part of an opponent's weight resting on them. The exposed arm of the lifter is its most vulnerable part. A severe collision or strike by a spinner can bend the arm, making it useless. On better-defended lifters, the arm retracts into an armored wedge front when completely lowered, exposing the arm only to lift when the wedge has already gotten under the opponent.

Lifters rarely damage the opponent by themselves; instead, the lifter strategy is to take advantage by getting the opponent's drive wheels off the ground. Many lifter matches are won with no damage being inflicted to either robot, instead leaving the losing bot flipped over or the match being decided by judges rather than by a disabled losing bot.

Strategy

The lifter is strongest against opponents that rely on traction to fight, such as wedges and rammers. Robots with overhanging enclosed shells will be easy targets for a lifter because it can immobilize them by simply tilting one side up enough to lift their wheels off the ground. A lifter relies on being able to get the arm seated under an opponent firmly enough to lift, and it is against spinning robots that lifters have their hardest time winning. Many spinners have enough kinetic energy in their shells or spinning appendages to knock a lifter aside on contact, and unless the lifter can somehow stop a spinner's rotation, the lifter will simply take blow after blow until one robot breaks. Thwack bots are also tough opponents for lifters, as their wild spinning and invertable, open-wheeled design make it difficult for a lifter to get into a position to knock the thwack bot's wheels off the ground.

Launchers

This design was first used on *Recyclopse* (*Robot Wars UK*, 1997). Other bots using this design include *Toro, T-Minus, Hexadecimator,* and *Chaos II*. The launcher features an actuated arm that's powered by extremely high-flow-rate pneumatics, capable of launching the unlucky opposing robot high into the air.

Launcher Design

The launcher is a specialized form of lifter, with an arm capable of not just lifting but hurling its opponent into the air. This attack will not only flip the opponent over, but quite possibly damage it from the impact, as well as make for a great show for the audience.

A launcher needs to release a tremendous amount of energy in a short period of time to work. Electric motors, linear actuators, and hydraulics are too slow for this kind of mechanism. Most launchers use specialized high-pressure pneumatics to get the impulse of force they need to fling the opponents, using modified hydraulic cylinders and high-pressure valves and hoses to run carbon dioxide or compressed air at 900 PSI or more. This is not a system that can be built with off-the-shelf parts—high-pressure pneumatic launchers take years of research and engineering to develop.

Another option is to use lower-pressure pneumatics and to engineer for a very high-volume flow, with large-bore tubing and valves, and either a high-rate pressure regulator or a large buffer tank. While the engineering of the pneumatics is simpler, large-bore, low-pressure pneumatics will take up a lot more room in your robot.

A third option is to use a spring mechanism. A powerful torsion spring or compression springs pushing the arm up, a powerful geared motor for re-cocking, and a latching mechanism to hold the arm down until remotely triggered should make a good spring mechanism. This type of weapon is heavier than a pneumatic system, and the cocking and latching mechanisms take some serious mechanical engineering skill to make. The time it will take to slowly re-cock the spring is also a significant disadvantage because until the arm is down, your robot will be helpless against an opponent. A single-shot launcher design that cannot be re-cocked should not even be considered.

Figure 10-4 shows how launcher robots work.

Whichever mechanism you use to power a launcher, your frame and drive system is going to be subjected to a tremendous jolt every time it fires its weapon. Your frame must provide a strong structural path between the launching mechanism and the drive wheels, as the arm is going to impart a massive downward force on the frame every time it fires. The entire robot should be built with major jolts in mind; be careful that nothing can shake loose and that all electrical components and connectors are solid.

Finally, this kind of robot can be dangerous to build and test. The forces involved in flipping several hundred pounds of robot through the air can kill you if the weapon misfires with part of your body in the path of the flipping mechanism. *Be careful.* Most competitions will require that you have some way of locking the mechanism when not in combat, usually with a pin or rod passing through a hole in the frame, which prevents the arm from moving.

A launcher does the most damage to an opponent when the hurled opponent strikes the ground after being flung into the air. The mass of the target bot and the

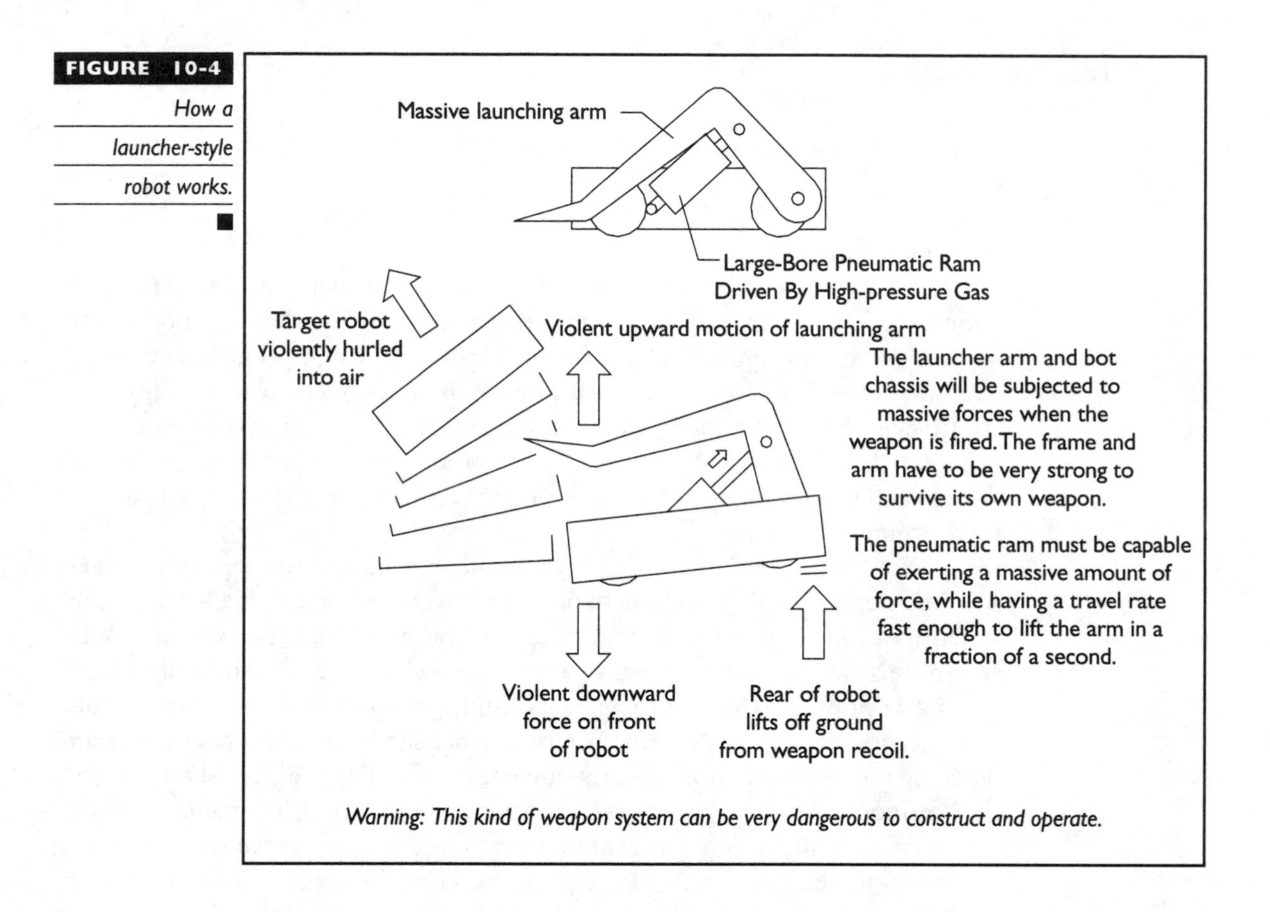

FIGURE 10-4 *How a launcher-style robot works.*

height to which it is thrown determine how hard it strikes the floor. The height at which the opposing bot is thrown is, in turn, determined by the energy developed by the launching mechanism. So, the more energy you can get into your launcher, the higher the opponent will reach, and the harder it will strike the ground.

Strategy

Like the lifter, the launcher works best against wedges and rams. It can also make a great weapon against slower lifters—even if a lifter can self-right, being hurled repeatedly into the air and slammed against the ground can eventually break it. Using a launcher against a spinner is tricky because, while most spinners will be broken or disabled after being flung into the air, getting a firm hit on the spinner with the flipper arm takes skilled driving, and surviving the hits takes a solidly built flipper arm. As with lifters, a thwack bot is a very difficult opponent; the launcher will have difficulty getting in a position to flip, and most thwack bots can run just as well upside down.

Clamp Bots

The clamp was first used on *Namreko 3000* (*Robot Wars*, 1996). Other bots using the clamp include *Complete Control*, *Tripulta Raptor*, *Spike IV*, and *Mantis*. A clamp bot features an actuated lifting arm, with an additional movable piece to act as a grabbing clamp that's capable of grasping the opposing robot and lifting it completely off the ground.

Clamp Design

The clamp bot takes the strategy of the lifter one step further, adding a second movable piece to the lifting arm to act as a clamp to solidly grasp the opponent robot. A well-balanced clamp bot can completely lift an opposing robot off the ground. As few robots can do anything when lifted off the ground, this places the match completely in the control of the clamp bot.

The clamping mechanism must open wide enough to grasp the largest opponent you are likely to face, and it should be designed to close in a second or less. A slower clamp risks the opponent getting free before the clamp is able to close. The grabbing mechanism should have a holding force at the tip at least equal to the target robot's weight, to prevent the claw from being forced open when the arm lifts. Pneumatics are a good choice for the closing mechanism, as they can provide both high closing speed and strong clamping force. Electric linear actuators or hydraulics will also work, providing superior closing force to pneumatics at the cost of a slower closing speed. Attaching the closing arm directly to the output shaft of a gearmotor is another possibility, although it's not recommended because it will not be as durable as driving the arm with a linear actuator. Figure 10-5 chows a clamp bot configuration.

Leverage is the key to a successful clamp bot. In most cases, your bot will be attempting to lift a target that weighs as much as it does. While a lifter usually has to lift up only one side of its opponent, a clamp bot must bear the entire weight of its opponent on the end of its arm. To avoid falling forward while lifting its opponent, the clamp bot will need frame extensions on either side of its arm extending forward as far as possible. Having a center of gravity as far back as possible will also help avoid tipping forward.

A successful clamp bot must not only be able to grab and lift its opponent, but it must be able to carry the opponent around the arena. This means having a drive train strong enough to carry twice the clamp bot's own weight, and the front end of the clamp bot's frame must be designed to ride smoothly on the ground. Ideally, a clamp bot would have drive wheels forward straddling the lifting fork, so that the opponent's weight is directly borne by the driving wheels. A clamp bot must also have the speed to catch fast opponents.

The need for a strong, well-balanced frame; a drive system having both great carrying power and high speed; and separately driven mechanism for grabbing and lifting make clamp bots one of the more challenging robot types to attempt.

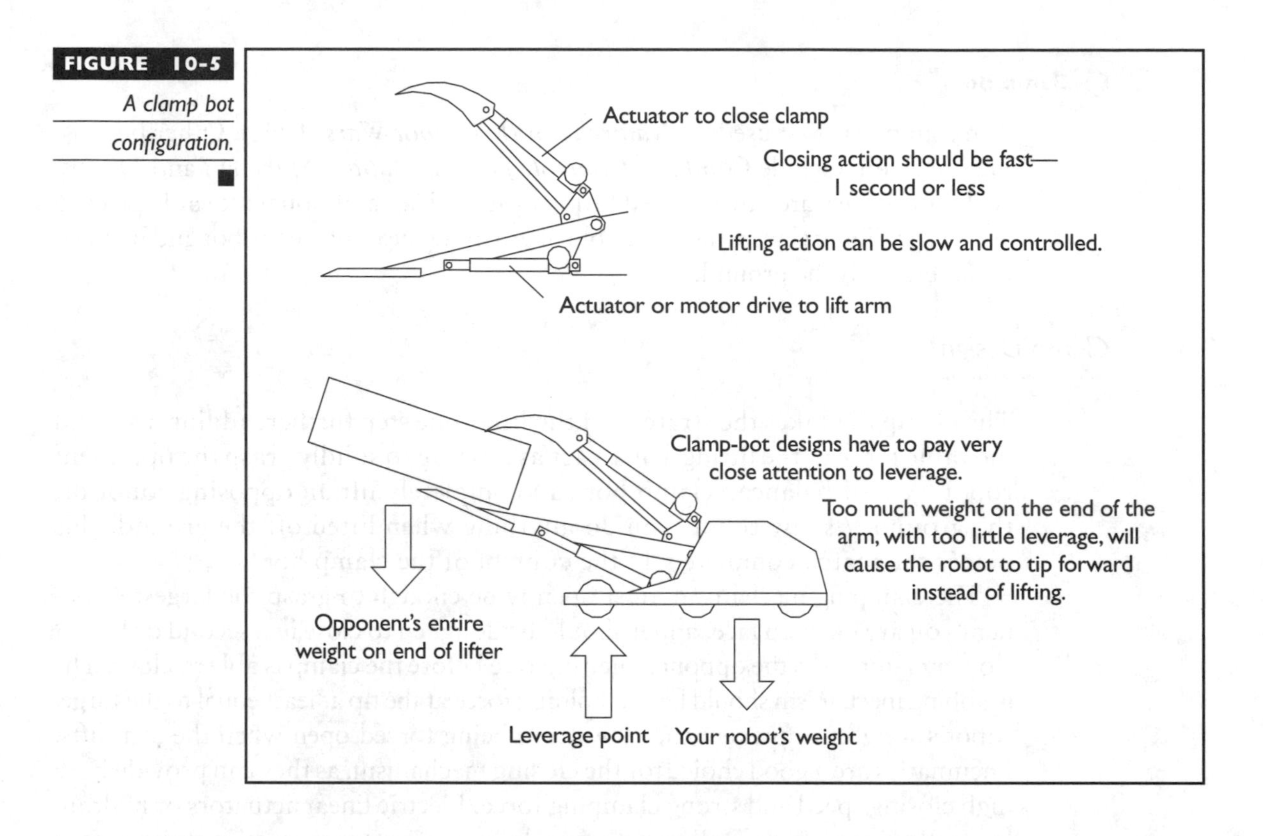

FIGURE 10-5 *A clamp bot configuration.*

Clamp bots use strategic designs and are intended to take control of the match by denying the opponent the ability to move. Usually their only option to inflict damage is to take the opponent over the arena hazards.

Strategy

Clamp bots work well against rams and wedges—these types of robots completely depend on their drive power for weapons, and once grabbed and lifted are completely helpless. Against a thwack bot, the challenge for a clamp bot will be in catching its opponent, because many thwack bots are very fast robots. Like the ram and wedge, once caught, a thwack bot can be rendered helpless by a firmly grasping and lifting clamp bot.

Spinners, particularly the completely enclosed shell-type spinners, are a tough opponent for a clamp bot. The spinner's weapon must be stopped before the clamp bot can grab it, but the only way the clamp bot has of stopping the shell is by repeatedly ramming it, taking punishing blows to the arm mechanism before the spinner is slow enough to be grabbed. With more working parts and typically lighter frames, clamp bots are more likely than most robot types to be damaged by this kind of punishment. A vertical spinner–or drum–type robot is an easier target, if the clamp bot can outmaneuver and grasp it without taking a hit.

Finally, care must be taken when grabbing hammer-wielding robots with a clamp bot, as a firm grasp can also give the hammer bot the leverage to repeatedly hit the clamp bot in the same spot. When attacking a hammer bot with a clamp bot, try to approach from the side so as not to be in the path of the hammer arm.

Thwack Bots

Spaz, Blade Runner, and *T-Rex* are thwack bots. Thwack bots feature a powerful, two-wheeled base, with a long-tail boom having an axe, pick, or hammer head on the end. They are capable of spinning in place at high speed.

Thwack Bot Design

Another design that uses only its drive motors for attack power, the thwack bot spins rapidly in place, whipping a weapon on a long tail about at high speed. Thwack bots are invariably two wheeled, as four- or six-wheeled designs cannot spin in place rapidly enough to make for a satisfying impact. Usually, this design—with exposed wheels and a symmetrical profile—allows them to run well when inverted, thus making them a difficult opponent for wedges or lifters.

Narrow wheels are key to a thwack bot because wide wheels will add scrub resistance and slow down the turning rate. Care must be taken to balance the robot so that as much of the weight as possible is resting on the main drive wheels; any weight resting on the tail or on any idler wheels is potential traction going to waste. The wheels should be soft rubber, high-traction types, and foam filled for survivability. Placing the wheels close together increases the top speed but will increase the time it takes to reach that top speed.

Figure 10-6 shows a thwack bot schematic.

Typically, ratio of wheel size to wheel spacing is between 2:1 and 4:1. Thwack bots typically have high driving speeds so that the high wheel speed can be turned into a high spin rate. The need for high wheel speed and spinning requirements can make this kind of robot hard to control.

The main design challenge with thwack bots is finding a balance between top speed and spin-up time. Ideally, a thwack bot should be able to reach top rotation speed in less than a single revolution, yet still have a top speed fast enough to do damage on impact. A thwack bot that takes too long to spin up will find itself helpless once an opponent has come within range to attack. Of course, more power makes for faster spinning (thus, less time to get up to full spinning speed) and higher top speed, so a thwack bot should have as much power as possible.

The primary weakness of the thwack bot concept is that it cannot move while spinning. This type of robot must either spin in place and hope its opponent drives into it, or charge to within spin radius and then spin—getting less than a full revolution before striking its opponent. Several attempts have been made to build a navigation system that allows a thwack bot to translate while spinning, by

FIGURE 10-6 *Thwack bot schematic of the rotational motion.*

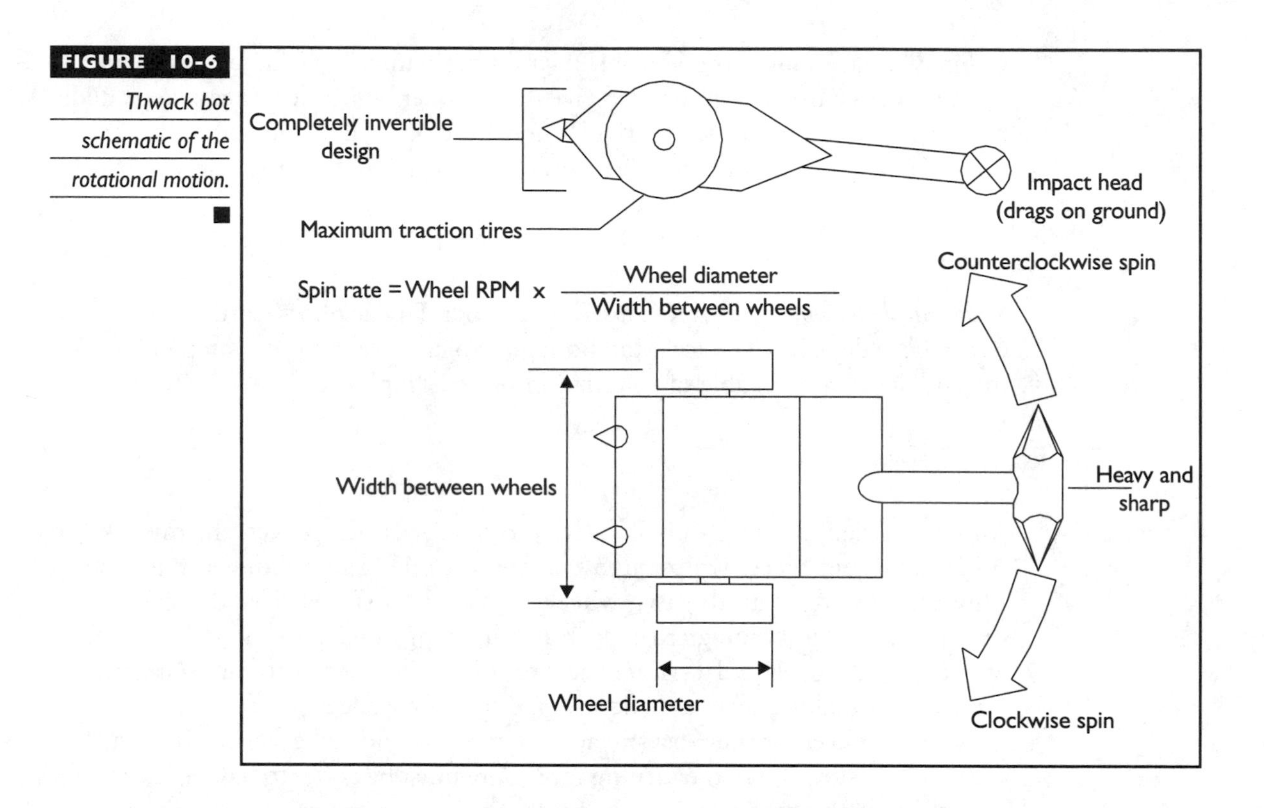

periodically varying the drive power in sync with the rotation, causing a slow wobble toward its opponent while spinning at nearly full speed.

A thwack bot's impact force comes from stored kinetic energy in the rotation of it's body. Its angular momentum is proportional to the body's moment of inertia times the speed of rotation. The faster it spins, the harder it hits. The robot's moment of inertia can be increased by moving its weight away from its center; however, this will also increase the time it takes to spin up.

Strategy

A powerful thwack bot has proven to be an effective robot against the lifter—a strong spinning attack can keep the lifter from getting its arm in a position to pick up the thwack bot, and the open-wheeled design and powerful drive of most thwack bots makes them difficult to keep a grip on. A clamp bot that gets a firm grasp on a thwack bot will render it helpless, but a powerful thwack bot can make it difficult and dangerous to get such a grasp. Wedge bots are difficult to fight against with a thwack bot, with the victory often coming down to speed and maneuverability.

Drums and vertical spinners can also be very dangerous customers for a thwack bot to fight, as the long weapon boom of a thwack bot can get hit and tossed upward violently, disrupting the thwack bot's spin as its wheels lose contact with the ground from the impact. The spinning, low-mobility attack of the thwack bot makes it impossible for it to choose its angle of attack, letting its opponent line up its attack strategy as it sees fit. A secondary attack mode of ramming and pushing can help in those cases.

Overhead Thwack Bots

This type of bot was first used in the *Spirit of Frank* (*Robot Wars,* 1995). Examples include *Toe Crusher, Over Kill,* and *Mjollnir.* The overhead thwack bot features a wide, two-wheeled base, with the main body being built entirely between the two wheels and fitting into their radius, and a long weapon–tipped boom such that the body flips over and brings the weapon down on the opponent whenever the robot reverses direction rapidly.

Thwack Mechanism Design

Like the thwack bot, the overhead thwack bot uses its motor torque to power an impact weapon. Unlike the conventional thwack bot, the overhead thwack bot attacks by reversing its drive power rapidly, the reaction torque from the drive motors swinging the entire body end over end and bringing the tail end down in front of it violently.

The challenge comes in getting enough inertia into the body of the robot, with significant force and accuracy to hit the target. The same rapid reversal of drive power that brings the weapon over will also drive the robot away from the target. Attacking with an overhead thwack bot is accomplished by charging at a target and then slamming itself into reverse just before impact. The entire robot has to be balanced just right, such that the robot flips over quickly before it starts to back up significantly. Insufficient or uneven wheel traction can cause the robot to veer to one side while flipping, causing the weapon to miss its intended target. Widely set wheels will help with accuracy.

Figure 10-7 shows an overhead thwack bot.

While a conventional thwack bot can take several revolutions to get up to speed, an overhead thwack bot must produce all its weapon power in less than one half of a full revolution of its drive wheels. The electrical and mechanical drive power components have to be optimized for a high rate of energy delivery—high current rate batteries, thick wiring, high-horsepower motors, and very rugged drive gearing are a must. All the main components must fit between the drive wheels for the robot to flip freely. Usually, these bots have large-diameter wheels set wide apart to allow sufficient room between them for the main body. Of course, large-diameter wheels usually means a high gear reduction to get the right speed and torque, and large wheels and a high gear reduction will make the wheels respond more slowly to rapid motor power reversal.

Optimizing an overhead thwack bot for maximum damage is difficult. The best tactic is to increase drive motor power as much as possible. Increasing the length of the tail and the weight of the mass at the end of the tail will increase

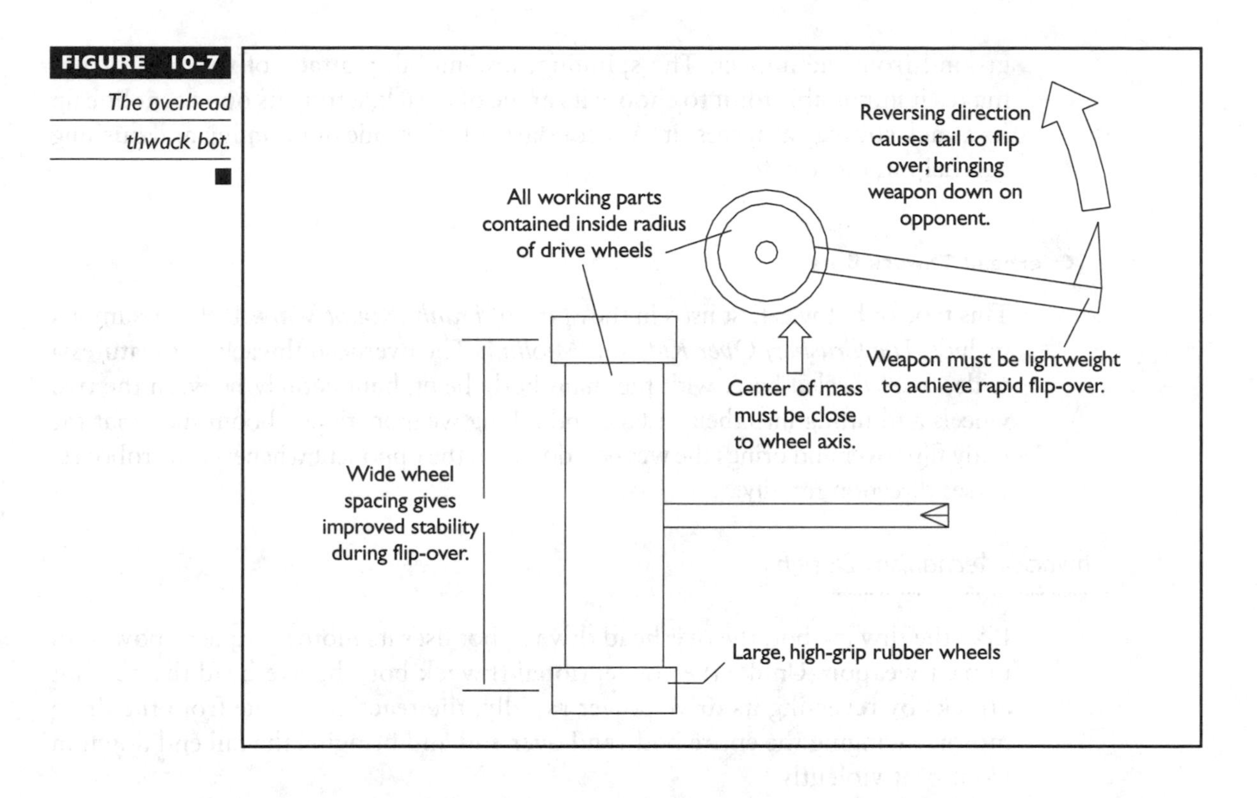

FIGURE 10-7 *The overhead thwack bot.*

the damage done to an opponent, but will also make it harder to strike an opponent accurately.

Strategy

The overhead thwack bot is a difficult design to make work successfully. The prime advantage of this design is its inability to be disabled by being flipped over. Lifters and wedges have a hard time getting a grip on this highly mobile design. However, even the best overhead thwack bots lack sufficient power to strike a killing blow, instead having to hit repeatedly and hope to win by judges' decision.

The most successful overhead thwack bot designs have been those that combined the conventional overhead hammer with a freely swinging wedge. The wedge must pivot on the axis of the wheels, a tricky mechanical bit to pull off, which can allow the overhead thwack bot to push an opponent around the arena or pin it in place before reversing to strike with the weapon.

Spinner Bots

A spinner bot was first used on *The South Bay Mauler* (*Robot Wars,* 1994). *Hazard, Odin*, *Ziggo*, *Tortise*, *Turbo*, and *Blendo* are spinner bots. These bots feature a heavy spinning bar or disk, possible with hammer heads, chisels, maces, or other protrusion pieces attached.

Spinner Design

The spinner uses the concept of a flywheel that stores the mechanical energy output of a motor in a spinning mass to be released in one massive blow to the opposing robot. The spinner was one of the first successful tactics for inflicting actual damage on the opposing robot; it remains one of the most dangerous—not only to the target robot, but also to the spinner robot, the arena, and the audience. Nearly all incidents of penetrated arena walls and injured audience members have been due to spinners with more energy than the arena could safely contain.

The earliest spinners were bar shaped, often with hammers or spiked balls on chains attached to the ends. While simple to build and lightweight, these designs don't store as much energy as disk- or ringed-shaped spinning weapons, though *Son of Whyachi* has mangled the very best opponents with its three flying spiked sledge hammers. Thin bar or tube spinners are also more susceptible to bending or breaking on impact. The ultimate form of the spinner is to enclose the robot completely in a spinning cone-, dome-, or cylinder-shaped outer shell. With this type of design, it will be impossible for an opponent to hit the spinner without being struck by the spinner's weapon.

Figure 10-8 shows a spinner.

Spinners allow the energy output of a motor to be stored over some time in a kinetic energy form, ready to be delivered into a target in a moment. This does not mean that you should use a small motor—the faster your spinner can get up to speed, the better your robot will fare against a determined and durable opponent. A spinner that takes more than 10 seconds to spin up may never get the opportunity to reach top speed; you should design for a spin-up time of 3 seconds or less.

FIGURE 10-8 A spinning weapon robot.

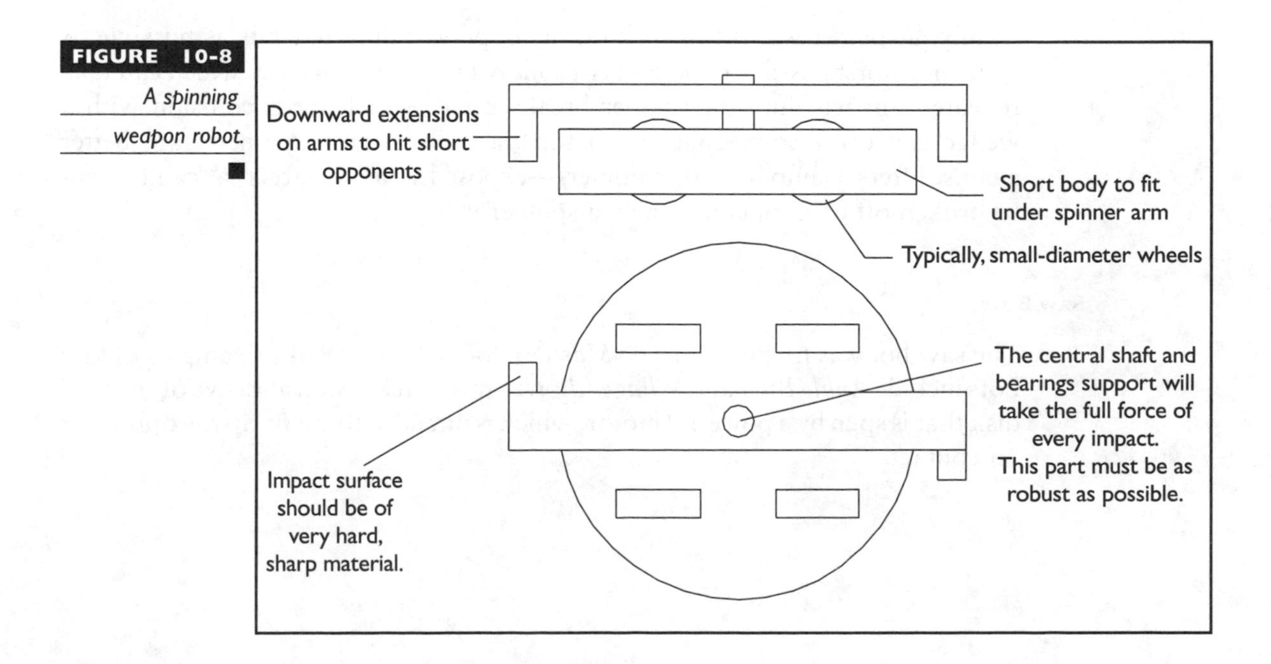

While a powerful spinner is the most destructive form of kinetic-energy weapon in the competition, this destructiveness comes with a price. The powerful kinetic impacts that the spinner delivers are felt as much by it as by its opponent; many spinners have crippled the opposing robot only to be themselves knocked out by the same impact. A spinner needs to be built as ruggedly as possible to avoid this fate. Many of the fully enclosed shell-type spinners use rings of rollers on the inner frame to allow the spinner to ride smoothly even if it becomes bent or dented.

A fully enclosed spinner has an additional difficulty not faced by other robots: when the weapon is running, it can be difficult for the robot's driver to see which way the base inside is facing! Methods of dealing with this include having a tail trailing out underneath the shell, having a non-rotating flag or arrow sticking up through the center of the shell, making part of the spinning shell out of transparent materials, or cutting windows in the shell to allow the interior to be partially visible.

The reaction torque of spinning the shell will produce a strong turning force on the base of the robot, which will make the bot want to swerve to the side when driving. A four-wheeled base is recommended to give some straight-line stability. Many spinner drivers also use R/C helicopter rate gyroscopes in their control electronics to compensate for the effects.

For optimum damage, the spinner weapon should be large and should have its mass concentrated as much as possible at the outside of its radius. Many spinner weapons are made of disks or domes with weights at the edges and holes in the middle, to maximize the rotary inertia of the weapon. Of course, more inertia in the weapon means a greater spin-up time.

Strategy

Ideally, a spinner wants to knock out its opponent in as few hits as possible. A spinner's worst possible opponent is a solidly built ram or wedge, which can take repeated impacts until the spinner breaks itself. A high-speed collision with a wedge can cause some spinners to flip themselves over. Spinners fare better against lifters, clamp bots, or hammers—exposed weapon parts that can be bent or broken off of an opponent help a spinner win.

Saw Bots

The saw bot was first used in *The Master* (*Robot Wars,* 1994). Examples of saw bots include *Ankle Biter* and *Village Idiot*. Saw bots feature an abrasive or toothed disk that is spun by a powerful motor, which is intended to cut or rip the opponent on contact.

Saw Design

Now increasingly rare, the saw was tried many times in the early days of robot combat, usually with little success. The idea of disabling the opponent by slicing it apart has proven to be a difficult challenge because the materials most modern combat robots are made of take too much time to cut, even under controlled circumstances, let alone when the target is actively trying to get away from the saw blade. The concept has been largely abandoned, aside from a few brave robots that use saws in combination with other attack styles.

Combat trials have shown that the best saw blades to use are the emergency rescue blades used to rescue accident and building collapse victims. Thick steel disks coated around the edge with hard abrasive make these blades able to cut a wide variety of metallic and non-metallic materials quickly—just the thing for a combat situation. They are, however, heavy, expensive, and available only through certain specialty dealers, and they require a seriously powerful motor to be used to full effect. Figure 10-9 shows some examples.

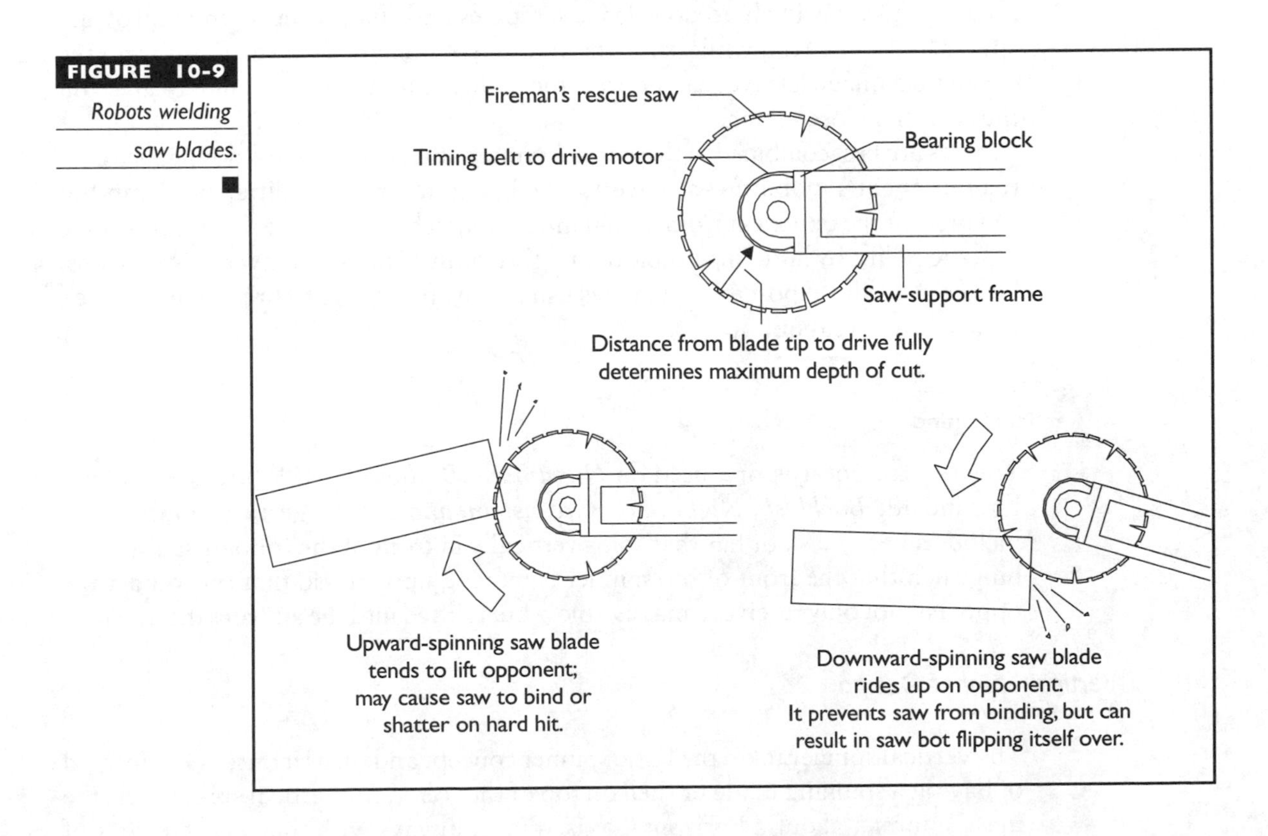

FIGURE 10-9 *Robots wielding saw blades.*

Saw blades, other than the emergency type, have not proven to be effective. Abrasive disks are nearly useless against soft materials like plastics, wood, or composites, and they easily shatter on impact. Toothed wood-cutting blades cut softer material nicely, but they stall on metals. Milling saws are heavy, can shatter on hard impacts, and usually knock the opponent away rather than cutting into it.

Damage from a saw does not come in the form of one or two big hits, but from many small gashes and cuts. The saw motor should have enough torque to keep the saw from stalling, and it should have speed of a few thousand RPM. More mass in the saw blades will help optimize damage on initial contact, keeping the weapon from stalling instantly. The best saw weapons act more like spinners than saws, storing up a lot of inertia in the weapon to deliver on contact with the opponent.

Strategy

The saw, by itself, is not an effective means of disabling an opponent. Unless already disabled, your target will not stand still and give your bot the time to cut into it, so the most a saw is likely to do is leave scratches and shallow cuts while throwing sparks and dust. Still, while rarely fatal to the opponent, a powerful saw and the cosmetic damage it leaves can impress the audience and judges enough to give you the win in a close match.

Saws are best combined with an attack strategy that gives you the dominance over the opponent's mobility—a powerful wedge, ram, or even a lifter or clamp bot can prevent the opponent from dominating the match and give the saw weapon time to score points by inflicting visible damage. Against a spinner, a saw may be useless, however, as the exposed saw blade is usually the first thing to break when struck by a serious weapon.

Vertical Spinner

This type of bot was first used on *Nightmare* (*BattleBots*, 1999). Other spinner bots include *Backlash, Nightmare, Greenspan, and Garm*. Vertical spinner bots include a heavy disk or bar that spins vertically in front of the robot, usually spinning such that the front of the spinner is moving upward, so that on contact the opponent not only receives a massive blow but is lifted into the air from the impact.

Vertical Spinner Design

The vertical spinner takes the basic spinner concept and turns it on its side. Instead of having a spinning blade or shell on top of the robot, the vertical spinner sets the mass spinning about a horizontal axis, almost always with the exposed front of the spinner moving upward. When it strikes an opponent, the impact force pushes the opposing robot upward, often flipping it over or subjecting it to a hard impact with the floor when it lands. The recoil force on the vertical spinner merely pushes it down against the floor, rather than flinging it sideways, as can happen with a conventional spinner.

Figure 10-10 shows a vertical spinning robot.

While the weapon can be much more effective than a standard horizontal spinner, the vertical spinner trades off improved offense with a greatly weakened defense. While a standard spinner can be built to cover the robot's body completely, such that an opponent cannot help but be hit by the weapon on any contact, the vertical spinner's narrow disk must be carefully lined up on its target. The large disk gives the vertical spinner a dangerously high center of gravity, requiring a large, wide body to support it, which makes the vertical spinner vulnerable to attacks from the sides or rear.

Spinning the disk will generate significant gyroscopic effects every time the robot turns, requiring widely set drive wheels and a slow turn speed to keep the robot from flipping itself over when turning. The vertical spinner also suffers the same self-inflicted impacts as the standard spinners. While the impacts are downward and the floor helps brace the robot in place, vertical spinners have been destroyed by their own weapon impacts.

As with the spinner, the optimum form of the vertical spinner will be a disk with the weight concentrated at the edges. Vertical spinners have the additional property of hurling their opponents into the air on solid hits, doing additional damage when the opposing robot crashes back into the floor.

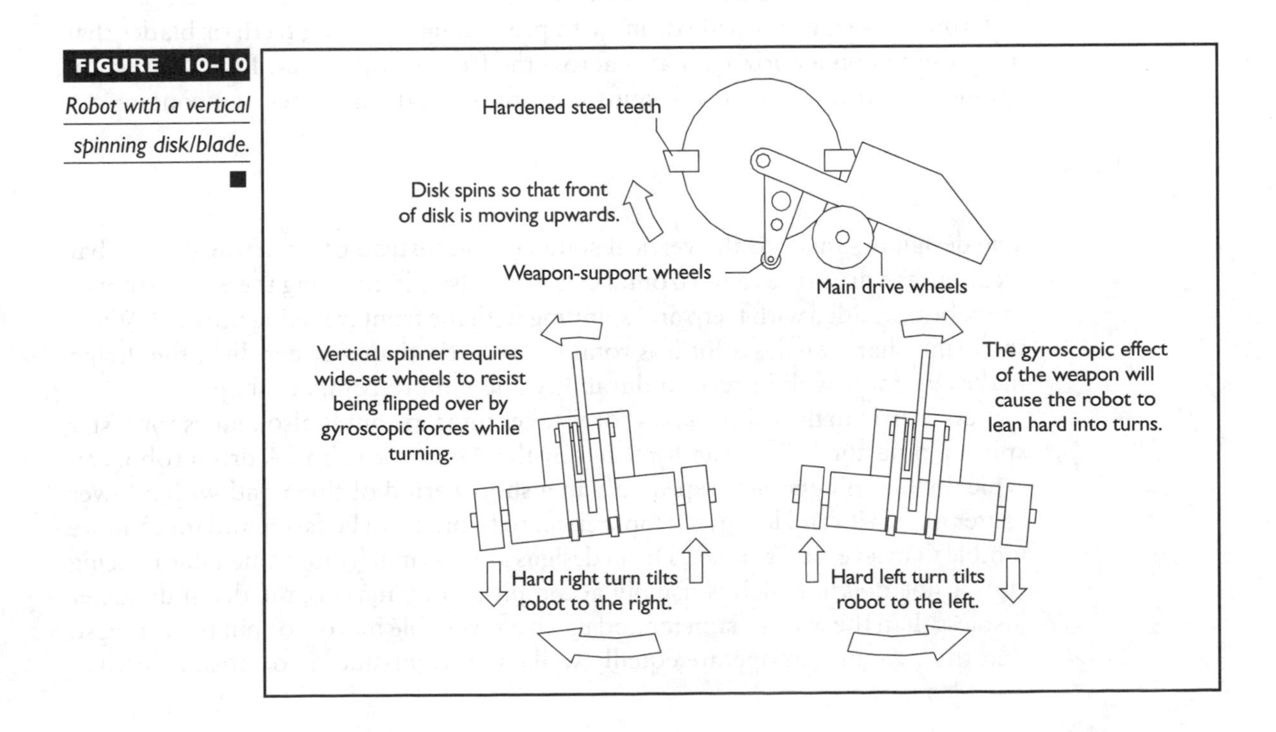

FIGURE 10-10 *Robot with a vertical spinning disk/blade.*

Strategy

Vertical spinners are good against any opponent that cannot disable them quickly or outmaneuver them to avoid being struck by their weapon. A slowly moving lifter, clamp bot, or rammer will be an easy target for a vertical spinner. A wedge may be a tricky target for a vertical spinner, especially if cone or pyramid shaped, because the spinner blade works best when it can catch on an edge on the target robot. A fast-moving wedge or lifter that outmaneuvers a vertical spinner can be a very difficult opponent.

A fight between a vertical and a horizontal spinner is usually short and violent, and can go either way. If the vertical spinner manages to bring its weapon into contact with the horizontal spinner's body, the resulting impact can damage the horizontal spinner's mechanism and disable it or even—in extreme cases—flip over the horizontal spinner. The vertical spinner can also take significant damage from the hit; and if the horizontal spinner is able to maneuver to strike at the vertical spinner's exposed drive wheels, it stands a chance of ripping them clean off and winning the fight without taking any direct hits.

Drum Bots

The drum was first used on *Gut Rip* (*Robot Wars* 1996). Other drum bots include *Little Drummer Boy* and *El Diablo*.

Drum bots feature a wide drum with protruding, spinning teeth or blades that are mounted on a horizontal axis across the front of the robot. Like the vertical spinner, the front of the drum spins upward to lift the opponent on contact.

Drum Design

The design is similar to the vertical spinner; but, instead of a narrow disk or bar weapon, the drum uses a horizontal cylinder—usually covering the entire front of the robot, studded with teeth and spinning with the front traveling upward. While the drum shape carries a lot less rotational inertia than a wider disk, the design makes up for it with improved durability and a more-compact shape.

Less inertia in the rotor makes for weaker impacts, but it also makes for faster spin-up time and less impact force felt by the rest of the robot. A drum robot can typically hit an opponent repeatedly in a short period of time; and with a lower center of gravity and less gyroscopic effect to fight, it can be faster and much more nimble than a vertical spinner. Drum designs are also much more amenable to being run upside down, which is usually accomplished by making the drum diameter just less than the wheel diameter and using a reversible motor to spin the drum, so that the weapon can operate equally well either right-side-up or upside down.

Drum robots are typically made in a four-wheeled configuration, with a roughly square overall shape. The wider weapon doesn't need much careful aiming to use effectively; and because the impacts of the weapon tend to lift the target robot into the air, the drum functions well in a ramming/pushing mode—repeatedly kicking its opponent across the arena with a combination of weapon hits and drive power.

Figure 10-11 shows a drum robot.

The vulnerable parts of the drum are the drive mechanism and support structure. The simplest and most common design is to support the drum with bearing blocks on either side and to use a chain drive to run the drum from a motor inside the main body of the bot. This method works until a strong blow to either front corner breaks a support arm, cracks a bearing block, or dislocates the chain. Hiding the drive motor inside the drum is a more durable but much trickier option.

Because the drum will be subjected to a major downward impact every time it strikes an opponent, support arms or wheels under the drum weapon to keep it from being driven into the arena floor are a good idea. Many drums also have some kind of ramp or scoop built into the drum supports, so that wedges will be fed up into the drum—rather than getting under it without being hit.

The drum doesn't pack nearly as much inertia in its weapon as the vertical spinner. What inertia it does have can be maximized by constructing the drum with as

FIGURE 10-11 *Robot with a spinning drum in front of the robot.*

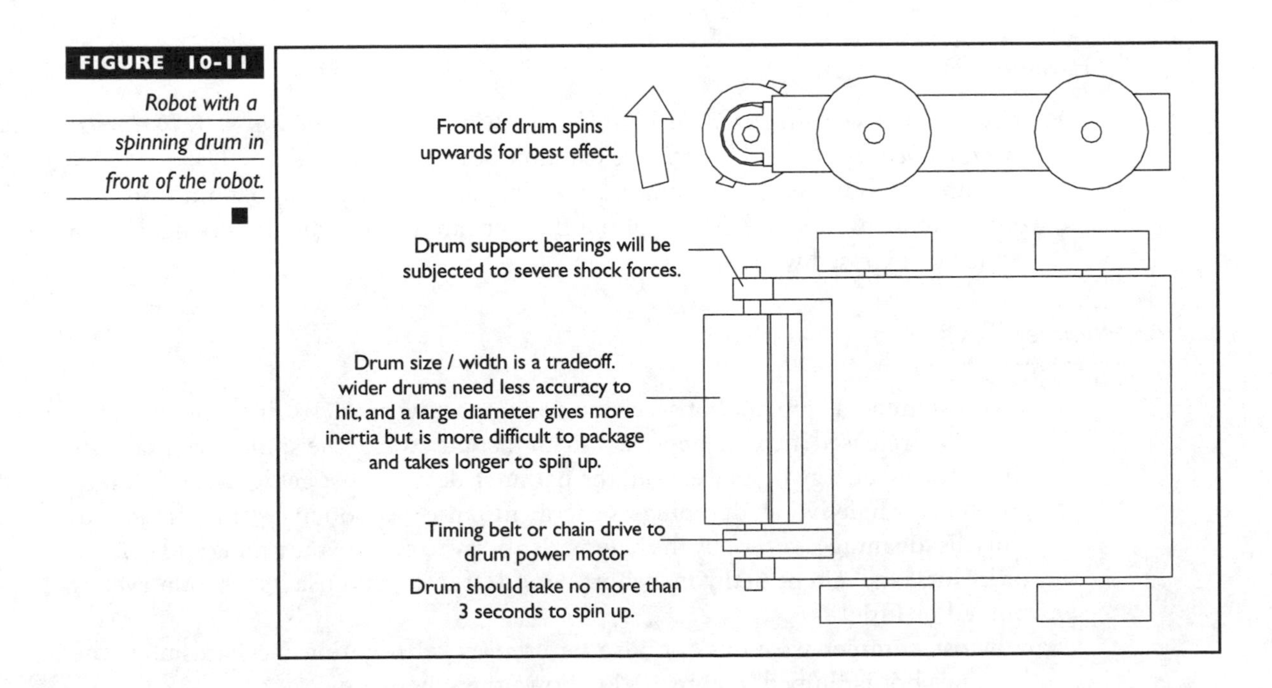

wide of a diameter as practical. A wide drum with short teeth welded to it will pack more of an impact than a thin shaft with larger blades.

Strategy

Drums lend themselves to an aggressive driving style; the fast weapon spin-up and ability to upset an opponent's footing on a good hit mean this style of robot can take control of the match and keep the opponent on the defensive. Robots that don't do much damage quickly or need time to set up a controlling move, such as thwack bots or lifters, can usually be beaten by a good drum.

The bane of the drum is the wedge. A wedge's sloped front and often sloped sides don't offer a good surface for the drum's weapon to catch. A well-designed, powerful wedge will have more of its weight budget devoted to drive power than the drum; and if the drum's weapon cannot catch on the wedge to damage and flip it up, the wedge will have the advantage.

In a fight between a drum and a spinner, the battle usually will hinge on whether the drum's weapon drive and support structure can hold together long enough for the spinner to be disabled. The drum's weapon can kick a spinner into the air, breaking its traction and spinning it around under the recoil of it's own weapon, but the drum weapon is going to take a significant impact from the force—possibly disabling it or even tearing it free from its mounts.

Hammer Bots

The hammer was first used on *Thor* (*Robot Wars,* 1995). *The Judge, Killerhurtz, Frenzy*, *Deadblow*, and *Mortis* are examples of hammer bots.

Hammer bots feature hammers, axes, picks, or mace weapons on powered overhead arms, and are designed to inflict repeated blows on an opponent's top armor or exposed wheels.

Hammer Design

Like a spinner, a hammer bot accelerates an impact weapon, storing kinetic energy that is all released into the opponent in an instant. While the spinner can take its time storing energy in its weapon, the hammer design must get its weapon up to speed in a single swing, dumping its energy into the weapon in less than a second. This disadvantage is offset by the hammer's ability to control the timing and placing of its hits, strike repeatedly in a short period of time, and use its weapon even if pinned or lifted.

Most hammer weapons can also be used as self-righting mechanisms if the hammer bot is flipped. Figure 10-12 shows the schematic.

Most hammer weapons are pneumatically driven. The most common and easiest method is to attach a pneumatic cylinder that pushes the hammer down from

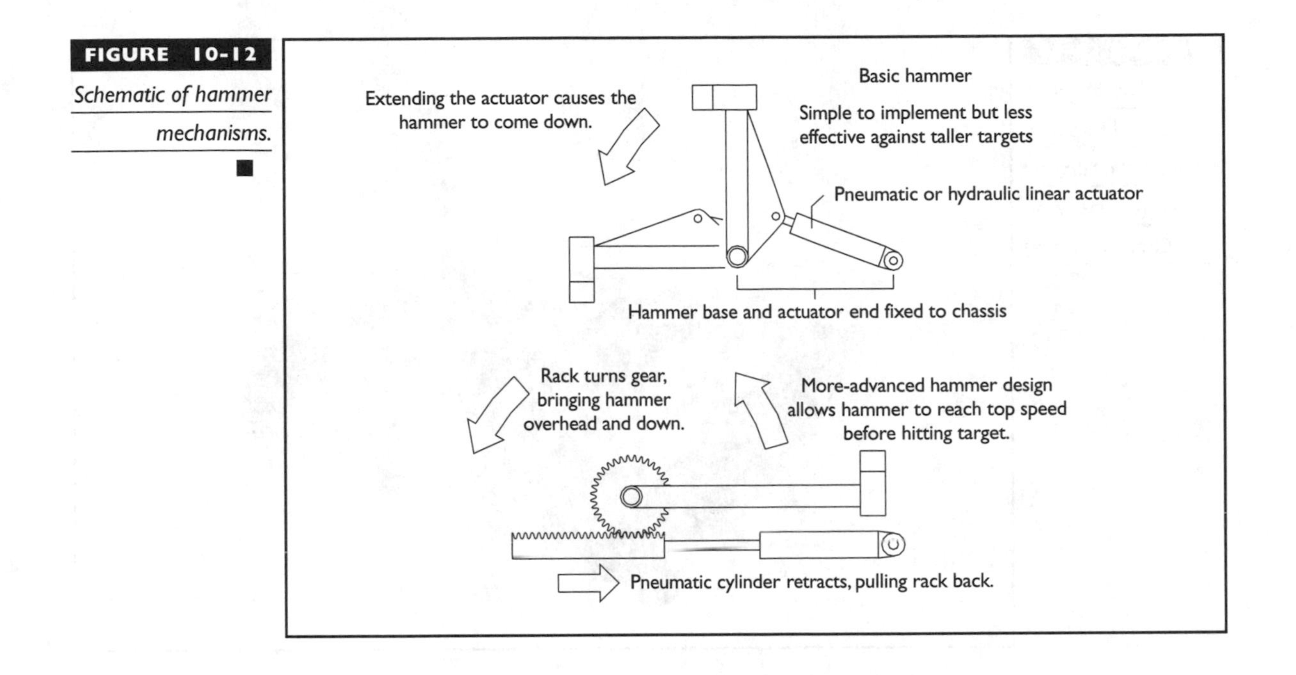

FIGURE 10-12 *Schematic of hammer mechanisms.*

behind. This limits the hammer's travel to at most 90 degrees, and less if you are striking a tall robot. This isn't much room to get the hammer up to full speed and will mean that your weapon will strike only flat robots with its full power. A better option is to use a mechanism that allows the hammer to travel a full 180 degrees, permitting it to get up to full speed before it impacts. This can be accomplished with a pneumatically driven rack-and-pinion mechanism driving the hammer arm, or by using a pneumatic cylinder to pull a chain wrapped around a sprocket connected to the hammer arm. Figure 12-13 shows a photo of *Deadblow,* one of the fastest rapid-firing hammer robots to compete in *BattleBots*.

Whichever mechanism is used, the limiting factor in a pneumatic hammer's speed will be the rate at which you can make the working gas flow from your storage tank into your driving cylinder. As the pressure regulator is a major bottleneck, some pneumatic hammer bots have huge low-pressure reservoirs downstream of the regulator to provide the high flow rates that the hammer needs. Other bots use massively large-bore tubing and valves to minimize flow resistance in the pneumatic lines. High-pressure systems that run gas straight out of a carbon dioxide tank with no pressure regulation can provide extremely high rates of force delivery, but these systems are expensive, dangerous, and difficult to build.

Carbon dioxide absorbs a lot of heat from its environment as it expands from liquid to gas, which means that a CO_2 tank called upon to provide gas for many hammer shots in a short period of time can freeze up and become too cold to deliver gas quickly enough to keep the weapon running. To get around this, some

FIGURE 10-13 Deadblow, a 114-pound pneumatic hammer bot. (courtesy of Grant Imahara)

builders use high-pressure air or nitrogen, which do not have to change state from liquid to gas. This gets around the problem of the tanks freezing up, but it doesn't store nearly as much energy in the same space and requires huge tanks to run a hammer for an entire match.

Another option is to drive the hammer with an electric motor. This makes it easy to give the weapon 180 degrees or more of travel, allowing it to reach full speed before hitting the target. Gearing should be optimized for maximum speed at impact, taking into account that with too low a gear ratio, the motor won't have enough torque to get up to speed, while too high a ratio will mean that your hammer will reach its top speed too early and not do as much damage as it should. Problems of both speed and torque can be solved by choosing the most powerful drive motor you can for the mechanism.

Some hammer robots have used a crankshaft mechanism to produce recipro cating hammer motion from a continuously turning drive motor. When considering this kind of mechanism, you should keep in mind two things: First, you want the hammer moving at maximum speed when it strikes the opponent; many simple crankshaft mechanisms will have the hammer traveling at top speed only in the middle of the stroke. Second, if the hammer's motion is interrupted mid-stroke, it should have some way of reversing and striking again without stalling or having to lift the entire robot off the ground.

Hydraulic-powered hammers have also been built. Hydraulics can provide tremendous force that can accelerate a hammer very quickly, but most hydraulic systems respond rather slowly and are not ideal for the high speeds required for rapid-fire striking a good hammer system needs. Building a hammer mechanism with a hydraulic drive will require a powerful motor and expensive, high flow-rate valves and tubing.

Some builders have experimented with using a large spring to power the hammer and a high-torque motor or linear actuator to crank the hammer back and latch it after firing. While this can give a powerful hammer action, the increased reload time makes the concept questionable. A hammer that takes more than 5 seconds between shots may never manage to hit its opponent more than once or twice in an entire match.

For optimum results, increase the hammer velocity as much as possible. Remember that your hammer may strike its opponent only partway through its stroke, so design for it to do most of its acceleration at the beginning of its travel.

Strategy

Even the strongest hammer bots have trouble consistently disabling opponents with their hammers. A hammer bot's best opponent is one with weak top armor or a fragile frame. Barring that scenario, a hammer bot should try to strike as many blows on the opponent as possible while avoiding being disabled. A hammer stands a good chance against a thwack bot, wedge, ram, or saw-wielding robot, because those designs won't be able to disable the hammer quickly and the hammer can get a lot of good hits in. Against a crusher, a hammer bot will have a hard time; the hammer may need to strike many blows to affect the crusher, but the crusher needs to get lucky only once.

Any good hammer bot should be able to self-right quickly with its weapon, which reduces the threat from lifters and launchers. Fighting a spinner with a hammer is often disastrous for the hammer, because the spinner's weapon will be nearly impossible for the hammer arm to avoid, and striking the active spinner with the hammer arm will likely result in a bent or even torn-off weapon!

Crusher Bots

The crusher was first used on *Munch* (*Robot Wars*, 1996). Some examples of crusher bots include *World Peace, Razer, Jaws of Death,* and *Fang*. Crushers feature a large, heavily reinforced claw, usually hydraulically powered and capable of closing with several tons of force to crush or pierce the opposing robot.

Crusher Design

Mechanically the most challenging concept to build, crushers use powerful claws to pierce and crush the opponent. Most crusher designs use hydraulics to achieve the incredibly high forces needed to pierce armor, although ball-screw linear actuator designs have also been used.

The challenge of a crusher design lies not only in achieving the force required, but in designing a claw structure strong enough to deliver the force without collapsing. Most crusher designs use claws that taper to narrow blades or spikes to focus the force on as small an area of the target's structure as possible. The claw not only needs to be designed to survive its own crushing force, but must be rigid enough to avoid bending on hits from spinners or off-center forces from closing onto a sloped surface.

Figure 10-14 shows a schematic.

Ideally, a crusher's claw should be large enough to bite into a sizable chunk of the opposing robot. A claw that's too small will not be able to damage much more than outer armor layers or small protruding pieces; and if used against a large target with curved surfaces, a small claw might simply slide off the target without digging in. Typically, you will want your claw to open as large as the height of the largest robot you expect to fight, and be long enough to get at least a third of the way into your opponent for maximum damage potential.

You also want the claw to close as quickly as possible. A claw that takes more than a few seconds to close will likely allow the opposing robot to escape before being crushed. A closing time of one second or less should prevent even an agile robot with high ground clearance from getting free. Of course, the combination of high force and high speed requires a powerful motor to drive the claw mechanism. A variable-displacement pump on a hydraulic-powered crusher will allow you to do both with less power—the hydraulic system can run in high-speed, low-pressure mode until the claw makes first contact, and then switch to high-pressure mode for the main crushing action.

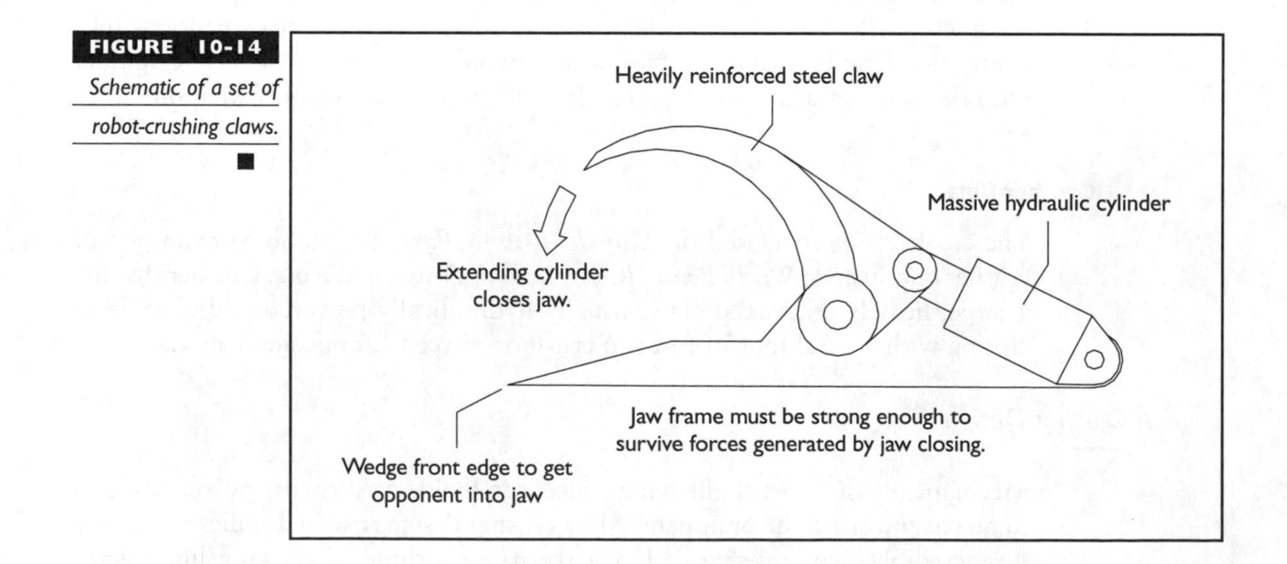

FIGURE 10-14 *Schematic of a set of robot-crushing claws.*

A well-executed crusher is one of the few designs with the potential to inflict significant internal damage to its opponent. While a powerful spinner might break up a robot's frame and rip off external parts, a crusher that hits the right spot on an opponent can punch holes through radio gear, batteries, or other electronic parts, decisively disabling its opponent. A crusher also has the advantage that once its claw has grasped an opponent, that opponent will find it impossible to escape. A crusher with a high-torque drive system can grasp, and then drag its opponent into arena hazards, or it can pin them against a wall before opening its claw and taking a second bite.

A crusher doesn't damage its opponent quickly, but the nature of its weapon is such that—once the crushing begins—there is no escape.

Strategy

The crusher mechanism will invariably take up a large part of the robot's weight, leaving little left over for armor and drive system. While they may need to score only one hit with their weapon, a crusher bot may be at a disadvantage when faced with a faster, more agile opponent, especially a wedge or lifter that might get to the crusher from the side and flip it over. Thwack bots typically have high mobility and good ground clearance, and they may be able to flip themselves free of the crusher before it closes. The easiest target for a crusher is a robot that doesn't have a method of taking control of the match or dealing a killing blow quickly—weaker rams and hammer bots are easy crusher prey.

A good spinner will be a challenging opponent for a crusher. While most spinners do not have strong drive systems, a spinner with a powerful weapon may be able to keep a crusher from ever getting in position to use its weapon by knocking it aside on every impact. Against a spinner, the crusher bot's best bet is to try and first knock the spinner into a wall to stop its weapon, and then rush in for a killing grab before the spinner can recover.

Spear Bots

The spear was first used in *Ramfire* 100 (*Robot Wars,* 1994). Some example spears include *Rammstein, DooAll,* and *Rhino.*

Spear robots feature a long metal rod, usually sharpened at the front, actuated by a powerful pneumatic or electric mechanism to fire at high speed at the other robot.

Spear Design

The spear design seeks to damage its opponent by firing a long thin rod, piercing the target's armor and impaling some sensitive internal component. Usually, a spear weapon is pneumatically powered, although other methods have been attempted.

The goal with a spear design is to maximize the impact when the weapon head hits the target. The force behind the weapon at the point of impact does not matter, because the effect on the target will be determined entirely by the kinetic energy of the spear at the moment of impact. The kinetic energy of the spear is proportional to its mass times the square of its velocity, so increasing the speed of the spear will do more to make it an effective weapon than increasing its mass. Excess force on the spear at the moment of impact will mainly have the effect of pushing the spear-armed bot and its target away from each other; the traction holding the bots in place on the floor is small compared to the forces required to punch through armor. Figure 10-15 shows such robots.

Ideally, your spear bot should strike the opponent near the end of its travel for maximum effect. In practice, however, this will be difficult if not impossible to arrange. In most cases, the spear will strike the target robot after only a fraction of its travel. If possible, you should design your spear to accelerate as much as possible early in its travel.

Most spear designs use a pneumatic cylinder to fire the weapon. With a pneumatic ram, the top speed of the weapon is limited by the rate of gas flow into the cylinder. All components on the gas flow path from the storage tank to the cylinder—regulator, valves, tubing, and fittings—should be made as large-bore (internal diameter) as possible for maximum flow rate.

FIGURE 10-15 *Robots carrying spears.*

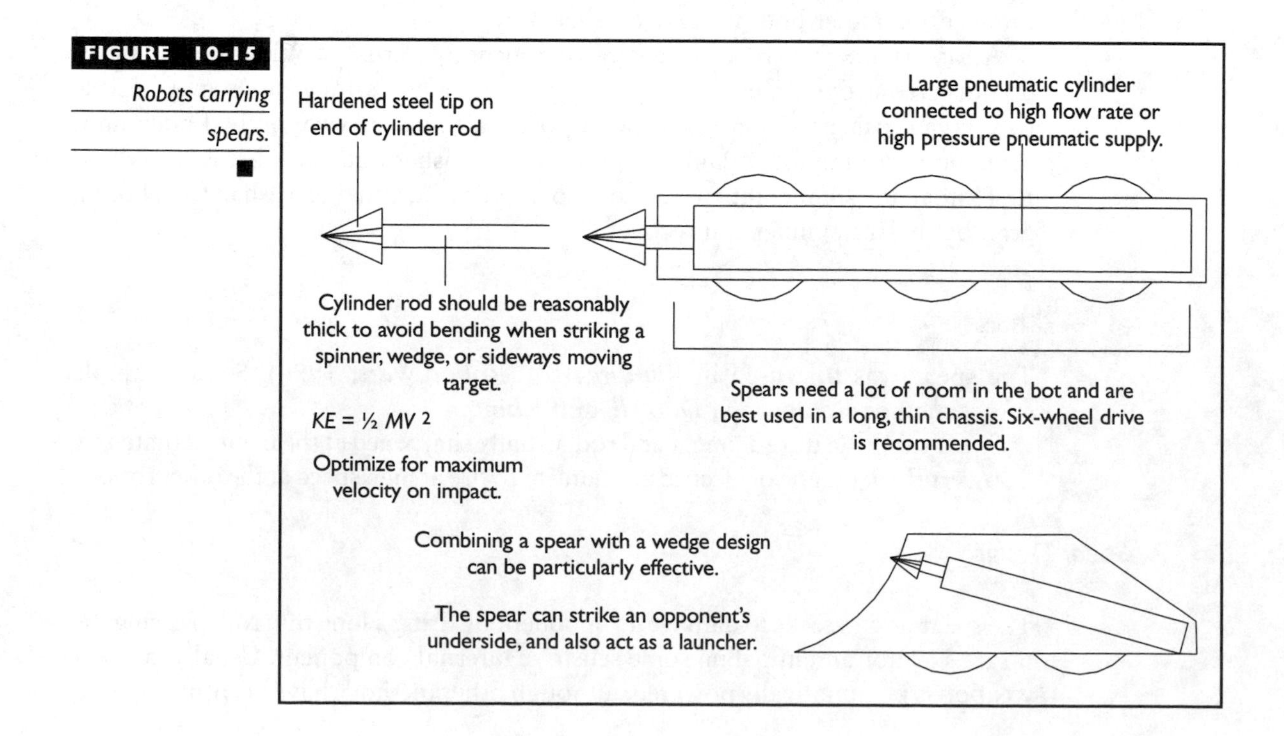

If a carbon dioxide tank is used for the gas source, consider using buffer tanks on the low-pressure side of the gas regulator to compensate for the limited conversion rate of the carbon dioxide from liquid to gas. A high-pressure air or nitrogen source will provide a greater air flow, at the expense of more room taken up by the gas tanks. The most powerful spear designs use no regulator at all, instead running full-pressure carbon dioxide straight from the storage tanks. Although this approach overcomes the gas flow problems by running at a much higher pressure, it is difficult and expensive to implement safely.

---

note *In the comparison of carbon dioxide and high-pressure air (HPA) (or nitrogen), it's true that HPA has an advantage in flow because there is no phase change from liquid to gas; but when using HPA, large bore tubing and valves and a downstream accumulator are still essential elements to achieving high flow in a system. Using HPA with small-diameter tubing will still have significant flow restrictions and less-than-optimal performance. This discussion also applies to the air flow discussion in the "Hammer Bots" section earlier in the chapter.*

Another approach is to use a powerful spring to accelerate the spear. This approach has the advantage of the spear doing most of its acceleration in the early part of its stroke. The disadvantage of this concept is the need for a complex mechanical re-cocking system to crank the spear back and latch it in place until it is needed again. A long re-cocking time on a weapon makes that weapon nearly useless, as the opponent can freely attack while the weapon is re-cocking itself. A third approach is to use a crankshaft to drive the spear to convert a constant motor rotation to reciprocating forward and backward motion of the spear. While it is a less-complex approach to the spear weapon, crankshaft drive spear weapons tend not to be effective in practice. The spear will reach its maximum speed only at the middle of its travel, and will actually be decelerating for the second half of its travel. Furthermore, on striking the opponent, the weapon will either stall and be unusable or push the other bot away and ensure that the next impact between the spear and the target bot will be near the end of the spear's travel—where it will be traveling slowly.

The best head design for penetrating armor is a three- or four-sided, thin, pyramid- or diamond-shaped head. Conical points are less effective at penetrating armor; the head should have sharp edges so it can cut open rather than force open the armor material. The downside of effective penetration is that the spear head may get stuck inside the target robot after being fired, jamming the two robots together and risking damaging the spear mechanism as the target bot struggles to get free. One possible way to minimize the potential to get stuck is to machine the entire shaft to slightly increase the diameter of the spear toward the robot's body. Some teams use deliberately blunt weapon heads, hoping to knock out the opponent through impact damage rather than penetrate armor.

Maximize the spear velocity to get the most effect. Mass of the weapon head is less important than the speed at which it travels.

Strategy

Barring a lucky hit against a thinly armored opponent, a spear is not going to disable an opponent in a single hit. A spear is best used on a fast, agile robot capable of avoiding its opponent's weapon while firing the spear at less well-armored spots. A wedge will be a difficult target for the spear because most wedges are well-armored; and if the spear strikes the wedge's sloped front, it will just slide up and lift the bot's front off the ground. A ram will also be a difficult target, again because most rams are well armored, as well as fast and agile.

A spinner can be a disastrous opponent for a spear, as the first hit on a spinning body will likely bend the spear, jamming it and making it unable to retract and fire again. Against opponents that need to place their weapons with some accuracy—clamp bots, launchers, crushers, or other spears—the fight will come down to maneuverability and driving skill as both bots try to place their weapons for best effect while avoiding the opponent's attacks.

Closing Remarks on Weapons

For most people, weapons selection is a matter of personal preference. This chapter has presented many of the different types of weapon systems that are currently being used in combat robot events, and lists their strengths and weaknesses. There are many different types of weapon systems that have yet to be seen in the world of combat robots. Use your creativity in coming up with a new weapons system! But remember, what ever weapon system you use, it must conform to the rules, regulations, and safety requirements of the event that your bot will enter.

The most-effective weapon that has not been discussed is driver control. One of the most-effective weapons you'll ever have is learning how to control your robot. A good driver can avoid the deadly blows of an opponent and then position himself or herself for the kill. Remember: there are more points awarded for strategy and aggression than for damage points.

chapter 11

Autonomous Robots

THE robots described in the book thus far are remote-controlled (R/C) robots, which are generally the easiest robot to build because all the traditional R/C equipment can be readily purchased at hobby stores and from the Internet.

The next level for the robotic evolution, however, is the *semiautonomous* robot. Including some semiautonomous features along with traditional features in your robots can simplify some of the work in controlling the robot, because such features mean that the robot will have some behaviors that will function on their own.

Autonomous control can range from little control to almost 100-percent control within the robot. Minor control could be in the form of a mixing circuit to help with tank driving, overload current sensors on the motors to reduce the power going into the motors automatically, or automatic weapon firing or driving mechanisms. Generally, a semiautonomous robot will have a sensor that can monitor its environment and some electronics that process the sensor data to make a decision and execute some action.

The next level for the robotic evolution is the *fully* autonomous robot. These robots act completely on their own in performing tasks, using microcontrollers or computers for brains and many different sensors that allow the robot to see its environment, hear its environment, and feel its environment. The robot's brain will interpret the sensor data, compare it to internal programming, and execute a series of actions based on the data. Various examples of autonomous robots are maze-solving robots, line-following robots, sumo robots, and soccer playing robots in the Robo Cup. Even NASA's Mars Sojourner has some autonomous features, also, that allow it to send images back to the engineers at NASA, who study the images and tell the robot to check out a particular rock or other interesting feature. The robot then determines how to get to its destination. If it senses an obstacle in the way, the robot figures out a path around it to continue its mission to the place of interest. Once the robot gets to its destination, it conducts a series of experiments and sends the data back to the engineers at NASA.

Most combat robots are either totally remote-controlled robots or semiautonomous robots. It is very difficult to make a fully autonomous combat robot, which needs a way to "see" its opponent and be able to distinguish it from its environment. Reliable robotic vision systems are difficult to develop. Consider the human brain, of which more than half is devoted to processing just what the eyes see. The rest of the brain does everything else.

The human eye-brain combination can easily spot a robot in a combat arena and know where it is, what direction it is going, how fast it is going, its motion relative to another robot's motion, where the hazards are, and where the perimeter of the arena is. The human eye-brain can do human *intuitive* things, such as calculate how heavy an opponent is and how dangerous it might be, and determine the weak points to attack and when to retreat and regroup. The human brain can do this all at once—plus throw out any information that is not needed for the task at hand. The eyes of a robot break the image it sees into picture elements, or *pixels*. A robotic vision systems has to interpret everything it sees, pixel by pixel, on the vision camera and then make decisions on it. Teaching a robot how to distinguish the difference between a steel box and an enemy robot is a challenging task. Research scientists around the world are still getting PhD's trying to figure out how to implement a reliable vision system in robots.

Though vision systems are rather complex to implement, autonomous robots are still possible. For example, longtime combat robot builder Bob Gross built a beacon system that can be placed on a robot allowing it to see where the opponent is in the arena. (See the sidebar "Bob Gross and Thumper," later in this chapter, for more information.)

Because every robot design and function is different, this chapter cannot provide the exact details on how to implement a sensor into your robot. What this chapter *will* cover, however, is the basic functionality of how various sensors work.

Because a specific sensor may have performance characteristics based on how it is implemented and the environment in which it is being operated, the robot builder should build a prototype sensor system and fully test it before implementing it into the robot. When using semiautonomous to full-autonomous components in your robot, testing is critical. It is best to build small-scale prototype models to test the various components and all of the failsafe features before implementing them in the final robot. All the bugs need to be worked out prior to a combat event. When at an event, you will have to demonstrate these features and the corresponding safety features to the safety inspectors; and you'll need to convince them that these features are reliable, or you won't be allowed to compete. Because of this, more time before the contest is required to test the robot and the advanced controls.

Using Sensors to Allow Your Robot to See, Hear, and Feel

Before implementing semiautonomous features in a robot, you need an understanding of how sensors work so that the appropriate sensor can be selected for the application. A multitude of sensors can be used by a robot to react to its environment. This chapter will cover some of the most common sensors used in robots. Most of these sensors break into two categories: *passive* and *active* sensors.

Passive Sensors

Passive sensors monitor some condition in the environment. They don't introduce anything into the environment; they simply sense what is happening around them. A thermometer and a photocell are everyday examples of passive sensors. If connected to a household heating system, a thermometer's findings are reported to a simple circuit in a household thermostat to tell the heater when to turn on or off. Similar circuits are used to control air conditioners in warm climates. Photocells monitor ambient light to sense how bright it is. These are used in street lights to sense when the lights are no longer needed—a circuit turns off the lights when the sun comes up. Similarly, when the sun goes down at night, the light level drops to a predetermined level and a circuit turns the lights back on.

Another type of passive sensor is the *passive infrared* (*PIR*) sensor, sometimes called a *pyroelectric* sensor. These sensors are commonly used to detect the presence of a person and activate a circuit. They can control lights within a room or outside a house, or they can be used as a burglar alarm. The sensor is a small crystal mounted within the housing that can sense the infrared radiation emitted by a person. The sensor has a circuit that charges the crystal, and the presence of the radiation discharges the crystal, which is detected by another circuit.This is called the *pyroelectric effect*. The radiation is focused upon the crystal by a row of Fresnel lenses that cause a series of signal peaks as a person moves by. Several autonomous robots have used these sensors to detect heat emissions from their surroundings.

For combat robots, an electronic thermometer can be used to monitor the internal temperature, the temperature of the motors, or batteries. If the temperature gets too high, cooling motors can turn on or the power requirements can be reduced to avoid overheating.

A tilt sensor can be used to monitor whether the robot gets flipped upside down. Once the sensor detects a flip, it can initiate an arm or piston that will flip the robot right side up. Or, if the robot was designed to run upside down, the tilt sensor can be used to reverse steering controls, since an upside-down robot will turn in opposite directions than a right-side-up robot. Another type of passive sensors that can be used are acoustic sensors that can listen for the motors of the opponent robot. These sensors can help guide your robot toward its opponent.

The most complex-passive sensor is a charged coupled device (CCD) camera that is used to "see" the environment. CCD cameras are part of a vision system. When used alone, they require advanced object-recognition software and usually a dedicated computer. They can also be used with active sensors to help simplify the computational software. Vision systems are most commonly found at robot soccer events. Recently, CCD cameras have been used to detect flames in the Trinity College Fire Fighting contest, and some members of the Seattle Robotics Society have developed methods to use CCD cameras and simple microcontrollers to see the lines in line-following contests.

Active Sensors

Active sensors often introduce a sound or light and look for how the introduced energy reacts with the environment. Some examples of this type of sensing are sonar, laser, and infrared reflective detectors.

Sonar detectors introduce a sound, typically higher in frequency than what humans can hear, and listen for the echo. The bounced-back echo is used by sonar range finders to send out a sound pulse and then compute the time it takes for the sound to return. This time is directly proportional to the distance the sound must travel to bounce off the nearest object and return. The speed of sound is around 770 mph, but it can easily be measured using a moderately fast microcontroller chip like those found in many robots.

Infrared (IR) reflective sensor systems emit a specific wavelength of light and look for a reflection of light. Since light travels so much faster than sound, it is difficult to measure the time it takes to receive the reflected light. Infrared detectors are typically used to detect whether an object is present within the range of the detector rather than how far the object is from the detector. Some clever infrared detectors use some simple geometry present in a triangle formed by the emitter that generates the light, the reflected object, and the detector that senses the emitted light. Everyday examples of infrared detectors can be found in modern bathroom stalls in public places. The mechanism that automatically flushes the toilet typically uses an infrared detector to detect the presence of a person using the toilet. The system is activated when a person is in the stall for a predetermined period of time. Many systems have a small flashing LED that speeds up its flashing when the time has elapsed. When the person leaves the presence of the IR sensor, the toilet flushes.

Lasers can be used to detect where an opponent is located. This type of system is fairly advanced and usually employs a CCD camera or a linear sensor array. A laser beam is emitted from the robot, and the CCD camera is used to see the laser spot—or line—on the opponent robot. Generally, a band pass filter is placed in front of the camera to filter out all wavelengths of light except for the laser beam wavelength. When the laser beam and the camera orientation is known, the range and location of an object can be determined through mathematical triangulation. This type of system is fairly complex, but not as complex as a true vision system, and it is beginning to be seen in robotic applications using simple microcontrollers. This type of system could be used in automatic weapons firing and assisted homing in on an opponent, and it can be placed in fully autonomous robots.

Devantech SRF04 Ultrasonic Range Finder

The Devantech SRF04 Ultrasonic Range Finder (shown in Figure 11-1) is a 40-kHz ultrasonic range finder that can be used to determine the range of objects from 1.2 inches to about 10 feet (or 3 cm to 3 meters).

FIGURE 11-1 *Devantech SRF04 Ultrasonic Range Finder. (courtesy of Aeroname, Inc.)*

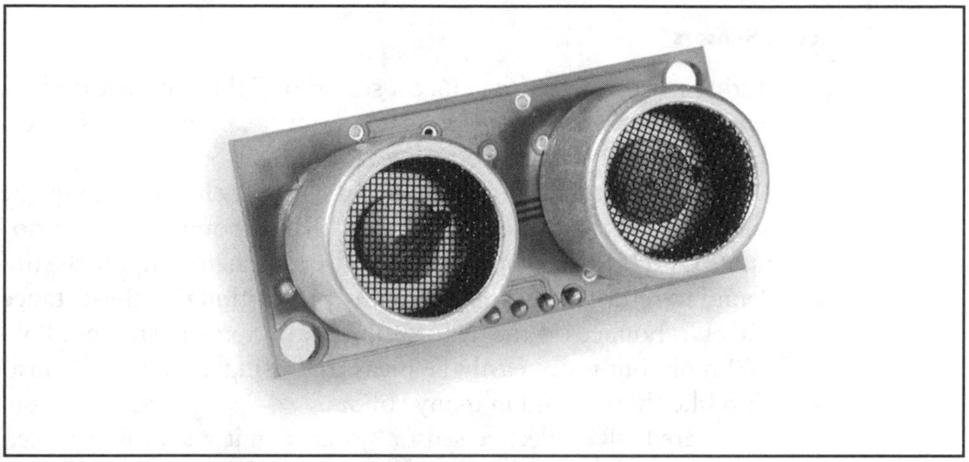

This sensor works by transmitting a pulse of sound and measuring the time it takes for the reflected signal to return to the sensor. The sensor then outputs the return time as a pulse. By measuring the pulse width and multiplying this value by the speed of sound, you can calculate the distance to the nearest object. Figure 11-2 shows how the timing pulse from this sensor is generated. The sensor can detect a 1-inch-diameter broom handle at 6-foot distance.

Polaroid 6500 Ultrasonic Range Finder

Another version of an ultrasonic range finder is the Polaroid 6500. Polaroid capitalized on the development of ultrasonic distance sensors designed for its instant cameras and made the technology available for other uses. This sensor can accurately measure the distances of objects from 6 inches to 35 feet. This sensor works similarly to the SRF04 sensors; the return echo pulse time must be measured. The distance is computed by multiplying the time by the speed of sound, which is approximately 1,130 feet per second. These sensors have found a lot of use in the autonomous robotics community.

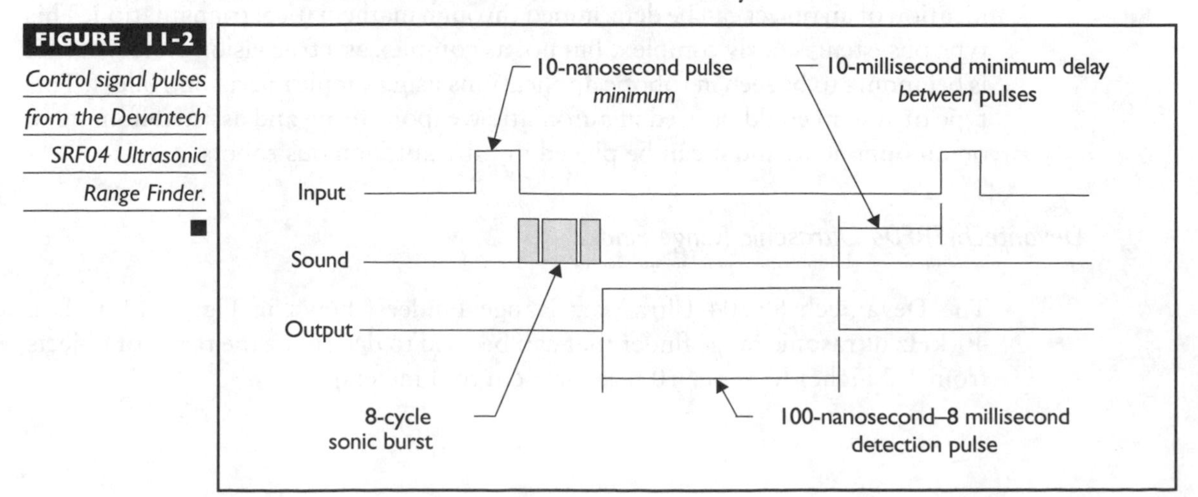

FIGURE 11-2 *Control signal pulses from the Devantech SRF04 Ultrasonic Range Finder.*

Sharp GP2D02 and GP2D12 Infrared Range Sensors

The GP2D02 and GP2D12 are infrared range finders. Shown in Figure 11-3, these sensors work by transmitting an infrared light and measuring the location of the reflected light on a position sensitive detector (PSD).

Next, you can see how this sensor works. By using a triangulation method, the range of an object can be determined by the location at which the reflected light hits the PSD. The detection range for these sensors is from 4 inches to 31 inches (10 cm to 80 cm). The GP2D02 outputs an 8-bit serial data set. As the object gets closer to the sensor, the output number gets larger. The maximum number occurs with a distance of about 4 inches, and the smallest number occurs out past 31 inches.

$$T_c = \frac{1}{A + B(\ln R_{sensor}) + C(\ln R_{sensor})^3} - 273.15$$

The GP2D12 works similar to the GP2D02, except the output value is an analog signal—in other words, it is a variable voltage that will range from 0 to 3 volts. The maximum voltage will occur at 4 inches, and the minimum voltage will occur out past 31 inches.

Sharp GP2D05 and GP2D15 Infrared Proximity Sensors

The GP2D05 and the GP2D15 are infrared proximity sensors that look physically identical to the GP2D02 and the GP2D12 sensors. The difference between these sensors is that the output signal changes when the object moves past a preset distance of 9.5 inches (or 24 cm). For any object that is between 4 inches and 9.5 inches (10 to 24 cm), the output signal is 0 volts, and any object that is past the 9.5-inch threshold will have a positive 5-volt output signal. The difference between the GP2D05 and the GP2D15 sensors is that the GP2D05 sensor requires an input trigger pulse to tell the sensor to make a measurement. The GP2D15 sensor continuously takes measurements.

note *The case of all four of the GP2Dxx sensors looks like normal black plastic, but it is actually a good electric shield when grounded. It is very important that you connect this shield to ground. This is mandatory for these sensors to work reliably!*

FIGURE 11-3 *Sharp PG2D02 Infrared range sensor.* (courtesy of Acroname, Inc.)

Thermal Sensors

One of the more popular types of sensors to measure temperature is the *thermistor*, a sensor whose internal resistance changes with temperature. By measuring the resistance of the sensor, the temperature can be calculated. Measuring the sensor's resistance is accomplished by using a voltage divider circuit. Figure 11-4 shows a simple schematic drawing of this type of sensor.

The voltage to be measured is the V^{ou}, which is defined in the following equation, where R^{sensor} is the thermistor resistance, R_I is some other resistor used in the circuit, and V^{in} is the input voltage:

$$V_{out} = \frac{R_{sensor}}{R_I + R_{sensor}} V_{in}$$

Because V_{out} is being measured and the thermistor's resistance, R_{sensor}, is unknown, equation 1 can be solved for the resistance of the sensor.

Equation 2 shows this new relationship:

$$R_{sensor} = \frac{V_{out}}{V_{in} - V_{out}} R_I$$

After the thermistor's resistance is measured, the temperature can be calculated using the Steinhart-Hart Equation, which describes how the resistance changes with temperature in semiconductor thermistors. The basic form of the equation is shown in equation 3, where constants *A*, *B*, and *C* are thermistor-specific constants that are obtained from the manufacturer of the thermistor, or they can be determined experimentally. T_K is the temperature in degrees Kelvin.

$$\frac{1}{T_K} = A + B(\ln R_{sensor}) + C(\ln R_{sensor})^3$$

A more useful Equation is shown in equation 4, where the temperature, T_C, is in degrees Celsius:

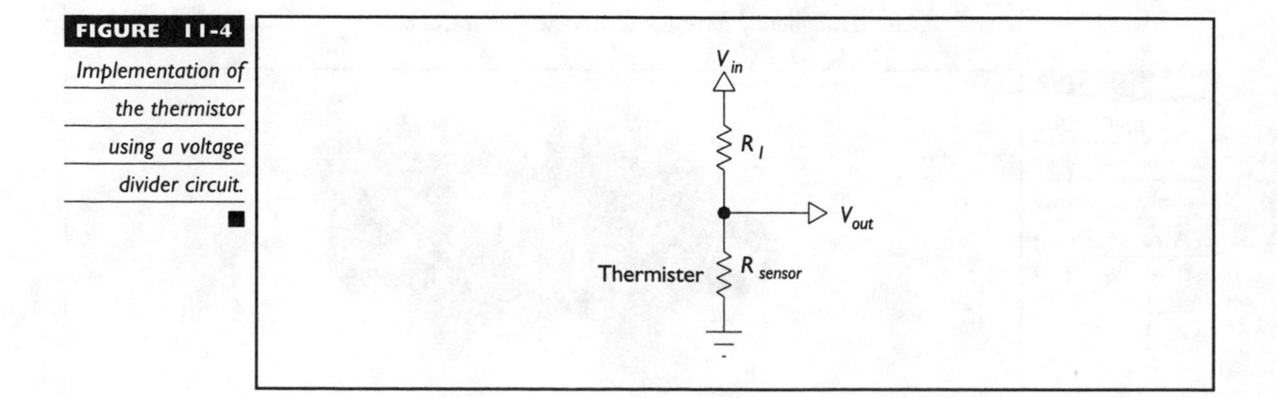

FIGURE 11-4 *Implementation of the thermistor using a voltage divider circuit.*

Tilt Sensors

A *tilt* sensor usually comes in three types: a conductive liquid tilt switch, a mechanical switch, and an accelerometer. Accelerometers can be used to measure the direction of gravity, which makes them a great sensor for determining whether your robot has been flipped on its back or on its side. Unfortunately, these sensors will detect every bump, slam, bash, and crash you robot will experience. Because of all of this extra activity, it will be difficult to implement accelerometers because a lot of filtering of the data will be required to differentiate between impacts and actually turning upside down. They are fun to play with, though. If you are interested in experimenting with accelerometers, check out Analog Devices' Web page at *www.analog.com*.

Conductive liquid switches are commonly used for tilt switches. The most common is the mercury switch, in which two electrical contacts are embedded inside a glass tube, along with a small amount of mercury. When the switch is held vertically, the mercury covers both contacts, which closes the circuit. When the glass tube is placed on its side or upside down, the mercury slides off both contacts, which opens the circuit. Mercury switches can be obtained at most electronics stores and some hardware stores. Mercury switches can be found in non-digital thermostats, and some companies sell a different version of this type of switch that uses a conductive electrolyte instead of mercury.

note *If you are going to use this type of switch, use the variety that uses the conductive electrolyte instead of Mercury. Mercury is a poisonous and an environmentally hazardous material. Most competitions have a rule clause that prohibits dangerous materials.*

The last type of a tilt switch is a mechanical tilt switch, which is basically a metal tube with a ball bearing inside it. Figure 11-5 shows a schematic of this type of switch. Gravity is used to hold the ball down on the bottom contact. When the

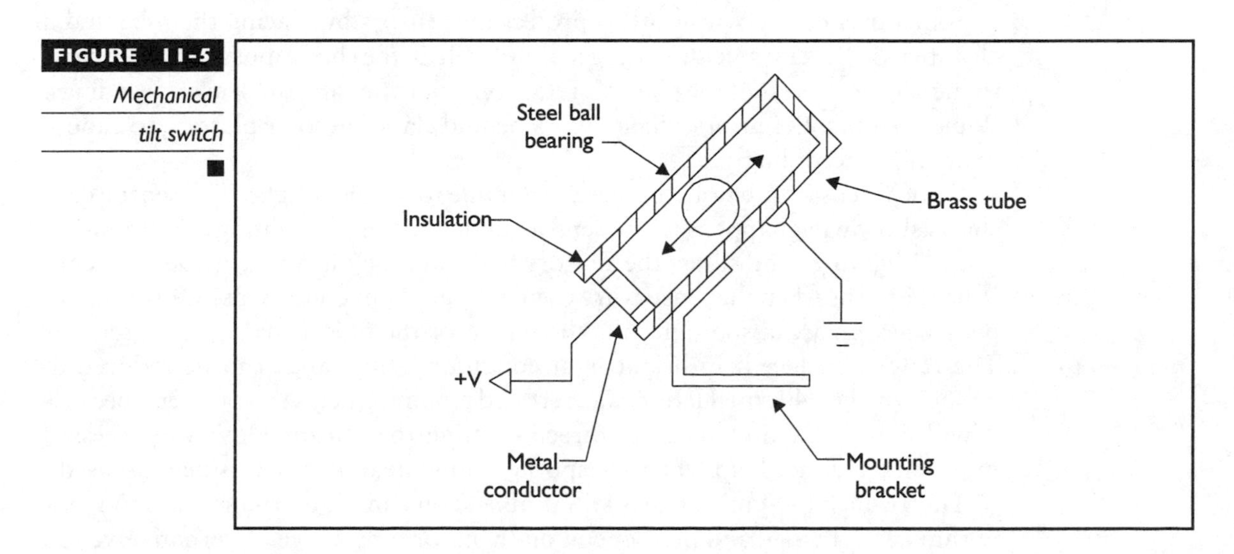

FIGURE 11-5 Mechanical tilt switch

tube rotates past horizontal, the ball will roll off the contact, thus opening the circuit. The figure shows a bracket at some angle. The smaller the angle becomes, the more sensitive the robot becomes to angular tilting.

Bump Sensors

A bump sensor is nothing more than a mechanical lever action switch that is attached to the underside of your robot's bumpers or armor. When another robot hits your robot, the bump switches will tell the robot that it was hit. One implementation of a bump switch is to place it on the sides and the back of your robot. When your robot is moving forward and the bump switches indicate that something is hitting the side of your robot, your robot can initiate an automatic spin move to face the attacker. To implement this type of sensor, the armor or bumpers must be semi-flexible so that when they get hit, they will move a little to trigger the switch.

Implementing Sensors in Combat Robots

Although many sensors have a few problems when used in the combat environment, the following techniques can help you overcome these.

Sensors obviously cannot be placed where they could be damaged by an opponent. In general, this means recessing them with the robot's structure. If you do recess the sensor, be aware that some varieties of optical, IR, or ultrasonic sensors will "detect" the sides of the exit hole, especially if this hole is too small. To help eliminate this problem, for optical or IR sensors, paint the inside area that faces the sensor flat black.

If you place the sensor too close to the floor, it is possible that the floor will return a distance measurement, which will be depend on the roughness of the floor. To help prevent this, try to mount the sensor so that it doesn't angle downward.

Sometimes people will want to protect the sensors by placing them behind a clear plastic (Lexan) shield. If using a plastic shield, the shield must be placed close to the sensor to prevent the plastic's refection from affecting the sensor's readings. Remember that IR sensors will not work behind glass and some plastics, so choose your shield accordingly.

Optical sensors are not completely immune to ambient light. The sensor data sheets show what happens to the sensors under "normal" lighting conditions. In normal lighting conditions, the sensors have a range out to 31 inches (80 cm). What happens when the sensors are used in bright light conditions? Or what happens when an arena spotlight hits the sensor or the object that is being sensed? The range is reduced! In bright light conditions, this range can be reduced to about 16 inches (40 cm). Here, again, recessed mounting helps this problem, because it will limit the spotlights from directly hitting the sensor. However, recessed mounting does not help when the spotlight hits the object that is being sensed.

The 16-inch (40-cm) range is still usable. You can set up the sensors to work within a short range and not depend on them for long range. Alternatively, you can help correct the problem by placing a small infrared filter in front of the receiving lens to block the bright light effects. A good one can be obtained from photography stores. One such filter is a Kodak Wratten #87 gelatin filter, or the #87C filter. Using this filter will yield normal distance measurements in bright light conditions.

Shock could damage the sensors. In the ring, robots can hit with such intense force that mechanical shock is a primary concern. Although the sensors are robustly built, if jarred hard enough, their precision optics can move enough to affect the sensor. To handle this, you should mount the sensor with rubber grommets.

Sensing: It's a Noisy World Out There

Often, when people first start using sensors in robots, they find the results are not quite what they expect. Most sensors used in robotics are subject to a great deal of interference, variation, and changing results due to the ever-changing environmental conditions. As a result, many people become rather frustrated that the results from the sensors change and give occasional false readings.

Consider an infrared sensor in a room full of infrared sensors. Because the sensor is looking for the light it generates with its infrared emitter, the light generated by other emitters in the room can confuse the detector and cause false readings. Similarly, a sonar detector in a noisy room may hear echos and sounds from itself or other sensors that cause false readings.

Humans suffer similar kinds of problems, but we have amazing abilities to correct the sensory input we obtain. When a sailor first walks on a ship, the rolling of the vessel in the waves can make him or her walk a crooked line or stumble around. Very quickly, typically in one or two days, the sailor's brain will adjust and compensate for the swaying ship so that the sailor doesn't even realize the boat is swaying after awhile.

This sophisticated adjustment and compensation is one of the most unique things about the human brain. The human brain also combines, or "fuses," the input from our vision, inner ear (balance), and pressure in our feet to keep us standing up. If one of these types of input changes, our brains can quickly adapt.

Robots need a similar ability both to combine the sensory input from several sensors and to adapt to changes in the function of the sensors. This is done in sophisticated autonomous robots using neural nets, Bayesian networks, genetic algorithms, and other complex computation. Your robot need not be this sophisticated to take advantage of sensors, however.

Techniques for Improving Sensor Input

Some sensors have built-in techniques that clean up the signal they create. Sonar detectors, for example, emit a "ping" in specific sound frequency ranges and ignore input from other frequencies. This helps filter noise and avoid interference from other sounds in the sensor's environment. Similar approaches are used with infrared detectors using filters and lenses to avoid unwanted wavelengths of light.

One of the simplest techniques you can employ for improving sensor input is based on simple statistics. If the sensor has an occasional bad reading, try averaging several readings, and perhaps toss out the high- and low-value ranges; then adjust within the microcontroller as part of the software or firmware driving the sensor on the robot.

Another simple and effective technique is to use hoods or shades over light detectors to avoid bright directional lights. Just about every robot competition has problems with lighting because the light in the arena is not identical to the light in the robot development environment.

Having to clean up sensor data when another robot is using the same sensor that your robot is using can be tricky. For example, your sensor might pick up the transmitted infrared light from your opponent's sensors. This may give false distance or proximity readings to your bot. To overcome such a problem, you can use an infrared receiver sensor, such as an infrared phototransistor, and take a measurement just before using the GP2D*xx* sensor. If the transistor detects an infrared signal, there is a chance that another robot is transmitting a signal toward your robot. If the transistor doesn't detect the presence of any infrared light, you can safely turn on your GP2D*xx* sensor. To add more reliability, you could use the phototransistor to take another measurement just after the GP2D*xx* measurement reading has been completed. The second reading will be used to determine whether your opponent has turned on his robot's sensor while you were using your sensor.

As you can see, the overall sensor package becomes more complicated as you attempt to improve the reliability of the sensors. Most important, don't assume that a sensor is perfect or that its output is perfect. Figure out a way to observe the output from the sensors directly while operating in the competition environment for test runs. You can often adjust to the output of a sensor after you know how the sensor is behaving.

Sensors can create much more sophisticated robotics behaviors that don't rely on constant human input to keep the robot going. The most robust robots typically have the most robust sensor input dictating the behavior of the robot.

Semiautonomous Target and Weapon Tracking

When you begin competing in robot combat matches, you will discover that it is a lot harder to get your robot positioned to deliver the deadly blow than it was when you were at home beating up garbage cans. This is because the garbage cans are not attacking you, there are no screaming crowds to distract you, and there is no 3-minute time limit to win. With all of this excitement happening during a match, when you finally get your robot positioned and the opponent is in the sweet spot for the attack, you could miss an opportunity because it took you too long to flip the attack trigger on your remote control. This sort of predicament is frustrating to the beginning combat warrior. If you look at videos of past combat events, you will notice that missing the opponent is a common problem for many beginning robot combatants. The experienced veterans always seem to hit their mark.

Semiautonomous Weapons

A semiautomatic weapon system is a valuable method that can be used to overcome this distraction and experience problem. Figure 11-6 shows a simplified schematic that demonstrates how to implement an automatic weapon system, such as a hammer or a spike. The system uses a proximity or range sensor such as the Sharp GP2D05 range detector. This sensor is designed to trigger a signal when the opponent gets within 24 inches of your robot. The output from this sensor is fed into a microcontroller that turns on the H-bridge that drives the weapon's motor.

A limit switch on the robot tells the microcontroller that the weapon completed its range of motion and that the motor needs to be reversed to retract the weapon. For safety purposes, the microcontroller must be connected to the radio control (R/C) equipment's receiver. The microcontroller must shut off the automatic weapon feature if it loses a command signal from the receiver. To enable a manual weapons control, the microcontroller can be used to control a single-pole double-throw (SPDT) relay that can bypass command signals between the receiver and microcontroller to the weapons motor controller.

With the automatic weapon system activated, all you have to concentrate on is positioning your robot against your opponent, and you can let the internal robot brain control the weapon for precise attacks. When you run up against a wall, you can quickly disable the automatic weapon system so that your robot doesn't attack the walls. And when the time arises, you can still manually attack your opponent.

Implementing Semiautonomous Target Tracking

The next level of semiautonomous control is to implement semiautonomous target tracking. With this type of system, you can simply drive your robot close to

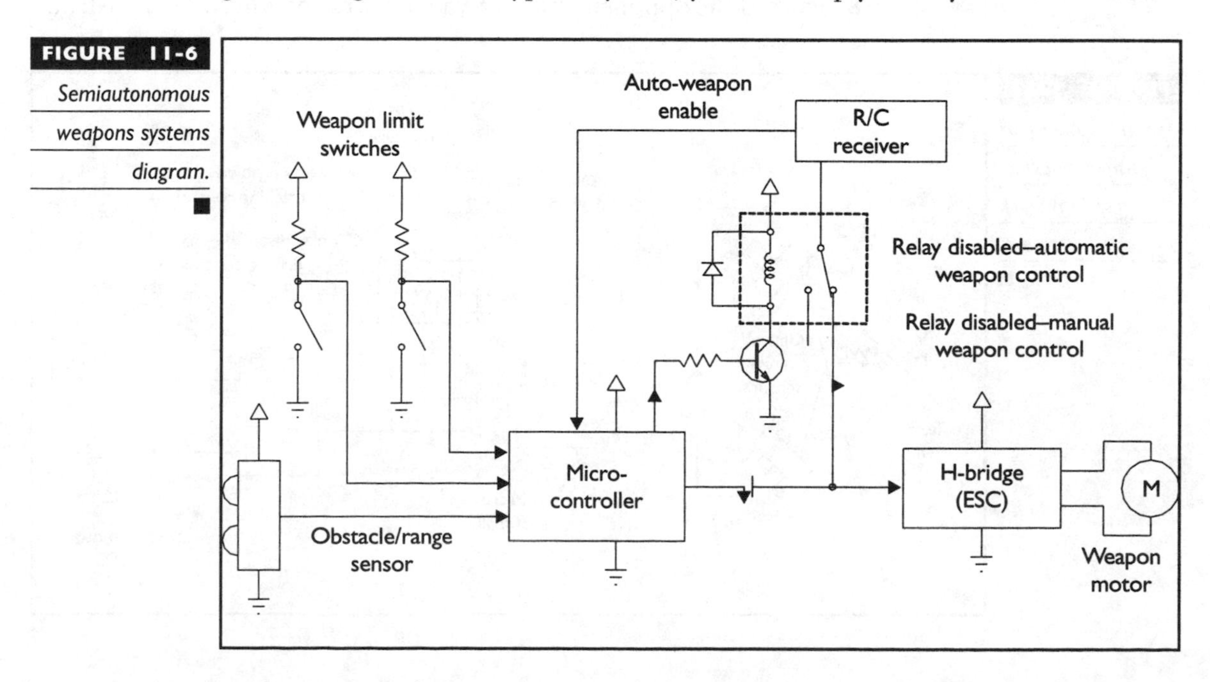

FIGURE 11-6 *Semiautonomous weapons systems diagram.*

your opponent, and your robot's sensors will lock onto the opponent and take over the driving. You maintain complete control of the weapon and let your robot push the opponent around the ring, you can have all the fun smashing its opponent to pieces with its weapon.

This type of a system needs at least two range detectors, such as the Sharp GP2D05 or the Devantech SRF04. Place both of these in front of your robot with the detection beams crossing each other. A microcontroller is used to monitor both sensors and to control the motor controllers. With this sensor configuration, the logic for driving the robot is relatively simple. If the left sensor detects the opponent, turn your robot to the right. If the right sensor detects the opponent, turn your robot to the left. If both sensors detect the opponent or both detectors do not detect the opponent, drive forward. You manually drive your robot up to your opponent until it is within your robot's crossing beams' reach, and then you can enable the semiautonomous tracking system and your robot will close in on your opponent on its own. Figure 11-7 shows a simplified schematic of this type of control system.

As with the semiautonomous weapons system, an active link must exist between the radio receiver and the semiautonomous target-tracking system's microcontroller. If the microcontroller loses contact with the radio receiver, the semiautonomous target-tracking system must shut down and enable manual control of the robot.

Semiautonomous Target Tracking with Constant Standoff Distances

The next level of control is to use the range-finding sensors such as the GP2D02, GP0D12, SRF04, or the Panasonic 6500. With these sensors, the microcontroller can be programmed to keep your robot a specific distance from your opponent, say 12 to 18 inches. If the opponent moves away, your robot will close in on it; and

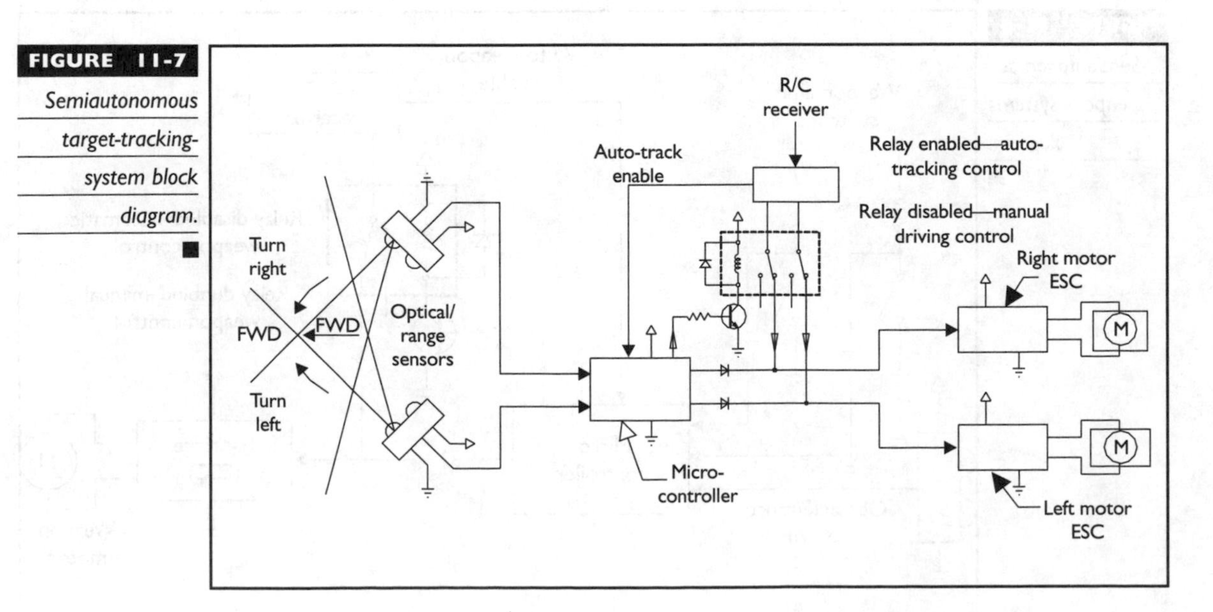

FIGURE 11-7 *Semiautonomous target-tracking-system block diagram.*

if your opponent moves too close, your robot will back away from it. Using this type of system, you can keep your opponent inside the "sweet spot" of your robot's weapon's strike zone. This type of system can be advantageous against the aggressive spinning robots. You can automatically keep your distance from the dangerous spinning weapons and focus your efforts on hitting the top of the spinning robot with your bot's axe or hammer.

Autonomous Target Tracking

As mentioned at the beginning of this chapter, fully autonomous robots are not easy to build with vision capabilities—the most difficult aspect of such system design. In the semiautonomous section, you learned about a few simple methods for a robot to "see" an opponent when it is close to your robot. But this robot still needed the human operator's eyes to get the job done.

Fully Autonomous Robot Class

In the early years of robot combat at *Robot Wars*, a fully autonomous class of combat robots existed. To account for safety, in 1996, specific rules were written about autonomous robots by Bob Gross. The key element to these rules is the use of an *infrared beacon*. The robots must be programmed to attack the beacon only, and they must ignore everything else. This way, the robot won't attack a person.

These beacons were issued to the robots by the event coordinators prior to the event. The dimensions of the beacons were 3.5 inches in diameter and 6.5 inches tall. The beacons were made of durable ABS plastic. Inside the beacon were 20,880-nanometer, infrared, light emitting diodes (LEDs) that provided infrared light 360 degrees around the beacon in the horizontal plane and 18 degrees in the vertical plane. The infrared light had a carrier frequency of 40 kHz with a superimposed modulation frequency. Each beacon had its own modulation frequency,

> **Safety First**
>
> Before we discuss how to get two robots to "see" each other, we must talk about safety. In all robot combat events, safety is the number-one concern. Most combat rules and regulations are written to protect humans from getting injured by a robot. Things like failsafes, automatic shutoffs, and manual kill switches come into play. Imagine a robot that is programmed to attack anything that comes close to it. After the match is over, who is going to walk up to the robot to shut it off to take it into the pits for repairs? If the robot is programmed to attack any robot that gets near it, how will it tell the difference between a human and another robot? It probably won't, and it will attack any human, or robot, that approaches. Because of this potential danger, some contests prohibit fully autonomous robots.

so different beacons could be distinguished between each other. The four different modulation frequencies were 550 Hz, 700 Hz, 850 Hz, and 1000 Hz.

The reason for using the 40-kHz carrier frequency was so that standard infrared remote control receiver modules could be used to detect the infrared light from the beacons. A set of these sensors could be placed around a robot to look for the beacon, and once it detected the beacon, the robot homed in on the beacon to initiate the attack. The infrared receiver modules were the same type of receiver module found inside television sets and video cassette recorders. Most electronic component stores sell them. Some models that work well with the 40-kHz signal are the Sharp GP1U58X, the Sharp GP1U59Y, or the Liton LTM97AS-40. These sensors specifically look for a 40-kHz signal, and they will ignore signals outside +/– 5-kHz tolerance band.

With this type of system, a beacon was placed on top of each robot in the match and the robots tried to find each other. The robot builder was responsible for developing the electronics and software for detecting and decoding the infrared signal from the beacons. Each robot was not allowed to use its own beacon design in combat, since the event coordinators provided them, or they were not allowed to transmit false infrared signals to confuse the opponent.

Figure 11-8 is a schematic drawing showing how to build a simple test beacon circuit. This circuit will generate the 40-kHz modulation signal and the 550- to 1000-Hz carrier frequencies. Resister R2 controls the carrier frequency, and resistor R6 con-

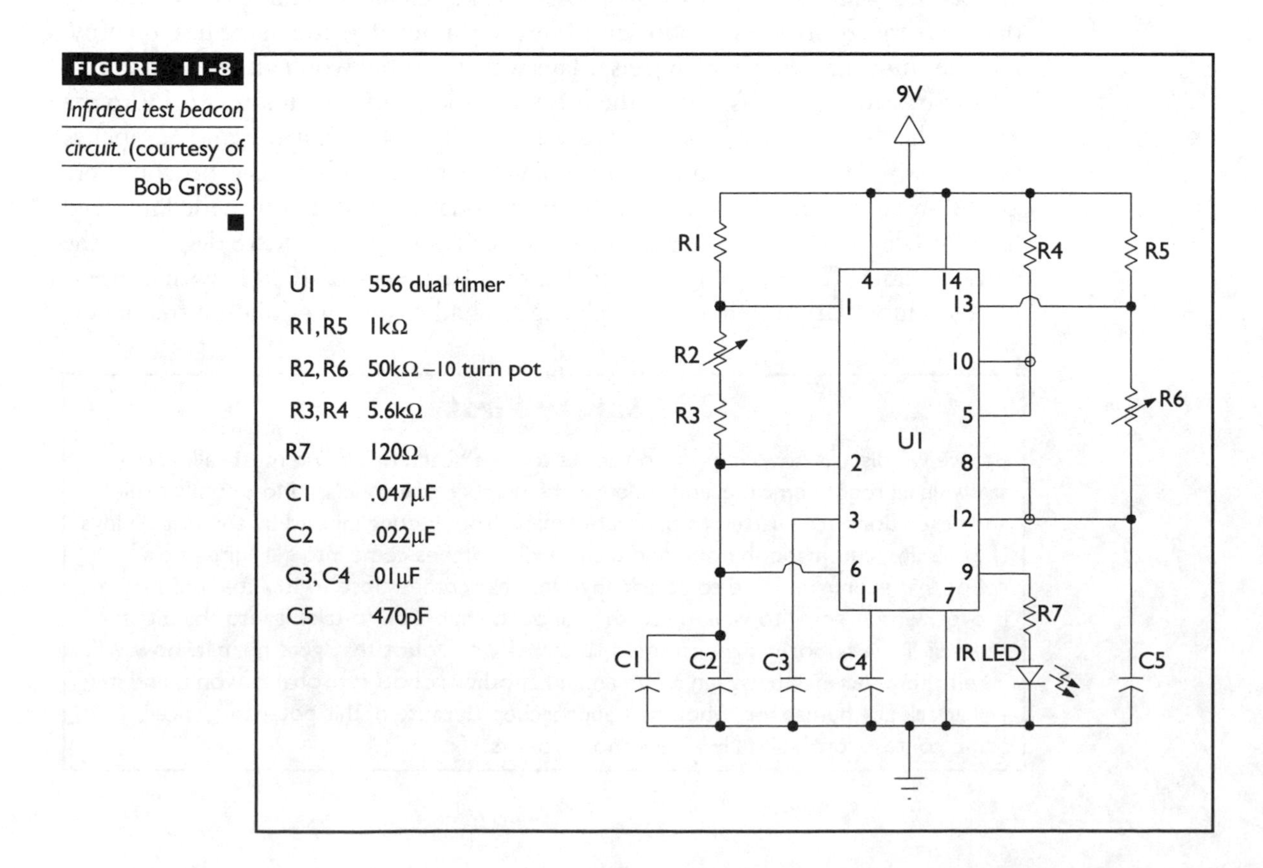

FIGURE 11-8 *Infrared test beacon circuit.* (courtesy of Bob Gross)

trols the 40-kHz modulation frequency. Using a 10-turn potentiometer will give you the best sensitivity control. To adjust this circuit, you first adjust R2 to 550 Hz, 700 Hz, 850 Hz, or 1000 Hz. You will need an oscilloscope or a multimeter that can measure frequencies. You will measure the carrier frequency from pin number 5. After the carrier frequency is set, then temporarily ground pin number 6 and adjust R6 until you get 40 kHz. You will monitor the 40-kHz frequency from pin 9. When you are done, remove the temporary ground from pin number 6.

For those of you who are mathematically inclined, the frequency carrier frequency is shown in equation 5, and the modulation frequency (the 40-kHz frequency) is shown in equation 6.

$$f_{carrier} = \frac{1.49}{(R_1 + 2(R_2 + R_3)(C_1 + C_2))}$$

$$f_{modulation} = \frac{1.49}{(R_5 + 2R_6)C_5}$$

To set up an autonomous system on your robot, you will have to build a circuit to decode the infrared signals your receiver unit detects from the beacons. You can use either hardware or software to decode the signals. With either method, you will need a microcontroller to interpret the results and plan the attack. A software method would measure the pulse length out of the receiver unit. Total pulse length is calculated from the modulation frequency, as shown in equation 7. The microcontroller will look for one-half the total pulse length, either the positive or negative portion.

$$f_0 = \frac{1}{1.1(R_1 + 2R_2)C_1}$$

A simple logic statement for detecting a 700-Hz signal might look like this:

```
IF [(Pulse_Width is greater than 650 microseconds)
and (Pulse_Width is less than 750 microseconds)]
THEN Beacon_Frequency is 700 Hz
```

With a Basic Stamp, to measure the pulse width can easily be accomplished using the *Pulsin* command. Using software to analyze the infrared frequencies can simplify the number of components that go into the robot controller and can give you more options in configuring your robot to attack. Programs can be changed between matches to account for conditions not originally accounted for. But software solutions are sometimes complicated to implement, depending on your programming skills.

A hardware solution can be simple. Figure 11-9 shows a schematic drawing of a circuit that uses the 567-tone decoder to interpret the carrier frequency from the infrared beacon. Potentiometer R1 is used to adjust the frequency this circuit will detect. A 10-turn potentiometer will give you the greatest sensitivity control in adjusting the desired frequency. To adjust this circuit, place the test infrared beacon in front of the receiver module and measure the voltage from pin number 8. Adjust

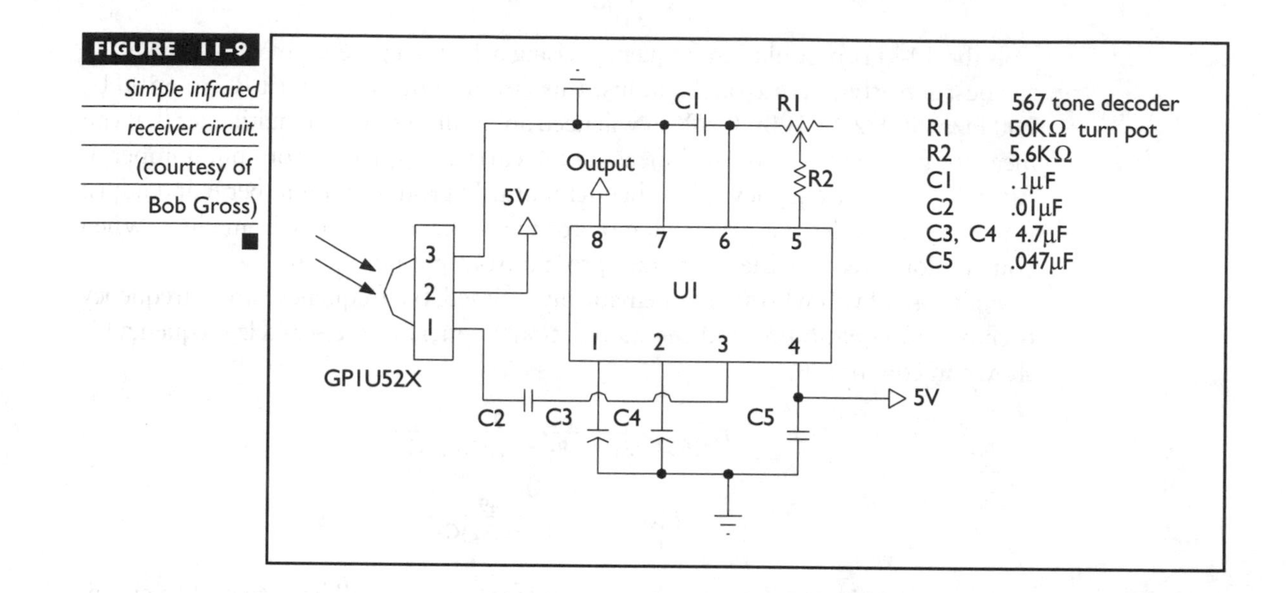

FIGURE 11-9 *Simple infrared receiver circuit.* (courtesy of Bob Gross)

R1 until the voltage drops to zero, and then remember the turn position. Continue turning the potentiometer in the same direction until the voltage jumps back up to 5 volts. At this point, you have found the sensitivity band with of this detector. Now back off the potentiometer position to someplace between the two positions you have observed. The voltage should be back to zero. Here you should be at the center frequency at which the test infrared beacon is transmitting.

The last feature that must be included in an autonomous robot is an actual R/C receiver. For safety purposes, you will want to be able to remotely shut down the robot. Even remotely turning on the robot is a good idea. The R/C receiver can be hooked up to a switch that turns power on and off to the main microcontroller in this robot.

Bob Gross and Thumper

Bob Gross implemented many of the features discussed in this chapter while building his champion robot *Thumper* (which won the autonomous class competition at *Robot Wars* 1997). To give you an idea of how effective a good autonomous robot can be, Thumper took on Jim Smentowski's R/C robot, *Hercules*, who weighed in 70 pounds heavier than *Thumper*. Through most of the match, *Thumper* was in the lead, chasing Hercules around the ring repeatedly and even pinning him against the wall twice. In the end, however, *Thumper's* drive motors burned out because of the extended pins. At that point, the heavier—and by then, stronger—*Hercules* was able to knock *Thumper* over and pin him against the wall as time ran out. Although *Thumper* didn't win that match, the crowd went wild seeing a fully autonomous bot give a remote-controlled machine a run for its money.

More Information

You can implement semiautonomous to fully autonomous features in a robot in many ways. A search on the Internet will yield thousands of pages of information on how to build different types of circuits. One of the appendixes in this book lists some good references for autonomous robots and sensors. The Seattle Robotics Society (*www.seattlerobotics.org*) has a Web-based magazine called *The Encoder* that has hundreds of tutorials that explain how to use different types of sensors and microcontrollers. With the correct types of sensors, microcontrollers, and software, you can develop a "turn-on-and-forget" type of combat robot and sit back to watch your creation single-handedly and autonomously destroy its opponent.

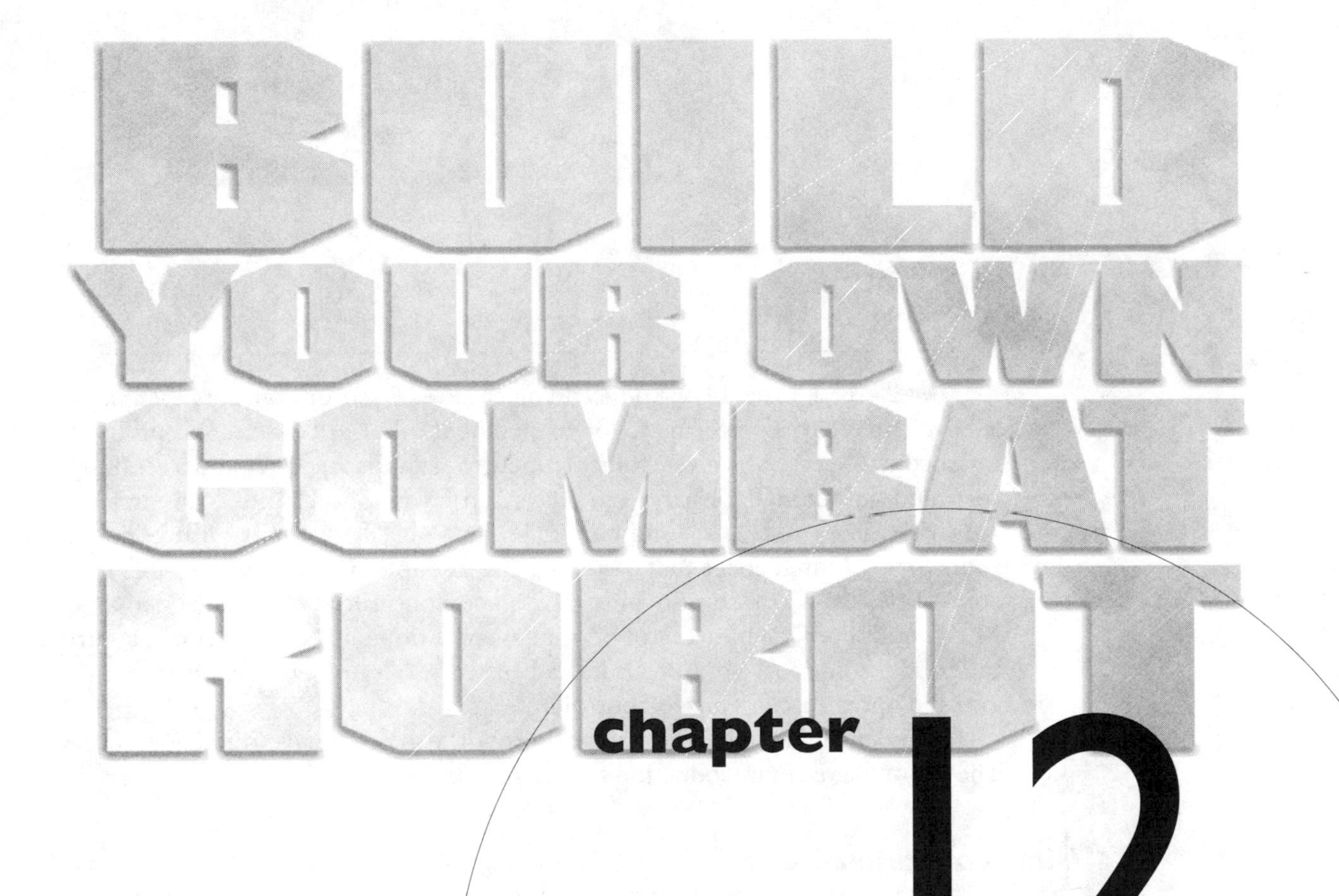

chapter 12

Robot Brains

CHAPTER 11 introduced you to several sensor concepts that can be used to enhance the performance of your combat robot. All of the concepts used a microcontroller that received control input from an R/C receiver and a set of sensors, and output control signals to relays and ESCs. This chapter will introduce what microcontrollers are, and how they can be used. Virtually all robots have some form of intelligence that can range from simple switches, to a simple radio control (R/C) system, to fully integrated microcontrollers with neuro-networks. Most of this book focused on robots that use traditional R/C equipment to control the robot. Some R/C equipment has advanced programmable features that can give the operator customized control options.

To implement advanced controls on a robot, you need to use microcontrollers. The following is an introduction to microcontrollers.

Microcontroller Basics

Most people who develop robots use the term "microcontroller" as a generic term to refer to a small control system with input and output control capabilities. Microcontrollers are not computers or microcomputers. Simply put, microcontrollers are designed to accept input from a set of electrical signals and output other electrical signals in response to commands programmed into the device. Computers are designed to accept input from humans and output the results back to the humans. A computer will include several microcontrollers, but a microcontroller will not have a computer inside. Microcontrollers, which interpret a human interface and send electrical signals to the rest of an electronic device, are often implemented as small, embedded processors found in many modern electronic gadgets from Furbies to watches, from thermostats to microwave ovens, from radios to television sets, and from cell phones to electronic ignition systems found in cars. They are found in many electronic devices made today.

Many controllers are designed specifically for robotic and similar applications, including Basic Stamp, Handy Board, BrainStem, OOPics, BotBoards, and countless other controllers. Figure 12-1 shows a photograph of several of the more-popular microcontrollers.

Some controllers are *slave* controllers in which commands are given and executed. An R/C transmitter/receiver pair is a form of slave controller: the input from an operator using the transmitter is executed and transmitted to the receiver,

FIGURE 12-1 *From top left to bottom right, LEGO RCX brick, Handy Board, Basic Stamp 2, BrainStem, Lineo, OOPic, traditional R/C receiver, and the Pontech servo controller.*

which then converts this data into commands the servos and speed controllers understand. Other controllers, such as the Handy Board, take programs that can be used to alter output results based on input results. Users download a piece of code to the controller, and the code then runs on the Handy Board to control the robot.

Still other microcontrollers offer reactive mechanisms that automatically manage outputs based on inputs—such as the thermostat in your house, which senses its environment and controls the furnace to keep your house at a comfortable temperature, and sometimes adjusting for times of the day when you are not home or asleep, to conserve energy. The OOPic and BrainStem controllers affect this type of control, called *virtual circuits* and *reflexes,* respectively. Some controllers can exhibit more than one type of control, and some can even perform multiple tasks at the same time.

You may have robots in your lives that you may not have thought of as robots. Consider a bread maker that knows the time, can mix various bread recipes, can sense heat, and can create auditory output with beeps and displays to inform you what is happening at all times. Some cars have anti-lock brake systems (ABS) that can sense each wheel's rotation and adjust the braking pressure so the wheels don't lock and skid, even "pumping" the brakes for maximum stopping power on wet and slippery roads.

The details of how to design electronic circuits using microcontrollers, how to write programs, and how to implement the microcontroller is beyond the scope of this book. You'll find many different books that have been written about various types of microcontrollers and programming techniques to help you. Appendix B lists some excellent books on microcontrollers. Some microcontrollers are simple to get started with, and some are so powerful that they require prior microcontroller experience to use them properly.

Table 12-1 shows a list of some specifications to several different types of microcontrollers. The number of input and output (I/O) lines represent the total number of individual control lines a microcontroller can have. This list combines both digital and analog I/O together. Digital I/O represents a data line where the input and output values are either 5 volts or 0 volts. This is to represent a binary 1 or a 0—or, in other words, and *on* or an *off* state. An analog I/O signal line represents a line that can interpret a variable anywhere between 0 and 5 volts.

A microcontroller's processor speed is the actual clock speed. Some microcontrollers require the four clock cycles to execute a single command, while other microcontrollers can execute a command in a single clock cycle. The time required to execute a command doesn't represent the time required to execute a line of programming code. When you write a program, each line will use many different internal commands that the microcontroller understands; thus, program speeds are always slower than clock speeds of a microcontroller. The specification that is really important is the execution time, which is the number of program instructions executed per second. Notice in Table 12-1 the difference in execution times when compared to the clock speeds of the microcontroller.

For programming space, the common term that is used to represent how much "memory" a microcontroller has is electrically *erasible programmable read only memory* (*EEPROM*), which is the number of - kilobytes of programming memory available on the microcontroller. In the microcontroller world, *memory* represents how much variable space the program can keep track of, not the amount of

Feature	Basic Stamp 2	Basic Stamp 2SX	Basic Stamp 2P	Basic Stamp I	BasicX-24	OOPic	BrainStem	Handy Board	Bot Board
I/O Lines	18	18	12	8	16	31	25	30	38
Processor speed, MHz	20	50	20	4	8	20	40	2	8
Execution time	4000	10000	12000	2000	65000	2000	9000	N/A	N/A
EEPRPOM Kbytes	2	16	16	256 bytes	32	4	16	32	2
Multitasking	No	No	No	No	Yes	Yes	Yes	Yes	Yes
Package, inches	24-pin DIP	24-pin DIP	24-pin DIP	14-pin SIP	24-pin DIP	2×3.5	2.5×2.5	4.25×3.15	2.2×3.2
Language*	Basic	Basic	Basic	Basic	Basic	Basic, C, and Java	TEA	C	Basic, C

TABLE 12-1 *Microcontroller Comparison* ■

space in which programs can fit. The concept is different than how regular PCs refer to memory. In Table 12-1, all the values shown are in kilobytes, except for the Basic Stamp 1—which has only 256 bytes of programming space. To some people, this doesn't sound like a lot, but 256 bytes represents quite a lot of programming space in a microcontroller.

Some microcontrollers execute one command at a time, and some can execute multiple commands at the same time. For some applications, such as controlling 16 different R/C servos in an animatronics movie puppet, being able to execute multiple commands simultaneously, or *multitasking*, can be helpful.

The microcontroller used in your bot can be either a small circuit board that connectors plug into, or a large integrated circuit. One of the common sizes for the microcontrollers is the 24-pin *dual inline pin* (DIP) socket. Basic Stamp started with this size, and several different companies have made Basic Stamp variants that are pin-for-pin, identical.

Unfortunately, no one programming language can be used to program all microcontrollers. Many of the languages are based on the popular Basic programming language or the C programming language. If you know how to program in either of these languages, you should be able to program one of these microcontrollers.

Basic and C are called high-level languages, and they are easy to learn and understand when compared to using the assembly language. A compiler compiles (or converts) the high-level language into a low level language that the microcontroller actually understands. For example, here is a simple instruction written in Basic that is easy to understand:

```
X = Y + Z
```

If this is written in assembly language, it would look like this:

```
MOVF Y,0
ADDWF Z,0
MOVWF X,0
```

This isn't easy to understand. The preceding assembly language example will be different from microcontroller to microcontroller, but the Basic language will be the same regardless of the microcontroller.

When you get started in the world of microcontroller programming—or, as the electrical engineers like to call it, programming embedded controllers—pick something you like and stick with it until you master it. Interfacing a microcontroller with the outside world is the same regardless of which microcontroller you choose. Master the interfacing techniques on one microcontroller before you move on to another type of microcontroller.

If you ever want to start a microcontroller "war," log onto one of the robot clubs' e-mail list servers and ask the question "What is the best microcontroller?"—and watch what happens. Many people think the microcontroller they use is the best, but there's really only one correct answer to this question: the best microcontroller is the one that you know how to use and program.

Every microcontroller has its advantages and disadvantages. Some microcontrollers have features that make certain tasks easier than other microcontrollers. For example, a number of microcontrollers have a built-in feature that can directly read in an analog voltage, and other microcontrollers have multitasking capabilities. Although users of these types of microcontrollers may claim they are better than other types of microcontrollers, that's not necessarily true. You can always find a way to make a microcontroller work to meet your specific needs, particularly if you're handy with electronics and/or programming. A "weak" microcontroller with good programming can outperform a "good" microcontroller with bad programming.

A search of the internet will yield dozens of companies that sell different types of microcontrollers. All of the different manufacturers have documentation that explains the capabilities of their products, an explanation of the programming language, and sample programs that illustrate the microcontrollers' capabilities. When selecting a microcontroller, keep in mind what you want it to do, and compare it with the literature you have collected. Then choose the microcontroller based on how well it can fit your needs and how well you understand its programming language.

The next few sections offer a short introduction to several of the popular available microcontrollers, and at the end of this chapter is a short discussion of microcontroller applications.

Basic Stamp

Throughout this book are many references to the Basic Stamp from Parallax, Inc. Basic Stamp applications include servo mixing—reading R/C servo signals to operate switches to turn on weapons.

For the beginner getting started with microcontrollers, the Basic Stamp is probably the best unit to start with. Parallax has created a rather extensive set of tutorials on how to use microcontrollers, basic programming, electronics, sensor integration, and actuator applications. All of its easy-to-understand tutorials can be downloaded from its Web site for free.

Probably the best place to learn about microcontrollers is to purchase one of Parallax's Board of Education Robotic (*BoeBot*) Kits and go through all of their experiments—see Figure 12-2. After you have worked through the tutorials, you should have a pretty good understanding of how to use a Basic Stamp inside combat robots. An excellent book on the subject is *Programming and Customizing the Basic Stamp* by Scott Edwards.

FIGURE 12-2 *The BoeBot from Parallax, Inc.* (courtesy of Parallax, Inc.)

Most Basic Stamp units come in 24-pin, dual-inline packages (see Figure 12-3). They can be plugged into a prototyping board, and a 9-volt battery is all that is needed to supply power to the unit. With some wire and a few resistors and capacitors, you can be up and running with your first Basic Stamp application.

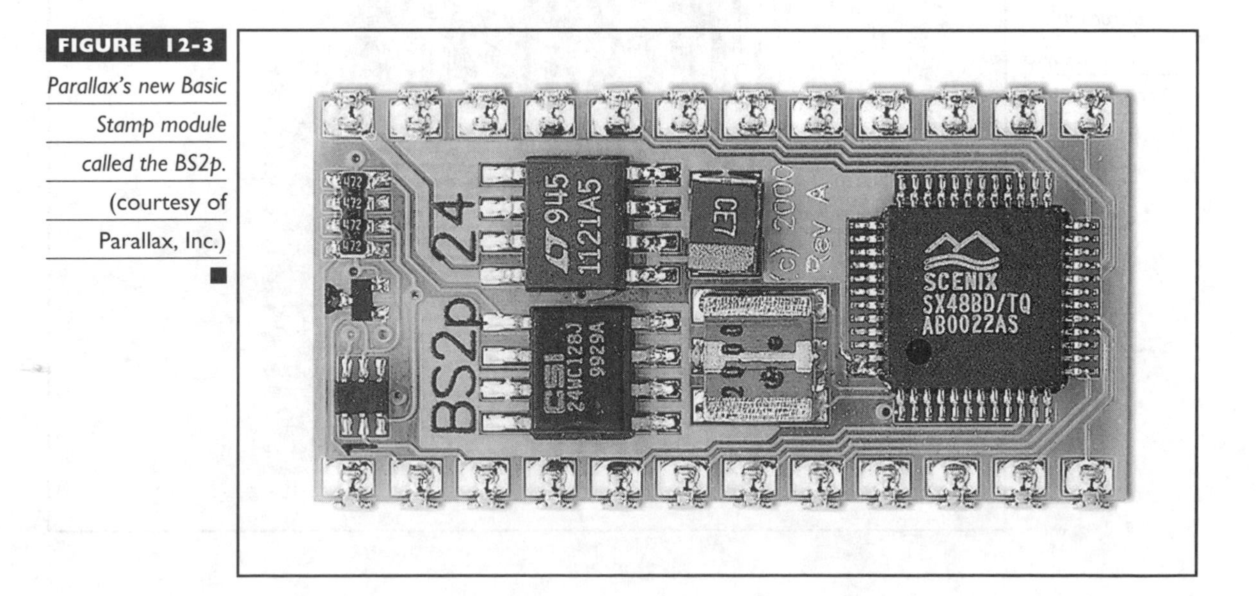

FIGURE 12-3 *Parallax's new Basic Stamp module called the BS2p.* (courtesy of Parallax, Inc.)

To program a Basic Stamp microcontroller, you will need a PC that runs Windows or DOS. The language is relatively simple for most of us to learn, because it is based on the BASIC computer language. Parallax had to make a few modifications to the language to make it work with Parallax products, but it is quite easy to learn and get up to speed with.

BrainStem

The BrainStem, by Acroname, Inc., is a new microcontroller board that has entered into the robotics community. This miniature microcontroller has been showing some really unique capabilities. Table 12-1 lists some of its specifications. The programming language used is called TEA, or Tiny Embedded Application, which is almost identical to the industry-standard ANSI C. This microcontroller is shown in Figure 12-4. It has some interesting features that are not found on other microcontrollers, including four dedicated radio controlled (R/C) servo ports. Thus, without any special programs, you can control four different servos, or four different electronics speed controller (ESCs). It also has a built-in port for controlling the Sharp GP2D02 Infrared range sensor. The BrainStem has software library support for Java, C, and C++ on Microsoft's Windows systems, and the PalmOS, MacOS, and Linux computer operating systems.

FIGURE 12-4 *BrainStem microcontroller* (courtesy of Acroname, Inc.)

Handy Board

The Handy Board is a powerful veteran microcontroller board that has been around for a long time. First developed at MIT (Massachusetts Institute of Technology) by Fred Martin, this microcontroller board uses the popular 68HC11 microcontroller from Motorola. The programming environment is called Interactive C, which is similar to the traditional ANSI C. This microcontroller has four built-in motor controllers for directly driving four different very-low-current (< 1.0 amps) motors, and it has a built-in liquid crystal display (LCD) screen for displaying information.

BotBoard

The BotBoard was developed by Kevin Ross and Marvin Green using the same 68HC11 microcontroller used by the Handy Board.The size of this board is significantly smaller, however, and it doesn't have the built-in features of the Handy Board. Because many people didn't want those extra features, this board offers a smaller and lower-cost solution to obtain the same level of power of the Handy Board. Karl Lunt has developed a version of the Basic programming language for the 68HC11 microcontrollers, which is called Sbasic. You can download it from Karl's Web site at *www.seanet.com/~karllunt/*. Karl is also the author of an excellent book about robots called *Build Your Own Robot* (see Appendix B).

Other Microcontrollers

Many other microcontrollers are out there. The OOPic uses an object-oriented programming language. The BasicX-24 and Basic Micro's Atom look almost like the Basic Stamp and are pin-for-pin compatible, but are faster, have more programming space, and uses a multitasking operating system. These microcontrollers are starting to gain a lot of popularity. A high-end microcontroller is the Robominds microcontroller, which uses the Motorola 68332, 32-bit microcontroller. It's very fast and very powerful.

Most of the microcontroller boards described here use either the Microchip PICs, the Atmel AVR chips, or the Motorola 68HC11 or 68HC12 chips as the core microcontroller. All of these microcontroller board companies have added some components to their boards to make their microcontrollers easy to use. When you get more experienced with microcontrollers, try experimenting directly with the PICs and the AVR chips. They are the microcontrollers found in most electronic appliances and systems.

Following is a short list of some of the most-popular microcontroller Web sites:

- **Basic Stamps** *www.parallaxinc.com*
- **BrainStem** *www.acroname.com*
- **BasicX** *www.basicx.com*
- **OOPic** *www.oopic.com*
- **Handy Board** *www.handyboard.com*
- **BotBoard** *www.kevinro.com*
- **PIC** *www.microchip.com*
- **Basic Micro, Atom Chip** *www.basicmicro.com*
- **68HC11 and 68HC12** *www.motorola.com*
- **Robominds** *www.robominds.com*
- **AVR** *www.atmel.com*

Microcontroller Applications

The following discussion offers several examples of the various applications for which microcontrollers can be used. Although they are not directly associated with combat robots, these features can be adapted to building combat robots. All of these examples are based on the BrainStem microcontroller from Acroname. Keep in mind when reading the following examples that virtually any microcontroller can be used to accomplish these applications.

The Robo-Goose

The *Robo-Goose* is a robot that can be driven by a human operator via remote control. The operator drives the robot using a standard R/C-type transmitter (much like a combat robot). What is different here is that the receiver sends the control commands into a BrainStem microcontroller module that manipulates the input and translates it into meaningful output for the motors on the goose. One input determines the steering and the other the speed of the goose. The BrainStem is performing a servo mixing function. Figure 12-5 shows a photograph of the Robo-Goose.

FIGURE 12-5 *Robo-Goose, a robotic goose that is controlled by a traditional R/C system and a BrainStem microcontroller to perform a servo mixing function.*

The mechanics for the Robo-Goose are two thruster motors lying below the surface in the water that can run to create forward or reverse thrust in the goose, as shown in Figure 12-6.

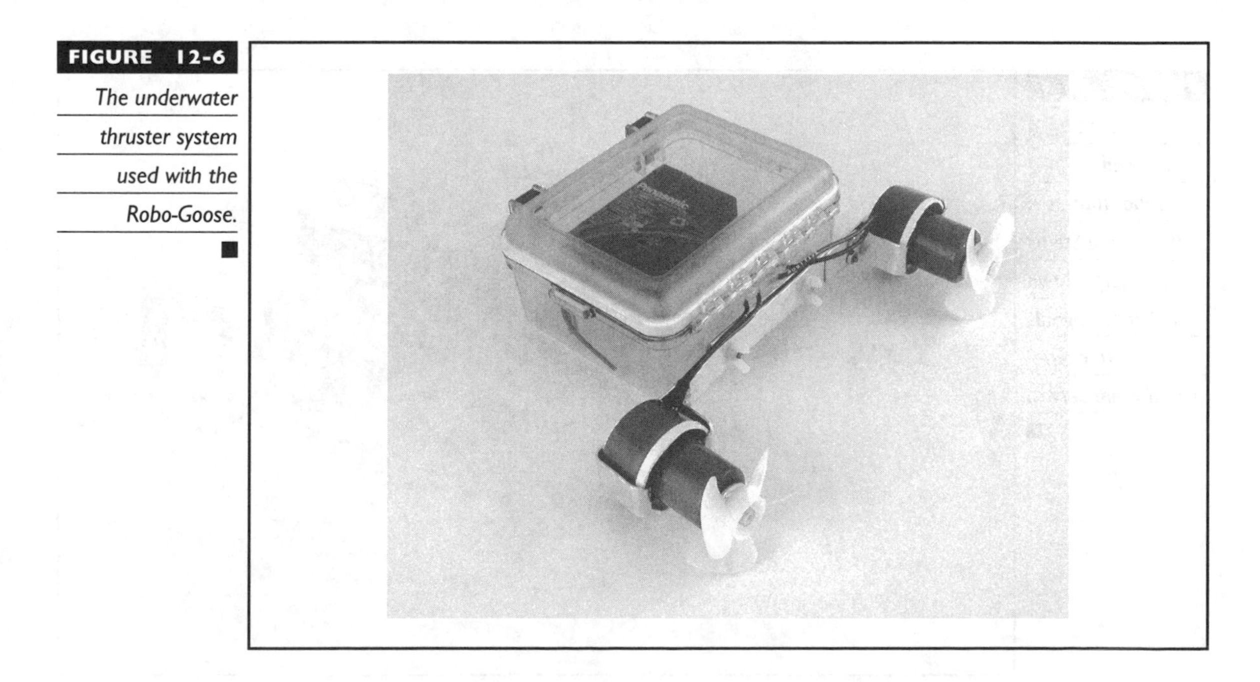

FIGURE 12-6 *The underwater thruster system used with the Robo-Goose.*

The Robo-Goose demonstrates an important concept in robotics control that we will call *microcontroller assisted control.* The inputs coming from the operator are translated into commands that affect certain motions on the robot. In mathematics, this is called a *mapping;* and in the case of the goose, two inputs (steering and forward motion) are translated, or mapped, into forward and reverse commands for the right and left thruster motors on the goose.

The BrainStem Bug

The BrainStem bug also uses microcontroller-assisted control to manipulate many different outputs from two simple inputs. The two outputs from the R/C receiver are fed into a small parallel microcontroller core consisting of three networked BrainStem controllers. Each BrainStem controls two legs, one for the front pair, one for the middle, and one for the back pair of legs. Figure 12-7 shows a photograph of the walking robot.

Simple forward and backward commands from the transmitter are translated into complex walking patterns with six servo actuators controlling the left legs and six more controlling the right legs of the robot. In this case, the assistance of the computer becomes crucial to the operation of the robot. Twelve servo actuators control the robot, and complex patterns are used to make the robot walk forward and backward, turn right and left, and even spin right or left while stepping in place.

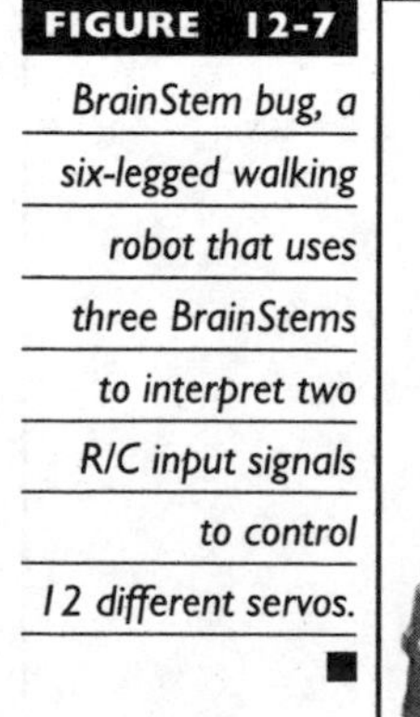

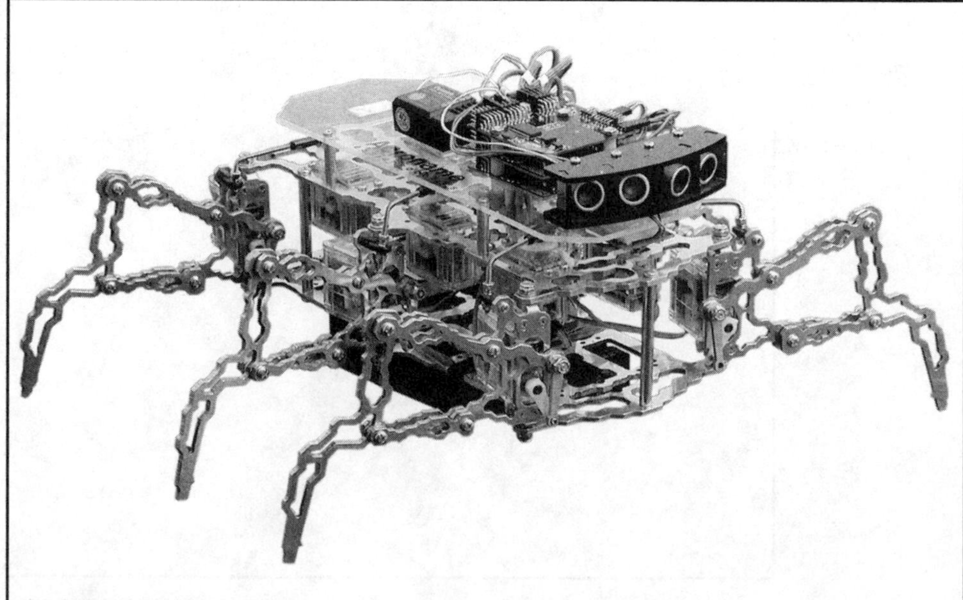

FIGURE 12-7 BrainStem bug, a six-legged walking robot that uses three BrainStems to interpret two R/C input signals to control 12 different servos.

Imagine trying to control the same robot with 12 sticks on the R/C transmitter while trying to do battle with another robot that is speeding toward you. The computer-enhanced R/C is crucial to sophisticated mechanical designs.

1BDI, an Autonomous Robot

1BDI takes the microcontroller control to the limit by completely controlling the robot without a human operator. This robot was designed to find a lit candle in a maze using vision, put out the candle using a fan, and then find its way out of the maze using its memory of how it got to the candle in the first place. Figure 12-8 shows a photograph of this fire-fighting robot.

The heart of *1BDI's* control is a BrainStem controller that is running a TEA program to read input from the sensors and to control the motors. The robot has various sensors to find walls using infrared light, to find lines on the floor using reflected light, and to sense whether the wheels have stopped spinning or not. 1BDI also has a secondary microcontroller system driven by a BSX-24 microcontroller that does vision processing from a charge-coupled device (CCD) camera similar to what you might find in a hand-held commercial digital camcorder.

The CCD array takes an image, and the BSX-24 processes the data to seek out the distinct shape of the candle. The BSX-24 can also distinguish yellow tubes placed in the robot's path that are meant to be color-keyed furniture for the robot to avoid. The programming for autonomous robots is typically much more sophisticated than that of microcontroller-assisted robots. Every possibility the robot may encounter must be handled so that the robot is not easily disabled. Building a robust autonomous robot is at the forefront of today's research in both robotics and artificial intelligence.

FIGURE 12-8 *A fully autonomous robot, named* 1BDI, *built to compete in the fire-fighting contest.*

The *Rover*, Teleoperated with Feedback

The Robo-Goose uses one-way communications to control the robot. If you drive the robot out of view around a clump of trees, you will have little luck in driving the robot back into view because you have no feedback from the robot. The *Rover* was designed to give more feedback to the controller both visually and through force feedback. The *Rover* uses a variety of controls to not only convert the inputs from the controller into the actual motion commands, but also to provide important feedback information to the operator. This allows the *Rover* to drive completely out of view from the operator at great distances. The feedback the operator gets allows the *Rover* to be quite robust in operation, even in confusing and difficult-to-navigate environments. Figure 12-9 shows a photograph of this robot.

Rover is manipulated via a traditional computer game controller (joystick). The commands given by the operator as she manipulates the joystick are translated in software into the commands for the motors that operate the four-wheel-drive arrangement of the rover's wheels. These translated commands are passed over a wireless computer network to a small hand-held personal data assistant (PDA) situated on the robot, where more processing takes place. The commands are then sent via serial communication to two networked BrainStem controller modules that control the motors.

FIGURE 12-9 *A tele-operated robot named* Rover *provides video feedback to the operator as to its actual position.*

What sets the *Rover* apart is that information can flow back to the operator from the robot along the same path in reverse. This information is in the form of a color video image from a camera mounted on the front of the robot, sounds coming from the vicinity of the robot, and sensor input from infrared proximity sensors mounted on the robot. The sensor input returned from the proximity sensors is manipulated in software and fed back into the joystick held by the operator. In this way, the operator can see what the robot sees, hear what it hears, and feel what it feels.

Each sense the operator can have from the robot makes for better teleoperation. Because the robot can only see forward, at times the operator may have to "feel" an obstruction as the robot backs up during navigation. By adding the sense of touch, the operator could "feel" the obstruction behind it before it even hits it. Since the infrared detector can detect the object from a distance of 6 inches, the software can make the joystick provide increasing resistance to moving back as the obstruction approaches—that is, it gets harder and harder to drive the robot back into the obstruction as the robot gets closer to the obstruction. You could call this "driving by Braille," as the sense of touch is being simulated and vision is not being used.

Summary

In a combat robot, you will be able to see the environment around the robot, but what about what is happening inside the robot? Is a motor overheating, are the batteries going dead, did one of your drive chains break? It would be nice to know if your robot is about to have an internal failure before it happens so you can initiate corrective actions during the match. Or, if your robot isn't moving correctly, you might be able to remotely fix the problem if you knew its cause, or alter the driving of the robot to protect a weak side. Without feedback, you can easily turn a minor problem into a major problem.

This chapter, and the previous chapter, presented some ideas about how you could use a microcontroller to enhance your robot-controlling efforts. Chapter 13 will show a simple implementation of the Basic Stamp 1 in a mini sumo robot. You will see some of the wiring requirements, and you can read the source code for two of the programs that make the robot work. They are written in PBasic so they should be easy to understand.

Have fun learning the world of microcontrollers. They can really help turn your robot into a super robot.

chapter 13

Robot Sumo

THE referee signals, and my heartbeat increases as I press my bot's start button. I stand back to mentally count down the 5 seconds that my bot must remain still before it can move. In my excitement, I mentally reach the 5 seconds before my machine starts to move. I panic, thinking he must be broken, but then both bots start moving forward. As my bot approaches his victim, I smile. The crowd cheers... I'm thinking, "I've got him now!"

But the bots just pass each other as if each one is the only player in the ring, and the crowd goes silent. My bot is now heading full-steam ahead toward the edge of the ring, and I suddenly think, "What if the edge sensors aren't working?!" As soon as my machine gets to the white edge of the sumo ring, it stops, backs up a bit, then spins around, and I breathe a sigh of relief that all seems to be working correctly. This time, my bot is heading right for his opponent, and nothing will stop him this time.

As my bot approaches his foe, he makes a couple of quick course corrections in order to zero in on the enemy. I'm thrilled that my object detection sensors are working. My bot closes in on his adversary, and the crowd starts cheering again. Just as he's about to hit his opponent, the rival bot suddenly turns toward mine. The crowd cheers louder. The bots crash into each other. My breathing almost stops as both machines halt in the center of the ring. The wheels of my bot are spinning on the ring surface.

My robot starts pushing his challenger steadily backward. Just as I think I've won the match for sure, the other bot gets better wheel traction and starts pushing mine backward. The crowd goes wild. I bite my lower lip. Thankfully, my bot's traction improves and he begins pushing his foe backward. As the backward and forward motions continue inside a one-inch area, our bots both start slipping sideways. As soon as they come apart, they shoot toward each other again—but their wheels get caught, and they start the classic "spinning dance." The crowd quiets down. This is turning out to be a much tougher battle than I'd anticipated.

The referee stops the match to separate the bots. I take a couple of deep breaths, and we restart. This time, when my bot reaches the edge of the sumo ring and backs up, he only turns 90 degrees. I'm glad I used a random turning method in the software! This time my machine approaches the rear of his rival. I smile to myself, because I can see he's going for the vulnerable spot. The crowd goes wild. I look at my human opponent's face, and I can see in his eyes that he knows his bot will lose.

With my bot right behind his adversary, I am sure this will be "game over." The other bot stops at the white edge of the sumo ring, and my bot runs right into him. The crowd screams as my bot pushes his enemy out of the ring. As the crowd continues to cheer, I pick my bot up from the ring, and marvel that this was only the first battle of three, and only 30 seconds have ticked by. It seemed like hours!

This is what you experience when you compete in one of the fastest-growing and most popular robot contests in the world. Robot sumo was originally started in Japan in the late 1980s by Hiroshi Nozawa, and was later introduced to United States robotics clubs by Dr. Mato Hattori. Robot sumo is a robotic version of one of Japan's most popular sports, sumo wrestling. Instead of two humans trying to push each other out of a sumo ring, two robots attempt the same feat. Since its creation, robot sumo has found its way into many robotic clubs, universities, high schools, and elementary schools throughout the world. There are even regional and national championship contests now being held in several countries, and some bots even go on international tours.

Robot sumo's growing popularity is due to a number of factors. First, the sport is relatively simple compared with other forms of robotic competition. Take, for example, *Robot Wars U.K.,* where robots are required to fight with not just the primary opponent but also with a number of house bots. Sumo fighting, where robots are only required to push one opponent out of the ring, seems pretty easy by comparison. Because the rules of the event are uncomplicated, bot builders are freed up to use any number of unique designs to give their bots a competitive edge over rivals. And because the bots come in a variety of designs, spectators can easily pick out their favorite bots to root for during the contests. Some bots become more popular than their builders.

One of the other factors making robot sumo so attractive to builders is the low cost of constructing this kind of machine. Often, sumo bots are made from parts scavenged out of old broken toys or household electronic products. Thanks to their small size, they can be easily carried around, and they do not require any significant repair costs after a contest.

Recent years have seen the growth in popularity of a more aggressive form of robot combat—the kind of contests fought on *BattleBots, Robot Wars, Bot Bash,* and *Robotica.* As exciting as these contests may be to watch and participate in, the costs to build these bots are significantly higher than those in sumo robotics. Most of the *BattleBots*-type robots cost at least $3,000 to build, and some of them cost more than $40,000.

On the other hand, it's rare to find someone who spent more than $1,000 on a sumo bot—in fact, most sumo bots cost less than $500 to construct, and some are virtually built for free if all of the parts can be scrounged out of junk equipment lying around the house. Because the rules of robot sumo prohibit bots from intentionally damaging one another, there are virtually no repair costs after a contest is over.

Robotic sumo rules vary in competitions throughout the world. The primary differences are in the size and weight of the bots. The basic rules of the game remain the same, where each bot must try to push its opposing bot out of the sumo ring. The first bot that touches the ground outside the sumo ring loses the round.

In robotic sumo, there are three rounds in a match, and the first bot to win two rounds wins that match.

In robotic sumo, there are two different general classifications: remote-controlled sumo bots, and fully autonomous sumo bots. The difference between the two, obviously, is that an autonomous sumo bot must operate completely on its own. No form of human control (except for turning the bot on) is allowed.

How a Sumo Match Proceeds

As stated earlier, a single robot sumo match consists of a best of two out of three individual sumo rounds. During a round, both bots are placed on the sumo ring. When the referee signals start, both bots are turned on, and the operators move away from the sumo ring. Each bot must try to find the other and push that other bot out of the ring. The first bot that touches anything outside the sumo ring boundary loses the round.

The other way to lose a round is to become disabled. For example, if a bot gets knocked onto its back and can no longer attack the opponent, the opponent wins the match. As with all contests, there is a time limit to each match. Each match has a total time limit of 3 minutes. There is no time limit to the individual rounds. This means that all three individual rounds must occur within the 3-minute time frame. If the score is tied after the 3-minute time limit has expired, the referee will award the match victory to the bot that appeared the most aggressive. If both bots appear to be equally aggressive to the referee, the referee may allow additional time for the bots to continue.

The contest coordinator will set the rules for determining the overall winner. The types of play include single, double, or round-robin elimination. This is usually determined based on how many bots are entered into the contest and the total available time to run the contest.

Robot sumo promotes sportsmanship and education. The rules of the event prohibit any action that will cause damage to the sumo ring, other sumo bots, or humans. Any bot that causes intentional injury or damage will be immediately disqualified from the competition. The exception to this rule is that any incidental damage caused by the bots running into each other is allowed. But if a bot has a feature with the primary purpose being, in the official's interpretation, to cause damage, that bot will be disqualified. For example, if a bot has a hammer that can swing down and hit its opponent, the bot with the hammer will be disqualified. Arms *are* allowed on the bots to try to help capture and confuse its opponent; but if the referee feels that the arm's primary purpose is to act as a weapon, then the bot will be disqualified.

The two most popular robot sumo classes are the international sumo class and the mini sumo class. The international class is also called the Japanese class (because this is the size class that is used in Japan), or sometimes it is called the 3kg class, indicating the maximum weight allowed for this kind of bot. Table 13-1 lists the specifications for these two bot classes. The mini sumo class was invented by Bill Harrison of SineRobotics. Except for the weight of the bot, every other specification is exactly half of the international sumo class.

	International Sumo Class	Mini Sumo Class
Length	20cm	10cm
Width	20cm	10cm
Height	Unlimited	Unlimited
Mass (maximum)	3kg	500g
Sumo Ring Diameter	154cm	77cm
Border Ring	5cm	2.5cm

TABLE 13-1 *Robot Sumo Specifications* ■

The size specifications of the bots only apply at the beginning of a competition round. Once the round has started, the bot can expand in size as long as its weight does not exceed the maximum, and all parts of the bot must remain attached together. This rule allows for some interesting design options. For example, a bot can have a pair of arms that deploy sideways to try to help capture its opponent. Since there is no height limitation, bots can have very long arms.

According to the rules, sumo bots must move continuously. Another rule states that the bot cannot be sucked down or stick to the sumo ring. This particular rule has resulted in many different interpretations. Basically, what it means is that builders can't use any adhesives to "glue" the bot to the surface of the ring, or use a vacuum suction cup to "suck" it to the ring. A literal interpretation of this rule states that if a bot is "glued" or vacuum-sucked onto the ring, then the bot is no longer moving continuously and will thus automatically lose.

But what if the robot can still move, despite being "glued" down? Because of the "continuous move" rule, some bots use vacuum systems to help pull the robot down to the sumo ring, and use sticky substances on the tires to increase traction. As long as these methods allow the bot to continuously move, and do not damage (or leave a residue on) the sumo ring, they *are* allowed. Some robot sumo contests have very specific rules that prohibit the use of sticky wheels and vacuum systems.

The official rules for international robot sumo are maintained by Fujisoft ABC, Inc., in Japan. The Web site for the rules can be found at *www.fsi.co.jp/sumo-e*.

The official rules for mini sumo are maintained by Bill Harrison of SineRobotics at *www.sinerobotics.com/sumo*.

Most robotic clubs have the same rules posted on their Web sites, along with any special amendments to the rules that are club specific. An excellent illustrated guide to American robot sumo, created by David Cook, is located at *www.robotroom.com/SumoRules.html*. This guide also lists several of the robot sumo clubs throughout the world.

The Sumo Ring Specification

The sumo ring is basically a large, smooth, flat disk made from solid black vinyl. Obtaining a 154cm-diameter piece of vinyl is often very difficult, so most sumo rings are made out of regular plywood. Figure 13-1 shows a drawing of the sumo ring. Note that all of the dimensions for the mini sumo ring are exactly half of the dimensions of the international sumo ring.

The sumo ring can be made out of virtually anything as long as the overall dimensions are maintained. Most sumo rings are made out of plywood. For a mini sumo ring, a 1-inch-thick piece of plywood will work. When building an international class sumo ring, it can be difficult to find a single piece of plywood that is 154cm wide. The easiest way to solve this problem is to make a set of four semicircles that have a 154cm radius. They should be glued together so that the seams between the two semicircles are at 90 degrees from each other. To make the sumo ring meet the 5cm height, a set of strips can be glued to the bottom of the sumo ring to form a spoked wagon-wheel pattern. The sumo ring can be made solid, but that will result in a very heavy sumo ring. For the large sumo ring, it is recommended to use screws in addition to the wood glue.

After the sumo ring has been assembled, the top surface needs to be sanded flat, and any depressions need to be filled in. Paint the finished top surface with a semigloss or flat black paint. Paint the outer ring gloss white, and the two starting lines brown. The side of the sumo ring can be any color, but white is usually the color of choice. The black-and-white color scheme for the ring's surface and borders

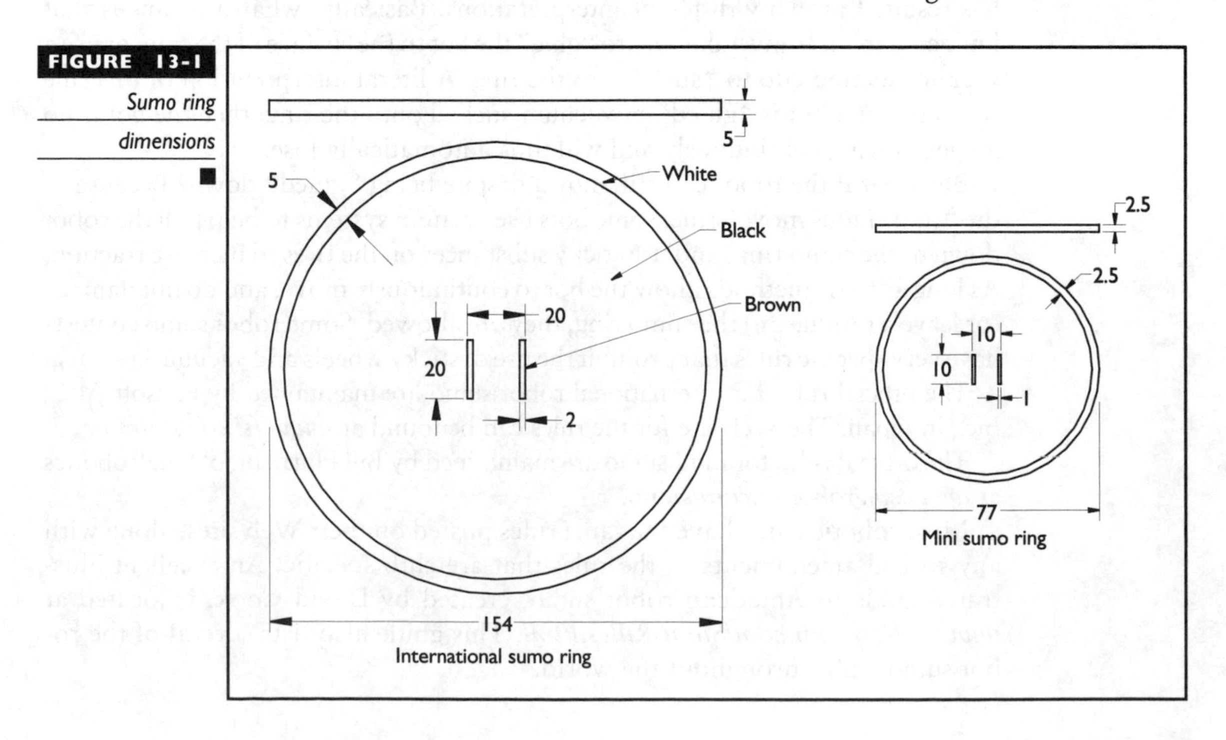

FIGURE 13-1 *Sumo ring dimensions*

were initially chosen so that bots could easily detect the color change and thus recognize the edge of the ring.

For most competitions, this type of ring is sufficient. The official international rules specify that the sumo rings be made from an aluminum cylinder with a height of 5cm and a diameter of 154cm. The top of the sumo ring will be covered with a hard black rubber surface. The official specification for the surface material is to use a long-type vinyl sheet NC, No. R289 made by Toyo Linoleum, Inc., in Japan. Unfortunately, this material is not available outside of Japan, and most vinyl sheet manufacturers in the United States do not make solid black vinyl sheets over 3 feet wide. Lonseal out of Carson, California, sells a solid black vinyl sheet that measures 6 feet wide. This material is called Lonstage, and is a flooring material. There are two different black color numbers to choose from: number 102 is for glossy black, and 101 is for flat black. Either one will work for the sumo ring surface. Lonseal recommends their adhesive number 555 to bond the vinyl to a plywood surface.

This material is generally not stocked in other flooring material warehouses, and you'll have to custom-order it. This material is fairly expensive, so only use it on official competition sumo rings. Regular painted plywood sumo rings will work for all other uses, including testing your sumo bot.

Mini Sumo

Mini sumo robots are becoming the most popular of the sumo classes because they're small, easy to build, and inexpensive, and you can easily carry their smaller sumo ring with one hand. This section will explain how to build a simple mini sumo bot that will be ready to compete in a contest or just show off to your friends.

Modifying an R/C Servo for Continuous Rotation

The first step in building a mini sumo bot is to modify two standard R/C servos so that they can rotate continuously around instead of having the normal 180 degrees of motion. This is a fairly simple modification to make. Use the Hitec HS-300, Futaba FP-S148, Tower Hobbies TS-53, or Airtronics 94102. If you use larger servos, then the completed bot will be wider than the 10cm specifications.

To modify the servos, remove the four screws from the bottom of the servo. Remove the servo horn so that only the small output shaft's spline is showing. With your thumb on the spline and your two forefingers under the front and real mounting tabs, push down on the spline. This will cause the top part of the case to come off. Figure 13-2 shows a servo with the top of the case removed. You'll then see a set of four gears on the top of the servo. Carefully lift the top middle gear off the center spindle shaft, and set down inside the top of the case. Then pull the output gear/shaft from the servo.

FIGURE 13-2 *Internal gears to a standard R/C servo.*

You'll notice a small brass shaft from a potentiometer. The potentiometer is used to monitor the actual position of the servo's output shaft. The next step is to cut the link between the potentiometer and the output shaft. By doing this, you can trick the servo into acting like a gearmotor.

First, you'll have to modify the output gear. There is a small, black tab on the top of the gear, as shown in Figure 13-3. Use a sharp knife to cut the tab off. Make sure you don't get any cuttings caught in the teeth of this gear. Now turn the gear upside-down, and look inside it. If you see a metal ring and a small removable elliptical retainer plate that grabbed onto the potentiometer's shaft, remove the metal ring and then remove the retainer. After the retainer has been removed, replace the ring back into the gear. This ring acts like a bearing, so be careful not to damage it. Figure 13-4 shows what this configuration looks like.

If your gear doesn't have this metal ring and elliptical retainer plate, then you'll need to cut off the output shaft of the brass potentiometer. Figure 13-5 shows how to do this with a pair of wire cutters. A Dremel cut-off wheel would work also here. Just make sure that no cuttings get inside the gearbox. Cut the shaft flush to the top of the gear support.

After both of these modifications, the output gear should freely rotate 360 degrees. Now it is time to calibrate the servo.

Remove the other two gears, and place them in the top of the servo case. Plug the servo into an R/C receiver, and turn everything on. You will probably notice the motor spinning in one direction. On the radio transmitter, move the stick to the center position. Then with a pair of needle-nose pliers, rotate the remaining output shaft from the potentiometer until the motor stops turning. At this point, you have calibrated the servo to *not* move when it sees an approximate 1.5 ms pulse width. Now if you move the stick on the radio transmitter, you will notice

FIGURE 13-3 *Removing the tab that prevents 360-degree rotation.*

that the motor spins one way when the stick is pushed in one direction, and the other way when the stick is pushed in the opposite direction.

If you want to use a microcontroller, like a Basic Stamp, then you can calibrate the servo by sending the servo a 1.5 ms pulse, then pausing for 15 ms, and then repeating this loop.

Once the servo has been calibrated, put all of the gears back in the servo in the opposite order in which you removed them; you might want to write that down ahead of time to help you remember! Place the case cover back on the servo, and then reattach the four screws. When the servo has been reassembled, test it again to make sure it was reassembled correctly. At this point, you have a miniature gearmotor for miniature robotics applications, such as a mini sumo bot.

FIGURE 13-4 *The internal potentiometer retaining plate found inside a Futaba FP-S148 server.*

FIGURE 13-5 *Cutting off the feedback shaft inside an R/C server.*

Building a Mini Sumo

The main components of a mini sumo are a body frame, motors, wheels, microcontroller, sensors, and batteries. For this project, we'll use two modified R/C servos for the motors for the mini sumo, as we just described. Figure 13-6 shows a drawing of a set of wheels, a body base, and a front scoop. Only use the servos that are listed in the table within Figure 13-6. Other servos will be too large. These parts can be made out of pretty much any material. Expanded foam PVC is an excellent material for bots. One of the common trade names for this material is Sintra. It is strong and very light. It can be easily cut with a coping saw and carved with a shape knife. This material can even be tapped with #4-40 threads. When using screws with this material, use only nylon screws, and only finger tighten them. You can also glue it with most superglues (cyanoacrylate glues). It is best to use thin aluminum for the front scoop.

The sumo wheels should be made out of a harder material such as plexiglass or aluminum. The hole spacing on the wheel should be selected based on the type of servo that you use, and it should align with existing holes in the servo horn. The existing holes should be redrilled to allow for either tapping a #4-40 thread or a 0.11in diameter clearance hole for a #4-40 screw.

Mini Sumo Body Assembly

Glue the two servo mounts to the base plate, as shown in Figure 13-7. Make sure that the mount with the 0.50-inch diameter hole is facing the rear (the hole is for routing the servo control wires). Feed the servo wires through the hole and screw the servos to the servo mounts as shown. It will be a tight fit when you're inserting

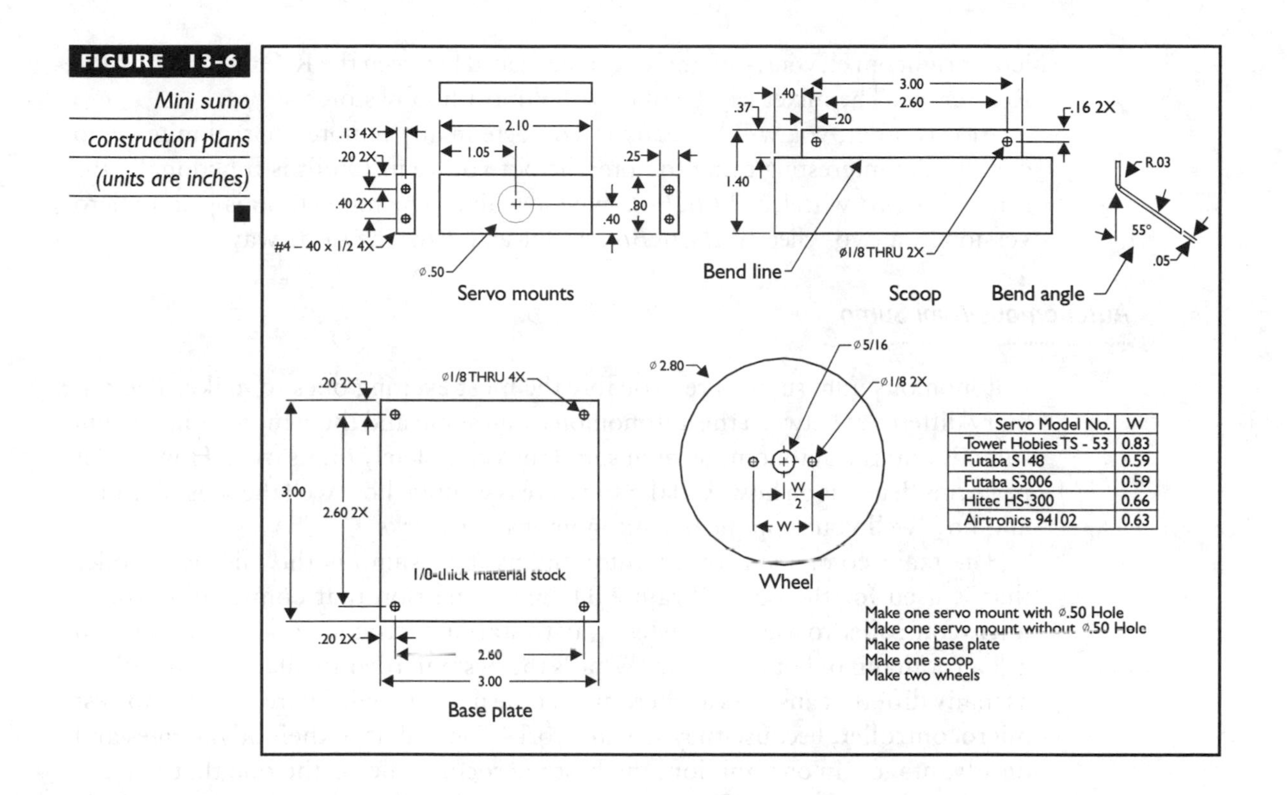

Servo Model No.	W
Tower Hobies TS - 53	0.83
Futaba S148	0.59
Futaba S3006	0.59
Hitec HS-300	0.66
Airtronics 94102	0.63

FIGURE 13-6 *Mini sumo construction plans (units are inches)*

the servos between the mounts. Now screw the front scoop to the bottom of the base plate as shown in Figure 13-7. Use the circuit board spaces as nuts for the front scoop. Screw the wheels to the servo horns using #4-40 screws, and then onto the servos using the screw that came with the servo.

The four circuit board spacers are used to mount control and sensor electronics to the top of the mini sumo. You can also add batteries and other electronic components to the bottom of the base plate, between the wheels. It is best to mount the batteries under the base plate to lower the sumo bot's center of gravity. At this point, you have a general-purpose mini robot sumo base that can be configured to your design ideas.

Remote-Control Mini Sumo

One of the most convenient features of the modified R/C servos is that they can still be controlled directly by a standard R/C receiver. The easiest way to make a remote-control mini sumo is use a two-stick R/C transmitter, and then attach the R/C receiver and the R/C battery to the top of the mini sumo. Turn the transmitter on, and adjust the trim settings to make sure the wheels are not moving when both sticks on the transmitter are centered. Then, drive the mini sumo around like a tank—each stick controls each wheel. To get better driving control where one stick is used for forward and reverse control, and the other stick is used for turning left or right, an elevon/v-tail mixer can be placed between the R/C receiver and the R/C servos. The mixer can be obtained at most hobby stores.

At this point, you will be ready to compete in any remote-control mini sumo contest. An interesting thing to note: the bot you've just built is functionally the same as a two-wheeled *BattleBots*-type machine. The mini sumo is just a micro version of a two-wheeled *BattleBot,* and it will drive the same way.

Autonomous Mini Sumo

Autonomous mini sumos are probably the most exciting ones to make. The primary difference between the autonomous mini sumo and the remote-control mini sumo is that the autonomous mini sumo runs completely on its own. How well it performs depends on how well the software is written, how well the sensors work, and how well your opponent's autonomous bot works.

The main component of an autonomous mini sumo is the microcontroller that is used for the bot's "brain." The next question that comes up is which microcontroller to use. The fastest way to start a microcontroller holy war is to ask a room full of bot builders, "What's the best microcontroller?" You will get as many different answers as there are people in the room. There really is no best microcontroller, because they will all work. They all have their advantages and disadvantages. In our opinion, the best microcontroller is the one that you are most comfortable with. The examples in the following sections will use a Basic Stamp 1 from Parallax, Inc. *(www.parallaxinc.com).* The Basic Stamp 1 was selected because it's a good microcontroller; it is relatively easy to learn how to use; and, most of all, it has been proven to be an effective microcontroller on champion mini sumos.

Edge Detector

The absolute minimum capability that an autonomous mini sumo needs is the ability to detect the edge of the sumo ring so that it doesn't run out of the ring on its own. There are many different ways to detect the edge of the sumo ring. The two more common ways are to use either mechanical contact switches or optical color-detection switches. If the switches and software work correctly, there really is no advantage to using one or the other. Some mini sumos use a combination of both mechanical and optical switches. This section will talk about how to implement an optical-edge-detection switch.

One method that can be used to detect the edge of a sumo ring is to use an infrared detector pair. This consists of using an infrared phototransistor and an infrared light emitting diode (LED). Because the edge of the sumo ring has a white band around the perimeter, the infrared detector pair can be used to detect the color change as the sensor passes over from the black surface to the white surface.

The basic theory behind this approach is that the amount of current that flows through an infrared phototransistor is a function of how much infrared light it receives, at least until it is fully saturated. Because different colors absorb different amounts of infrared light, different colors will reflect different amounts of infrared light (surface texture will affect the amount of reflected light, and some materials allow infrared light to pass through).

By placing an infrared detector pair near a surface, the infrared light from the infrared LED will reflect off the surface toward the infrared phototransistor. Because the amount of current that flows through the phototransistor is a function of the amount of infrared light it receives (reflected infrared light from the surface), this type of arrangement can be used to detect surface color changes. Figure 13-8 shows you a simple schematic of this type of sensor. This circuit was first demonstrated in a mini sumo by Bill Harrison of SineRobotics.

When the detector pair is over the black portion of the sumo ring, the signal out from the sensor is high. This is due to the 10 kΩ pull-up resistor and that the transistor is not conducting any current. When the sensor passes over the white sumo ring edge, the output signal from the detector pair will go low because the transistor is not conducting the current straight to ground. The potentiometer is used to adjust the intensity of the infrared LED, adjusting the sensitivity of the detector pair.

The relative distances between the infrared LED, the phototransistor, and the surface will have an effect on the sensitivity of this circuit. The reflective sensors from Optek P/N OPB706A and QT Optoelectronics P/N QRD-1114 have both the infrared LED and infrared phototransistor built into a single small package. Both of these sensors operate well at distances from 0.04 to 0.20 inches from the surface.

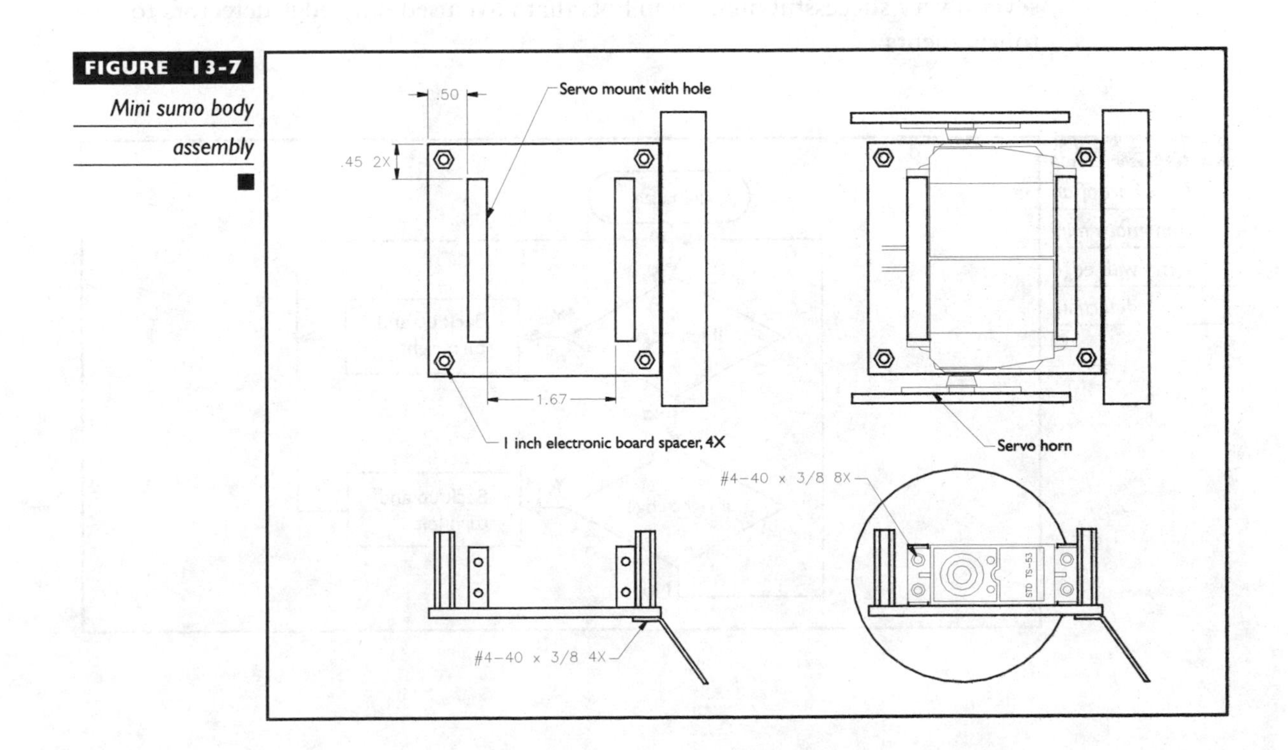

FIGURE 13-7 *Mini sumo body assembly*

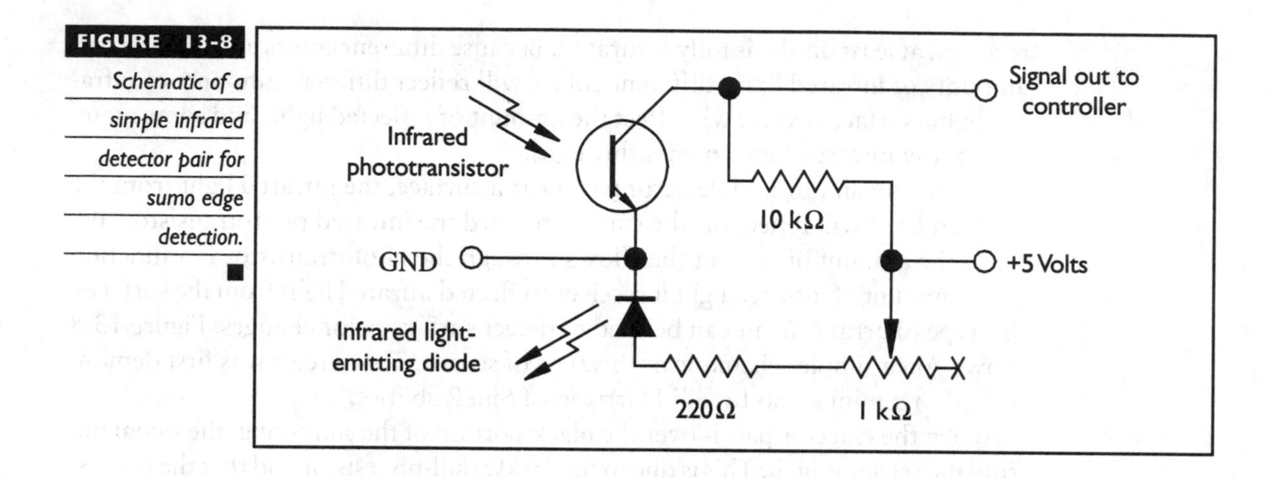

FIGURE 13-8 *Schematic of a simple infrared detector pair for sumo edge detection.*

Two different sensor packages should be used in a mini sumo. Each sensor should be mounted on the front corners of the mini sumo, just behind the front scoop. This way, the mini sumo will know which side of it approached the edge of the ring.

Figure 13-9 shows a flowchart of how to get a mini sumo to work with only its edge detectors functioning, and the Basic Stamp 1 source code shows an example of how to implement the sensors and modified R/C servo motors together into a working mini sumo. At the end of this chapter, there's an example program that uses two edge detectors to keep a mini sumo on the sumo ring. There have been several very successful mini sumo bots that have used only edge detectors to win tournaments.

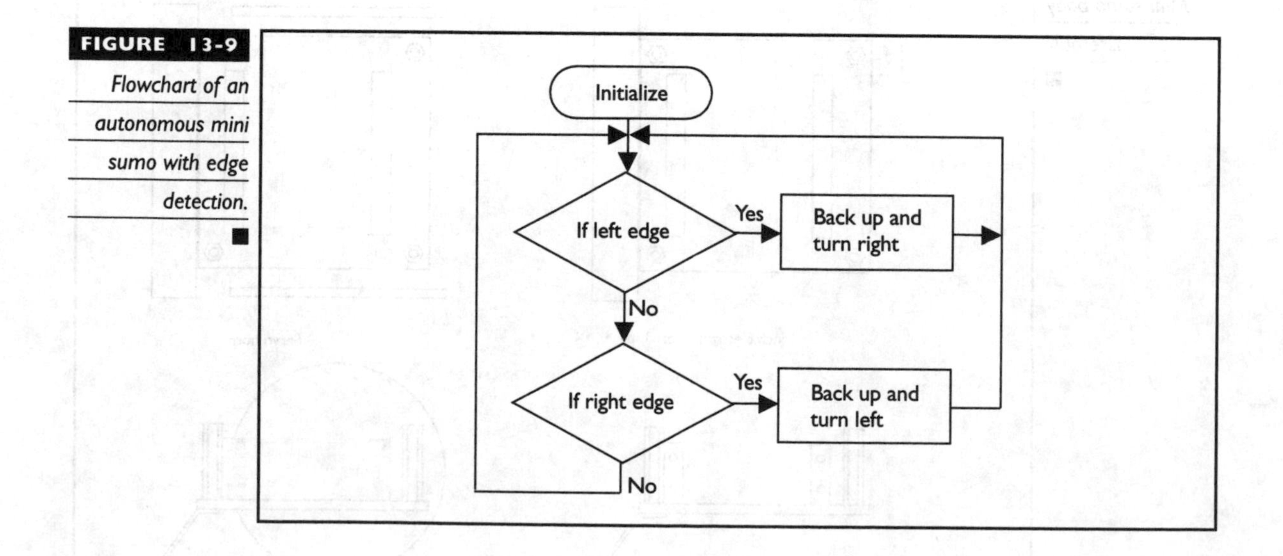

FIGURE 13-9 *Flowchart of an autonomous mini sumo with edge detection.*

The following is a sample Basic Stamp 1 program that uses two edge detectors to keep the bot on the sumo ring. This bot will more or less randomly run around the sumo ring. There are several mini sumos that have won competitions using this type of an approach.

```
'Mini Sumo Edge Detection, msedge.bas 9/27/2001
'This program is a sample program that uses the IR edge detectors to detect the
'white sumo ring edge. The mini sumo will move in a straight line until it hits
'the white edge. After the mini sumo hits the white edge, it will back up, turn
'the opposite direction of the edge detector, then moves forward again.

'pin0 = Right Servo:               These pin(s) are I/O pins, not
'pin1 = Left Servo                 physical pins on the Stamp 1
'pin2 = Right Edge Detector
'pin3 = Left Edge Detector

dirs=%00000011                'Initialize I/O pin directions
                              'pin0 and pin1 are outputs.
                              'pause 5000 'Pause 5000 ms (or 5 seconds)

main:                         'Main program loop
    if pin2 = 0 then Lturn 'Check right edge detector, if the detector detects
  ' the white line then go to the turn left routine.
    if pin3 = 0 then rturn 'Check left edge detector, if the detector detects
  ' the white line then go to the turn right routine.
    pulsout 0,100             'Send a 1 ms pulse to the right servo
    pulsout 1,200             'Send a 2 ms pulse to the left servo
    pause 15                  'Pause 15 ms. This delay sets of the
                              '~50 Hz pulse frequency to the servos
goto main

Rturn:                        'This is the Right Turn routine
    gosub back                'Call the back up routine
    for b2 = 1 to 30          'This loop determine how much the
        pulsout 0, 100        'mini sumo turns. Increasing the
        pulsout 1, 100        'value (30) causes the sumo to turn
        pause 15              'more to the right, decreasing the
    next                      'value causes the sumo to turn less.
goto main

Lturn:                        'This is the Left Turn routine
    gosub back
    for b2 = 1 to 30
```

```
            pulsout 0, 200    'Send a 2 ms pulse to the right servo
            pulsout 1, 200    'Send a 2 ms pulse to the left servo
            pause 15
      next
goto main

back:                         'This routing causes the mini sumo to back up.
      for b2=1 to 25          'This loop determines how far the mini
            pulsout 0, 200    'sumo backs up. Increasing the value
            pulsout 1, 100    '(25) will cause the robot to back up
            pause 15          'more, decreasing the value will cause
      next                          'the sumo the back up less.
return
```

Object Detector

The goal of the object detector is to enable your bot to detect or "see" your opponent while it is far away from your bot, so that your bot can position itself to push the opponent out of the sumo ring. There are many different ways to locate your opponent, including bump switches, infrared reflective sensors, ultrasonic sensors, laser range finders, and vision cameras. The most common are infrared reflective sensors.

An infrared reflective sensor consists of an infrared LED and phototransistor. They are placed next to each other, facing the same direction. When the LED turns on, infrared light is emitted forward. If an object gets in front of the infrared light, some of the light is reflected back toward the phototransistor. The transistor turns on when it detects the infrared light. This type of sensor will actually work with any type of light, as long as the phototransistor is sensitive to the same wavelength as the emitted light.

Because normal light usually contains all wavelengths in the visible light spectrum and light in the near infrared wavelength spectrum, it becomes difficult to distinguish the difference between natural light and the light we are trying to detect. One way to distinguish a man-made (or bot-made) light source from natural light is to modulate the light source at some frequency that is not found in nature. A sensor tuned to this frequency will ignore all of the light sources except for the light source of interest.

The easiest way to make this type of object detector is to use the same type of infrared sensor that is found inside a standard TV remote control. You probably already know that you can change the TV channel just by aiming the remote at a wall opposite the TV set. The TV detects the reflection of the infrared light off of the wall, which in essence is the same way your object detector should work. Inside the TV is a small sensor that contains all of the filters and amplifiers needed to act as a stand-alone infrared sensor. Most of these sensors are tuned to receive a modulated infrared light source operating at either 38 kHz or 40 kHz.

A simple infrared object detector consists of two infrared LEDs, a 40-kHz frequency generator, and a 40-kHz infrared receiver module. For robotic object detection applications, a modulated LED is mounted on both sides of the receiver module pointing slightly away from the receiver module. By alternating which side of the LED is active, you can determine which side the object is on.

The schematic shown in Figure 13-10 is for a simple infrared object detector using a few common components. This circuit uses a single 74HC04 CMOS hex inverter to generate the 40-kHz modulated signal, and act as switches to turn on/off the modulated infrared LEDs. The potentiometer R1 is used to adjust the modulated frequency. When selecting the infrared LEDs and the infrared receiver module, make sure that they are both sensitive to the same wavelength.

The two most common wavelengths are 880nm and 940nm. For the Sharp detectors that come inside a metal can, the metal case must be grounded to the rest of the circuit. Resistors R4 and R5 can be decreased in value to increase the range of the detector. To turn on the infrared LED, apply 5 volts to the particular LED signal line. To turn it off, ground the signal line. The output of the infrared receiver module is normally high at 5 volts. When it detects the proper modulated infrared light, the output voltage will drop to zero.

The Sharp G1U52X and GP1U581Y series infrared receiver modules are the most common, and the Panasonic PNA4602M series infrared receiver modules are becoming more popular since they are less sensitive to visible light than the Sharp detectors, and they are less than half the size of the Sharp detectors.

FIGURE 13-10 *Infrared object detector schematic drawing.*

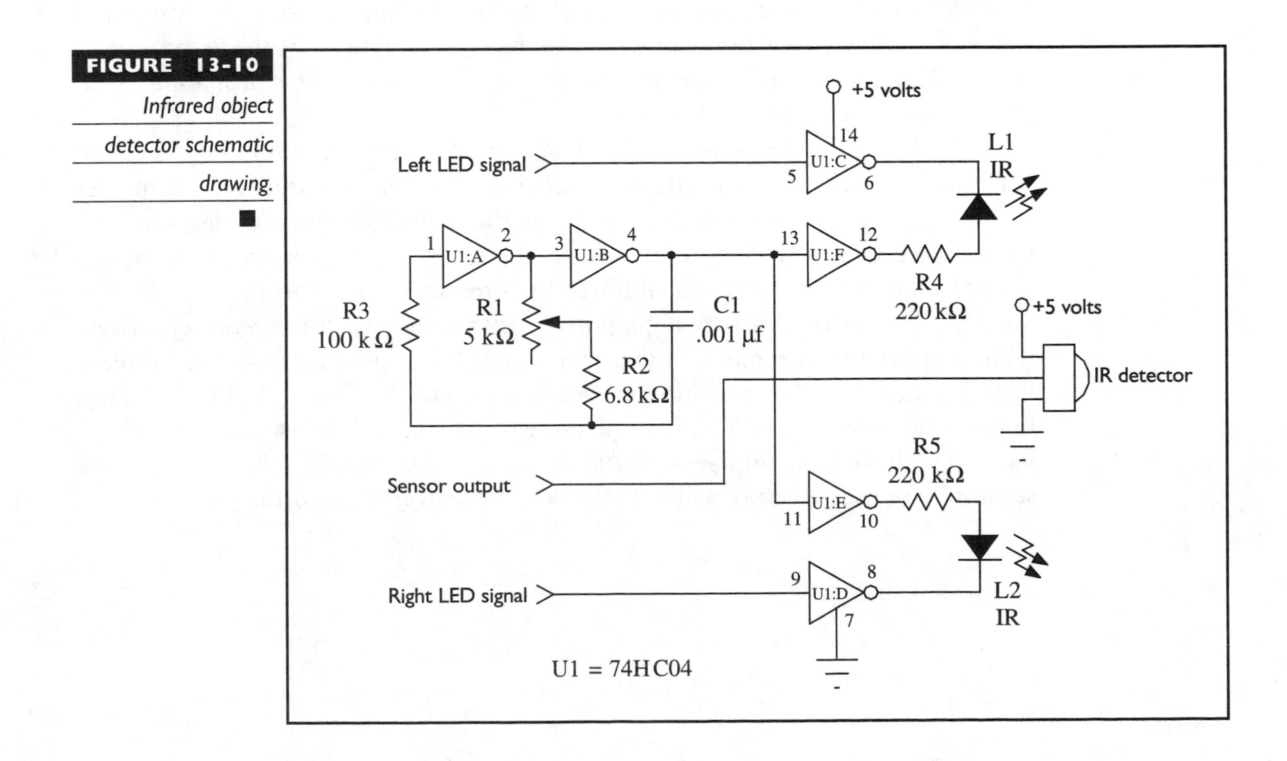

Once the circuit is built, put a small tube around the LEDs to help focus and collimate the infrared light. Although most of the light is projected in front of the LEDs, a small fraction of the light goes sideways and to the rear of the LED. This could interfere with the infrared receiver module, causing false readings. The tubes also help reduce this interference. The IR receiver modules that are not enclosed inside a metal case, such as the Panasonic PNA4602M, are very susceptible to this setback. To solve this problem, place a small piece of aluminum foil duct tape on the back and sides of the receiver module. Do not let the tape touch the wire leads. This will help prevent false readings from sideways and backward emitted IR light.

The basic operation of this circuit is to flash the left IR LED, then take a reading from the IR receiver, then flash the right IR LED, and then take another reading from the IR receiver. If the receiver detects something from the left IR LED, then there is something that is either to the left front or in front of the detector. If the receiver detects something from the right IR LED, then there is something either to right front or in front of the detector. If both left and right IR LEDs returned a signal, then there is something directly in front of the detector. Using this approach, a relative direction of the object can be detected. The left and right angular range detection can be adjusted by angling the IR LEDs toward or away from the IR receiver module.

Depending on which infrared receiver module you choose, the modulated infrared light must be on between 400–600μs to allow for the receiver module to stabilize. Otherwise, a false signal is more likely to occur. In the real world, there's a lot of "noise" in all signals, so it's better to take a sample of readings instead of relying on a single measurement. One method of sampling is to take five consecutive readings. If you get more than three hits, there is a greater probability that there is an object in front of the sensor.

A 40-kHz infrared receiver module has its peak sensitivity at 40 kHz. They are still functional when receiving light at +/- 5 kHz of the center frequency. The further away the actual modulated frequency is from the center frequency, the less sensitive the sensor becomes. With this knowledge, the sensitivity of this circuit can be adjusted by shifting the modulated infrared LED frequency away from the center frequency. The reason this may be important is that the detector circuit will detect white objects that are much farther away than black objects. Also, the ambient lighting at an actual competition is usually different that the ambient lighting at home or wherever you're building and testing your bot. Sensors usually respond differently in the different ambient lighting conditions. Having the ability to adjust the sensitivity of the detectors will improve your bot'soverall performance.

Sensor Integration

Integrating the object detector and two edge detector circuits along with a Basic Stamp 1 can be accomplished on a small prototyping board. Figure 13-11 shows a schematic of the entire circuit for an autonomous mini sumo.

In a remote-control mini sumo, the same battery pack powered both of the modified R/C servos and the receiver module. For the autonomous mini sumo, you need two different power supplies. A 4-cell AA (6-volt) battery pack will provide power to the modified R/C servos, and a 9-volt battery will provide power to the microcontroller and the sensors. A 4-cell AA battery box should be attached to the bottom of the mini sumo. Double-sided foam tape should be sufficient to attach the battery box to the bottom of the mini sumo.

All battery, microcontroller, and electronic circuit grounds must be tied together. If the grounds are not tied together, you'll see erratic performance in the bot. The reason for the two battery supplies is that the servos can momentarily draw up to 1 amp of current each. This could cause a voltage drop in the microcontroller, which will cause the microcontroller to reset. Using a separate power source for the microcontroller will help ensure a uniform voltage supply to the microcontroller.

The flowchart in Figure 13-12 shows how logic in this mini sumo should work. The following program example will make a fully functional mini sumo. This mini sumo will follow your hand as you move it in front of the mini sumo, and stay on the sumo ring. Using the information presented here, you will have a working autonomous mini sumo bot.

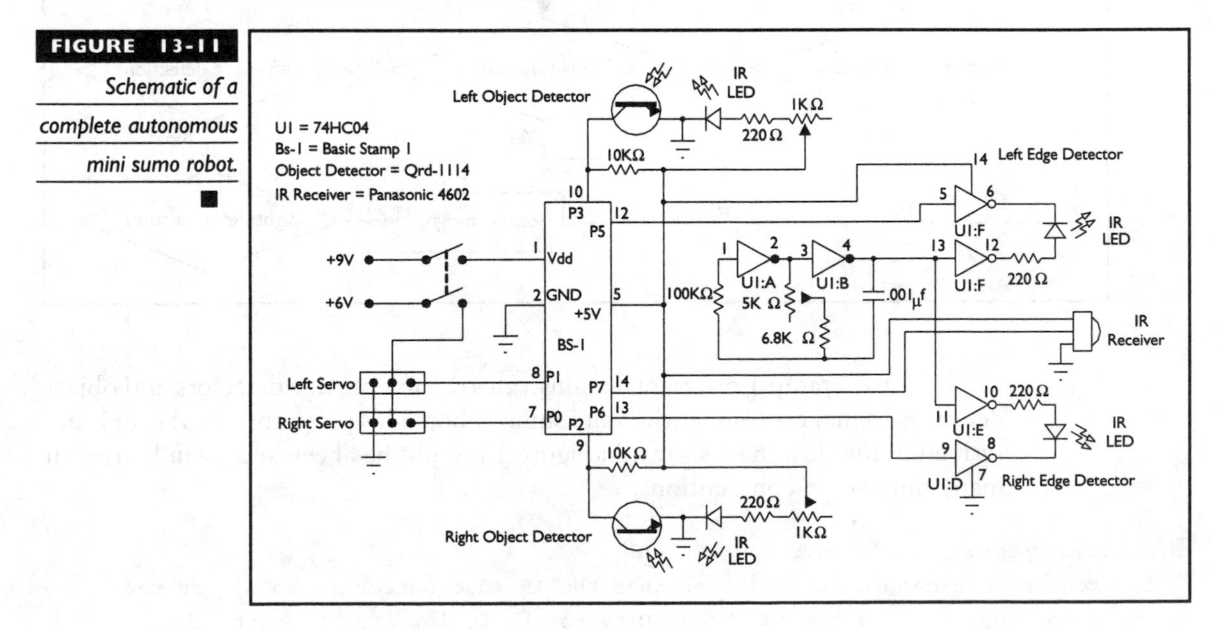

FIGURE 13-11 Schematic of a complete autonomous mini sumo robot.

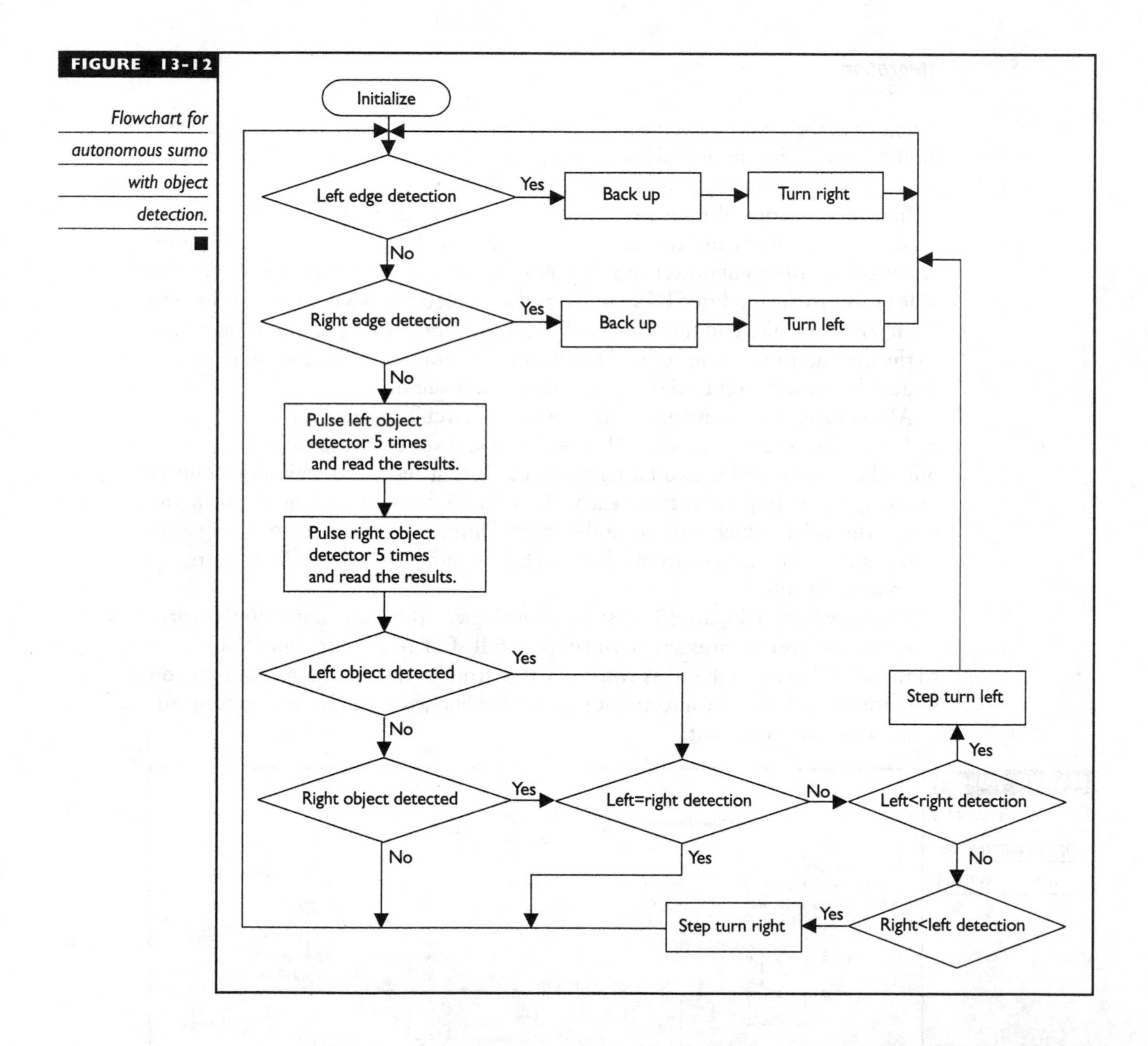

FIGURE 13-12 *Flowchart for autonomous sumo with object detection.*

This Basic Stamp 1 program demonstrates the use of edge detectors and object detectors to make a competitive mini sumo robot. This program uses the logic described in the flowchart shown in Figure 13-8 and has been successfully used in many mini sumo competitions.

```
'Mini Sumo Program, minisumo.bas 9/30/2001
'This program is a sample program that uses the IR edge detectors to detect the
'white sumo ring edge and the IR Object Detector to follow its opponent. The
'mini sumo will move in a straight line until it hits the white edge. After the
'mini sumo hits the white edge, it will back up, turn around, then move forward
```

```
'again. If the mini sumo sees an object in front of it, it will turn towards
'the object.

'pin0 = Right Servo                These pin(s) are I/O pins, not
'pin1 = Left Servo                 physical pins on the Stamp 1
'pin2 = Right Edge Detector
'pin3 = Left Edge Detector
'pin5 = Left Opponent Detector LED
'pin6 = Right Opponent Detector LED
'pin7 = IR Receiver Sensor

dirs=%01100011                  'Initialize the I/O pin directions pin0, pin1,
                                ' pin5, pin6 are outputs
pause 5000                      'Pause 5000 ms (or 5 seconds)

main:                           'Main Program Loop
      if pin2 = 0 then Lturn    'Check right edge detector, if the detector sees the
  ' white line, then goto the left turn routine.
if pin3 = 0 then rturn          'Check right edge detector, if the detector sees the
  ' white line, then goto the right turn routine.
      pulsout 0,100             'Send a 1 ms pulse to the right servo
      pulsout 1,200             'Send a 2 ms pulse to the left servo
      b0 = 0                    'Sample the left object detector for
      for b2 = 1 to 5           '5 times by toggling the IR LED on/off
            pin5 = 1            'The output pin will be high if there
            pin5 = 0            'is no reflected signal. If b0 (or b1)
            b0 = b0 + pin7      'is less than 3 then over 50% of the
      next                      'signals returned back to the receiver.
                                'This gives a good indication that
      pulsout 0, 100            'an object was detected, and a
      pulsout 1, 200            'less chance that the signals were
      b1 = 0                    'random noise or false signals
      for b2 = 1 to 5
            pin6 = 1
            pin6 = 0
            b1 = b1 + pin7
      next

      if b0 < 3 then turn       'If a positive object detection was obtained, then
      if b1 < 3 then turn       'goto the turn routine

goto main

turn:                           'This routine determines which direction
```

```
      b2 = b0 + b1             'to turn. If both detectors return
      if b2 < 5 then main      'equal values, then go straight,
      if b0 < b1 then left     'otherwise turn in the direction that
      if b1 < b0 then right    'had the stronger return probability.
goto main                      'i.e. a lower hit number.

left:                          'Make a small left turn move
      pulsout 0, 200           'Send a 2 ms pulse to the right servo
      pulsout 1, 200           'Send a 2 ms pulse to the left servo
      pause 15                 'Pause for 15 ms. This delay sets up
goto main                      'the ~50 Hz servo update frequency

right:                         'Make a small right turn move
      pulsout 0, 100
      pulsout 1, 100
      pause 15
goto main

Rturn:                         'This is the Right Turn Routine.
      gosub back               'Call the back up routine.
      for b2 = 1 to 30         'This loop determines how much the
            pulsout 0, 100     'mini sumo turns. Increasing the
            pulsout 1, 100     'loop value (30) causes the mini
            pause 15           'sumo to turn more to the right,
      next                     'decreasing this value decreases
goto main                      'the amount the mini sumo turns.

Lturn:                         'This is the Left Turn Routine.
      gosub back
      for b2 = 1 to 30
            pulsout 0, 200
            pulsout 1, 200
            pause 15
      next
goto main

back:                          'This is the back up routine
      for b2 = 1 to 25         'This loop determines how far the mini
            pulsout 0, 200     'sumo will back up. Increasing the
            pulsout 1, 100     'loop value (25) will increase the
            pause 15           'overall distance. Decreasing the
      next                     'value will cause the mini sumo to
return                         'back up less.
```

Performance Improvements

In sumo, two of the most important factors that make a winning bot are strength and technique. Simply having the strongest bot doesn't mean that you will have a winning bot; and having the smartest bot doesn't mean that you will have a winning bot, either. Your bot needs both of these skills.

Strength is related to pushing power. From physics, we know that pushing force is equal to the coefficient of friction between the bot wheels multiplied by the weight of the bot. This simple relationship pretty much tells you what you need to have in a strong bot: weight and traction. The higher the coefficient of friction, the better the traction the bot will have. The heavier the bot is, the greater the amount of force required to move it. It is best to make your bot as heavy as possible for it's weight class. For a mini sumo, this is 500 grams. As for traction, soft wheels usually have better traction than hard wheels. Some bots have placed rubber O-rings or rubber bands on the outside diameter of the wheel to improve traction, and others have used foam wheels like you see on model airplanes.

Weight and traction are the two most common ways to improve the performance of mini sumos. The other way to win is to use better strategy during the actual contest. This really comes down to the type of programs you use in your bot. Some bots spin more than they move in straight lines. Some bots use more sensors to improve vision capabilities, where others use a stealth approach to keep from being seen. Some bots even use arms to try to capture or corral their opponent. This is what makes robot sumo exciting, because it allows for many different types of bots to enter the competition. In fact, biped and hexapod bots have competed and have even won some matches. The Basic Stamp 1 microcontroller used in this example doesn't have the memory space for advanced software control. You will need to use a different microcontroller such as the Basic Stamp 2 or the BasicX-24 from NetMedia (*www.basicx.com*).

Various Mini Sumo Robots

Figure 13-13 shows a mini sumo named *Minimum Capacity* built by Pete Miles, one of this book's authors. This mini sumo uses the circuit shown in Figure 13-11 and the logic shown in Figure 13-12. The actual source code is shown at the end of this chapter. Although this mini sumo is not the best-looking bot on the block, it has placed in the top three positions in tournaments in Seattle, San Francisco, and Los Angeles, and the *All Japan Robot Sumo Tournament* in Tokyo.

One of the most exciting aspects of robot sumo is that any type of robot can be entered into the contests. Pete has also built biped and hexapod walking robots that are fully functional and have won several matches. These robots were built to demonstrate that walking robots can compete in robotic sumo contests. Figure 13-14 shows two photographs of these walking bots.

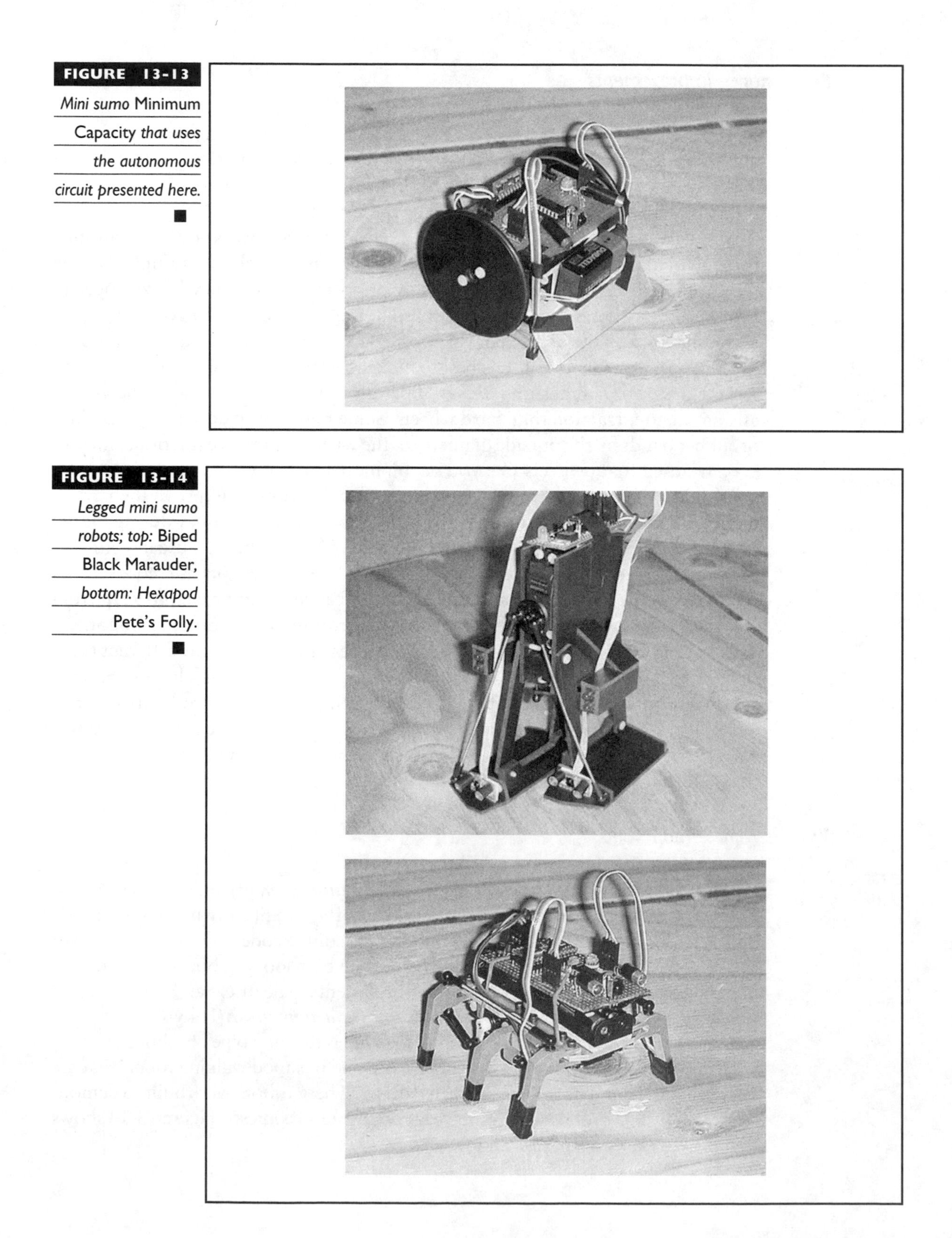

FIGURE 13-13 *Mini sumo* Minimum Capacity *that uses the autonomous circuit presented here.*

FIGURE 13-14 *Legged mini sumo robots; top:* Biped Black Marauder, *bottom: Hexapod* Pete's Folly.

International Robot Sumo Class

The general functionality of the international robot sumo class is basically the same as a mini sumo, except that they are heavier, faster, and smarter. The size of the international robot sumo class is 20cm square. This is a 4x increase in the area over mini sumos. The maximum weight increases by a factor of six, to 3kg. This allows for more powerful propulsion systems, more sensors, and improved microcontrollers, which increases the flexibility in the designs of the 3kg sumo robots.

Motors

Most 3kg sumo bots don't use modified R/C servos. This is because most R/C-style servos are not strong enough for the increased weight or fast enough to rapidly move the robot across the larger sumo ring. Typical motors include high-powered 12-volt and 24-volt gearhead motors from Pittman Motors, Barber-Coleman; planetary gearheads from cordless screwdrivers; and the electric motors from high-performance R/C racing cars. Most gearhead motors are purchased from surplus stores, because they usually cost 1/10th the cost of a new one purchased directly from the manufacturer.

When using stand-alone electric motors, you must build a custom gearbox. The advantage to this is that the gear reduction ratios can be set up to maximize the desired speed and torque range of the motors. Otherwise, you will have to use what is available in the regular gearhead motors. The drawback to this approach is that it requires custom machining of the gearboxes, which can be expensive. Because of this, most people use off-the-shelf gearhead motors and vary the motor voltage to get the performance they want.

Motor Controllers

Using high-powered motors requires high-powered motor controllers. The motor controllers are commonly called *electronic speed controllers (ESCs)*. In mini sumos, the peak motor current requirements are usually around 1 A. For the international robotic sumo class, peak motor current demands can exceed 100 A. This really depends on the type of motors selected for the sumo bot. Usually, higher-voltage motors require less current. The most cost-effective ESCs are the ones made for the R/C racing car industry. These controllers are designed to handle large amounts of current for short periods of time. They are also easy to integrate into a sumo bot.

When looking at electronic speed controllers, make sure that yours has a reversible speed controller. More than half of the electronic speed controllers made today are for forward use only. A sumo bot will be spending about half its time going backward as compared to going forward. The other factor to consider is the current handling capacity when operating in reverse. Of the ESCs that are reversible, about half of them have lower current ratings in reverse than in forward. You will need an ESC that has the same capabilities in forward and in reverse.

Most ESCs advertise the peak current capacity. This is a very misleading value. It is usually a theoretical value under ideal operating conditions, and not to exceed that value for more than one second. In reality, if the motors are drawing current near this advertised value for more than a few seconds, you will let the "magic smoke" out of the ESC, and it will stop working. Since sumos spend a lot of their time pushing other robots around, the motors will be drawing near maximum current for long periods of time. Because of this, you will need to look at the 30-second and 5-minute current ratings of the ESC. The ideal ESC will have a 30-second current rating greater than the stall current of the motor. Obtaining this information usually means contacting the manufacturer. One method to obtain a little performance improvements out of the ECS is to add a cooling fan above the heat sinks on the ESC.

Generally, the R/C style electronic speed controllers are the easiest and most cost-effective solution to driving the motors. These controllers can be found at most hobby stores. Another source for electronic speed controllers is using H-Bridge type controllers. There are many companies that sell a wide variety of these types of controllers. One of the big differences in these controllers is that they accept a true pulse-width modulation (PWM) signal to vary the motor speed, which can give you better speed, braking, and direction control resolution. Many bot builders build their own version of a high-powered speed controller using MOSFET power transistors. Although this can be done, it is generally a difficult task to produce a reliable controller. In the end, off-the-shelf speed controllers are less frustrating and cost less to implement.

Advanced Sensors Because the international robot sumo class is much larger than the mini sumo class, there's a lot of extra room for sensors. Most international sumo bots use more than one type of sensor. The edge-detection sensor is still used. Some bots use more than two sets of these, and some bots even have them on their backs to detect whether they are being pushed out to the sumo ring. The infrared object-detector circuit is very popular, and is used on the larger sumo bots. A new type of sensor that is used on the larger sumo bots is the range-detecting sensor. The two most common methods used are ultrasonic sensors and infrared range detectors.

Ultrasonic Range Detectors

Ultrasonic range detectors are becoming more popular because they are becoming more widely available. They work by measuring the time of flight from a sound signal being reflected off an object. The object's distance is computed by multiplying the measured time by the speed of sound in the current air conditions. For robotic sumo applications, any returned signal outside 5 feet can be ignored because it is outside the maximum diameter of the sumo ring. Ultrasonic sensors have a wide field of view, so it is difficult to obtain the opponent's direction with a stationary sensor. Because of this, multiple ultrasonic sensors are normally used.

One of the drawbacks to these sensors is that they have a minimum effective range. For example, the Polaroid 6500 sensor has a minimum distance of 6 inches (or nearly the width of the sumo robot). This can be dangerous because your bot may not see the opponent standing directly in front of it! Combining an infrared object detector with an ultrasonic sensor will give a good range of detection capabilities. A new ultrasonic sensor made by Devantech Ltd. (*www.robot-electronics.co.uk*) has a minimum sensing distance down to about 1 inch. The model number for this sensor is SRF04. It is small and compact, and has been successfully implemented on several sumo bots.

Infrared Range Detectors

Recently, Sharp started selling a set of infrared range detectors. Particular models include the Sharp GP2D02 and GP2D12. These sensors have both the infrared receiver and infrared emitter in the same package. The LED is positioned at a slight angle relative to the receiver to use an optical triangulation approach to determine range. The output from these sensors is either an analog voltage or a digital signal. As with the ultrasonic sensors, the drawback to these sensors is that they have a narrow field of view—thus multiple sensors must be used to obtain a wide field of view. Many bots have successfully used these sensors to detect objects and detect ranges for these objects.

Laser Range Finding and Vision Systems

Some advanced sumos can use laser range-finding systems and actual vision camera systems. These types of systems not only determine the range of the opponent, but they also provide positional information, which is very advantageous to finding your opponent quickly. These systems require powerful control systems to process all of the data in real time. They are also very expensive and fragile to implement. Currently, they are used more for experimental purposes; but as the microelectronic technology improves, these types of systems will become more widely used. The autonomous and semiautonomous robotics industry will drive the development of these types of sensors.

Advanced Software Algorithms

Most sumo bots collect data from the sensors and then plan a reaction based on the input. This type of approach is usually the easiest to program. Some sensors are given higher priority over other sensors. For example, an edge-detector result has higher priority than an object detector. The more sensors the bot has, the better the information it can process to determine a better reaction.

You can also collect a time history of the data in order to predict where the opponent will be, and then plan your attack based on the prediction. For example, if your bot detects that its opponent is off to one side, it can conduct a preplanned attack move, such as moving forward for 6 inches and then making a U-turn maneuver to get behind its opponent, instead of just turning toward the opponent. This generally requires a lot more processing power than a Basic Stamp. There are many microcontrollers available today that have this type of capability, such as the MIT Handyboard that uses the Motorola 68HC11 microcontroller, or the Robominds *(www.robominds.com)* board that uses the Motorola 68332 microcontroller.

Traction Improvements

As stated earlier, weight and traction are very important in a sumo bot. Most mini sumos are two-wheeled bots. In the international robot sumo class, there is a wide range of two-, four-, and six-wheeled bots. And most of them have a single motor directly driving each wheel. After the bot's wheels have been modified to have the highest possible coefficient of friction, and the bot is at its maximum weight, what is left to increase its pushing power? Increase the robot's apparent weight.

The way this is done is to add a vacuum system to the bottom of the bot. The vacuum system then sucks the bot down to the sumo ring, thus increasing the forces on the wheels, and increasing the pushing power of the bot (assuming the motors don't stall!). The rules of the contest prohibit sticking or sucking down to the sumo ring; but if the robot can continuously move while it's "stuck," then the vacuum system can be used because it doesn't interfere with motion.

The Japanese make the best vacuum-based sumo bots. These bots are so good that they can compete on a sumo ring that is *upside down* without falling off! One of the drawbacks to the vacuum-based bots is that they can generate so much vacuum that it literally tears the vinyl surface off the sumo rings. Under the rules of the contest, if a bot damages the sumo ring, it is disqualified. Unfortunately, once the ring is damaged, no other bot can use the ring. This is why most clubs specifically prohibit the use of vacuum systems.

Robot Part Suppliers

There are several companies that sell parts to build sumo bots. Lynxmotion *(www.lynxmotion.com)* sells enough parts to build complete and competitive sumo bots. Figure 13-15 shows a photograph of a six-wheel-drive international class sumo bot built by Jim Frye of Lynxmotion. This bot has a unique feature where the front scoop deploys forward after the match starts, which makes it easier for this bot to get underneath its opponent. This bot also uses a Basic

FIGURE 13-15 International class sumo robot named Overkill.

Stamp 1 for the microcontroller. Acroname *(www.acroname.com)* sells a wide variety of parts that can be used to build quality sumo robots. Mondo-tronics *(www.robotstore.com)* and HVW Technologies *(www.hvwtech.com)* also have a wide selection of robot parts.

Annual Robot Sumo Events

The following is a list of some of the largest annual robot sumo contests. This is not a complete list. There are many other contests held each year. This list only shows some of the largest events:

- All Japan Robot Sumo Tournament: *www.fsi.co.jp/sumo-e*
- Seattle Robotics Society Robothon: *www.seattlerobotics.org*
- Northwest Robot Sumo Tournament: *www.sinerobotics.com/sumo*
- Portland Area Robotics Society: *www.portlandrobotics.org*
- Western Canadian Robot Games: *www.robotgames.com*
- Central Illinois Robotics Club: *www.circ.mtco.com*
- San Francisco Robotics Society of America: *www.robots.org*

At this point, you should have enough information to get started in the exciting world of robotic sumo. As you gain more experience competing in sumo tournaments, you'll learn how to improve the designs of your bots, and help your competitors improve their designs, as well. Caution: robot sumo can be addictive!

BUILD YOUR OWN COMBAT ROBOT

chapter 14

Real-Life Robots: Lessons from Veteran Builders

IN this chapter, we'll conclude our discussion of building combat robots by offering two first-person accounts from veteran robot builders. Contributor Ronni Katz recounts her experience building *Chew Toy* for a past *Robot Wars* event, and co-author Pete Miles tells what it took to construct his machine *Live Wires* for a *Robotica* competition.

A lot of the technical details covered previously in the book will be addressed in some fashion in each builder's story. The steps they went through to build their machines are similar to what many builders go through constructing their robots, especially newer builders. Although their methods are not presented here as the only way to build a robot, they are intended to inform the reader as to the particular methods these builders chose to build their machines.

Anyone who builds a robot is going to do things in his or her own way; still, it's a good idea to keep in mind what methods others have used. When you begin your project, talk to others who have built robots and ask them about their experiences—what worked and what didn't. Learn from others' mistakes, and duplicate those efforts that worked well.

Ronni Katz—Building *Chew Toy*

I have competed in several *Robot Wars* competitions and have come up with three different designs. For this discussion, I will be using my lightweight design, *Chew Toy,* as the example model. Of the three possible entries, this one is the most basic robot that was actually a "garage-built" robot created using easily obtainable parts and tools that most builders either already own or can acquire.

First, I will cover the research and conception stage and the preconstruction phase. The latter phase comprises everything you do short of cutting the metal and welding it together. Figure 14-1 shows *Chew Toy.*

Step 1: Research

If your introduction to robot combat has come only from watching TV, you need to know much more before you begin building your first bot. First, it's a good idea to get familiar with the current rules for whatever competition you have in mind

FIGURE 14-1 Chew Toy

before you begin your design. The rules do change slightly from year to year, so it's best to make sure you're current.

Aspiring robot builders can obtain rules, information on robotic design and competitions, and building tips from many Web sites. On these sites, you can gather information on which engineering efforts have worked in the past and which efforts haven't. One of the best "unofficial" places to look is the *BattleBots* Builder's Forum at *www.delphi.com,* where you can read conversations between experienced builders and find other tidbits of information that should prove helpful to fledgling designers.

It is also worth sending e-mails to builders you might come across on the Internet, asking whether they're willing to share videos or other information with a newcomer. Many people in this community are open and welcome discussing ideas and questions with those interested in participating in robot competitions. More experienced builders can provide the names of reliable suppliers, and information about where to get good-quality, radio control (R/C) radios and speed controllers, and sometimes will even critique designs for a first-time competitor.

In addition to the Internet, other good sources of information are magazines such as *Robot Science and Technology* and other hobbyist magazines that deal with radio control and similar electronics scenarios. Ordering the parts catalogs advertised in these publications can be extremely useful. Some robot parts are just exotic enough that the average hobby, electronics, or hardware store won't carry them, but a larger catalog company might. If you have access to a university library, especially at a school with an engineering program, chances are it will have periodicals and books that may be of use.

Research what supplies you already have on hand to do your building. What tools do you own or have access to? Do you have space in which to build or have access to a place to do the construction and testing? Do you have access to a machine shop or know someone who does? How about a milling machine or lathe? Check out the availability of time on the milling machine in your friend's garage or the willingness of a local metal shop to cut aluminum or steel to your specifications; this will indicate what resources will be there when you need them. Local machine shops might want to be involved themselves, and you might wind up with a sponsor. (That happened with my team's robot, *Spike II*. The machine shop that did all the aluminum cutting and welding donated a portion of their services in exchange for advertising and help redesigning a printed circuit board. Yes, barter still exists today. If you have skills to trade for time on that milling machine or access to the heli-arcwelder, you should go for it. Bartering cut down on the expense of building our robot, and we made new friends and contacts.)

It definitely pays to look into the technical expertise that exists in your own neighborhood. Radio Shack can supply electronic bits and pieces at a decent price. Investigate what equipment—specifically, radio control parts—your local hobby store can get for you. Hobby stores that cater to model makers (especially model makers who build their own R/C planes, boats, and so on) often have a good selection of speed controllers and other essential equipment. Be sure that you purchase a speed controller that will handle the current you intend to pump through it. Many contestants at early *Robot Wars* competitions fried their speed controllers because they didn't check this detail. As far as R/C equipment goes, my advice is this: Don't get a cheap radio. It pays to invest in a good-quality PCM or FM aircraft radio set for ground frequencies. The aggravation you save will be well worth the money you spend.

Step 2: Conception

After you've done all your research—gone through those parts catalogs, know the rules, and are sure of the weight class you want your robot to compete in—the next phase is coming up with the design sketch. You don't need heavy-duty engineering computer aided design (CAD) software to create a basic design sketch. Our work was done on an artist's sketchpad and on notebook paper. The average builder won't have AutoCAD on his or her home PC, and it isn't necessary if you plan a simple robot design.

The photographs of my lightweight entry *Chew Toy* (Figures 14-1 and 14-2) show its simple design. *Chewie* is a basic robot—all the essential parts, such as the motors, batteries, and major weapons, were not that hard to lay out and assemble. The robot's conception came out of the hypothesis, "If I could use only a surplus store's catalog to get parts to build my robot, what would I design?" In reality, I use a lot more sources for parts. However, I was curious. Could I come up with an effective design by pretending I was limited in parts availability?

As you can see, *Chew Toy* has a simple structure. It relies heavily on its 3.5-hp, four-stroke motor and those rather evil sharp saws to do its battle damage. The

FIGURE 14-2 Chew Toy *with protective armor removed.*

body frame—the square steel tubing and the wire mesh used for the armor—came from Home Depot, another great inexpensive supplier. *Chew Toy* is something that all designers like—a cheap entry. The cost for this robot (everything *but* the speed controller) was about $500. (Instead of doing what I had initially conceived—create a simple relay system—I splurged on a Vantec speed controller for *Chew Toy*. It cost about as much as the *entire* rest of the robot, but, because the speed controller is an item that can be reused in future designs, I looked at my extravagance as an investment. In addition, it saved the time that it would have taken to construct and properly test the relay system I had devised in the early phase of *Chew Toy*'s development.)

Once you figure out what you want to build, the next step is building the mockup. I cut out a balsa wood frame and the parts into which the motor, the drive train, weapons system, and so on, will be fit. Balsa is easy to work with, and any hobbyist who has done original designs of model airplanes, boats, or the like has probably done mockups in balsa wood. Balsa wood is also cheap and readily available, and if you botch something in the mockup phase, you can redo it much more easily than if you were working in metal.

After your balsa wood mockup is within your parameters and everything looks workable, you are ready to spec out your final project. The balsa wood project can be broken down into the component parts and used as guides for cutting the metal for the final project. If you are doing your own metal cutting, you can take apart your mockup and use each piece as a template for your metal pieces. I laid the pieces on top of the metal, traced the shape onto the metal, and then cut out the shapes. That way I was sure all the metal shapes would be the exact size I specified, and when I cut and fit the bot together it would replicate the mockup.

Metal shops can also use your balsa template as a guide. If a shop is also going to be doing all your welding, it is a good idea to give these folks your design sketch and review it with them so they understand exactly what you want your finished piece to look like. Showing them the balsa mockup before you disassemble it for template parts is also useful, especially if you are working with people who have no prior experience with robotics.

Step 3: Building the Bot

I decided to use a surplus ammo box as part of *Chew Toy*'s structure because it was inexpensive, yet an effective way to house the electronics, but it wound up becoming the structural backbone of the robot. All the weapons systems and other features on *Chew Toy* are attached to the ammo box. The metal of the ammo box was not as tough as I'd originally hoped, but it provided adequate protection from impacts. All the electronics of the robot went inside, as well as the stationary axle that was a part of the robot's drive train. The axle—a long steel rod that goes lengthwise through the center of the ammo box—does double duty as part of the drive mechanism and as a means of holding the batteries securely in place.

The robot's motive power is supplied by a pair of kiddy-car motors (power wheel motors) that were inexpensive. I found them in the same surplus catalog where I found the ammo box. Because of their low price, I could purchase extra motors to use for experiments. When I tested these motors to achieve maximum performance, I found that when these 12-volt units are run at 24 volts, a good amount of power was produced. Subjecting motors to higher-than-rated voltage occurs frequently at robotic competitions. It's risky, though, so it requires a lot of trial-and-error testing to determine how much extra voltage the motors can handle. *Chew Toy*'s motors were broken in before being tested to their voltage limits.

It is also important to cool the motors properly. Breaking in the motors and cooling them well will prevent their melting. I learned this the hard way during the test phase. Knowing a few motors would fail during testing, we purchased extras to ensure an adequate supply.

My team chose motors that were easy to modify and that were designed to use a stationary axle. Working from the outside in, we attached the motor casing solidly to the chassis. The armature of the motor is mounted on a hollow shaft, or torque tube, that turns on the motor's stationary axle. Attached to this torque tube is a plate that transmits the motor's power to the gearbox input. The motors use a three-gear reduction system that gives a motor-to-wheel ratio of 110 to 1, greatly increasing the torque delivered to the drive wheels—no chains or belts here! The wheels are also designed to fit on a stationary axle and have bearings so all that was needed was to drill holes through the wheels and the drive plate of the gearbox and bolt them together. If you look at how a wheel is arranged on the axle (Figure 14-3), you can see a washer over the axle with a cotter pin securing the wheel in place. The point where the wheel is bolted to the drive plate of the gearbox is also visible. The wheels are decent sized with deep treads for added traction.

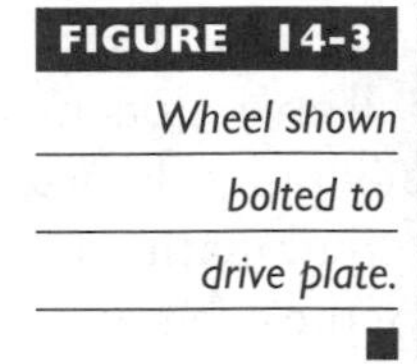

FIGURE 14-3 *Wheel shown bolted to drive plate.*

The ammo box was destined to receive all the electronics. It took time to determine the arrangement of all the items inside the limited space. Inside the ammo case are the Vantec speed controller, the radio and its battery pack, two Futaba servos driving standard microswitches to switch the weapons systems, and three relays for the weapons systems—two for the arm mechanism and one for the saw motors. An evening of careful planning and trial-and-error assembly found the configuration that worked best. They all fit, albeit in a densely packed configuration.

Between the axle and the rear of the box are the batteries—two high-rate-discharge Yuasa MPH1-12 batteries that can supply 100 amps or more. They were chosen for their high discharge rate, something many gel cell batteries are incapable of, as it was needed to run the saw motors. Quality varies widely among gel cell manufacturers. The Yuasas ran $26 each—not inexpensive, but battery quality is an area where you can't afford to scrimp. Everything was fitted in and tested; the robot was driven around as a mechanical ammo box to be certain the design worked. The axle through the center of the robot, the gearbox, and the wheels help to brace the batteries in place. The motors are held in place by hose clamps over PVC pipe. It may not have looked pretty, but the parts were inexpensive, effective, and easily obtainable. Most of this robot's parts were obtained from scrap yards, hardware stores, scavenged materials, and a surplus catalog or two. Although work on the basic drive box was completed and initial testing showed the design to be a solid one, there was still much more work to be done.

Step 4: Creating Weapons and Armor

Chew Toy's weapon is a rotary spinning mass. The design is simple: two milling saws on each side of the prow are driven by a chain sprocket mechanism. As you can see in Figure 14-1, a large chain sprocket was used; it takes chain reduction out of the system and in doing so transmits the maximum amount of torque. These saws were designed for low speed and high torque. The idea is to pull an opponent into the "mouth" area of the robot to "chew" on it and send many parts flying. *Chew Toy*'s weapons system and armor were constructed from a combination of surplus catalog goodies and scavenged parts. The prow (the arm) of the robot was fabricated of steel obtained from a rack-mounted computer system. A 1/4-inch aluminum plate, part of the support structure for the weapons systems, came out of a dumpster. Cut into the desired shape with a jigsaw, it was honed with a Dremel tool and welded to the main support structure (the ammo box).

The weapon support structure fits neatly between the two fan outlets. Attached to the front part of its underside is an inexpensive small furniture castor. When the prow is down, that foremost wheel is not visible, but in Figure 14-4 it can be clearly seen. It's bolted to the front of the machine and supports the two pillow boxes that hold the saw bearings.

The bearings used for the weapons system were designed for misalignment—the bearings are sitting in a rubber gasket, which can move around slightly. This way, we didn't have to be precise on alignment. We just stuck the bearings in there, slid the axle through them, and clamped it down to get a system that is reasonably strong and spins. The central theme of *Chew Toy* was building a robot cheaply and easily, and the *KISS (Keep It Simple Stupid)* weapons array helped us continue that theme.

The large rod you see mounted to the front of the robot in Figure 14-5 is the saw axle. The saws are milling tools that we picked up at a metal scrap yard. Berg sprockets and chains were used to construct the saw's drive. The shoulder on the sprocket was cut down with a lathe and the sprocket bolted to the saw, making one unit. Although combining the saw mechanism in this way made the unit heavier, it was desirable in this case because of the increased spinning momentum

FIGURE 14-4 *Front view with arm up; note the motor and chain drive for saws and caster under pillow boxes.*

FIGURE 14-5 *Front view with arm down showing 2-by-4 and nails.*

it offers. The design allows *Chew Toy*'s saws to strike an opponent and keep on spinning and doing damage instead of stopping abruptly.

The motors that power the saws are mounted on a support structure welded to the front of the robot. The saw motors also run on 24 volts instead of the recommended 12. When in battle, these motors get only intermittent use; thus, the reduction in life span from this hard usage should not pose a problem. If one motor should blow out during a competition, the second one will be able to power the saws. These motors were found through a surplus supply catalog. Although I had no specs on their design, and I knew nothing about who made them, they were inexpensive and testing proved they had the necessary torque and would work well for their intended purpose.

The arm was originally intended to right the robot if it were turned on its back, but then it became a weapon in its own right. The arm is made out of angle iron bought from a local hardware store. Welded onto the ammo box and attached to the front is a little bent piece of steel with a hook.

The initial welding on *Chew Toy* was farmed out, and one of my teammates who had welding equipment (and skill at using it) did later welds. The original arm conception has evolved considerably, and the appearance changed as we continued our improvisation. Things were added as the inspiration hit us. The old motherboard and perforated metal screening were attached as armor. The 2-by-4 with nails was incorporated to make sure the robot could right itself should it be flipped. The nails, and the reach they added, were necessary to accomplish the flipping. When the arm is lowered (Figure 14-5), the nailed 2-by-4 gives the robot additional protection. More of the armor in the form of circuit boards, perforated metal, and another 2-by-4 to protect the robot's rear was added when construction was nearing completion.

The arm actuators seen in Figure 14-6 were donated by Motion Systems. These actuators have 3 inches of throw, which gives us about 70 degrees of travel, enough to flip the robot upright. When the robot is flipped, it rests on the nails, and the process of raising the arm rolls the robot back onto its wheels.

When the arm is lowered, the hook part fits neatly between the saw blades, allowing the saws to do their work. Raising the arm provides 70 pounds of lifting force, which should be enough to pick up an opponent and allow the saws to cut away at its underside. The lifting arm can also be used as an "upper jaw." The pressing force of the motors of this upper jaw can trap an opponent between it and the "lower jaw" prow. Saw-like teeth welded to the underside of the arm and the top of the prow makes a "mouth," making *Chew Toy* live up to his name.

When we designed our armor, our focus was on our weight class and our potential opponents. We were influenced by other robot designs we saw online. One robot, *The Missing Link,* had a huge and nasty circular cutoff wheel on its front. These wheels, which were designed to cut through steel, could cut through *Chew Toy*'s frame without slowing. However, cutoff wheels bog down and get jammed when cutting through wood. So we attached thick pine 2-by-4s as part of *Chew Toy*'s armor. This would slow *The Missing Link* and any other robot using weapons designed to cut steel. Many builders don't perceive wood to be good armor. Actually, a thick piece of pine is hard to cut through, especially if it is attached to a robot that is fighting back. Robots mounting large-toothed, wood-cutting blades have a good chance against *Chew Toy*'s pine armor (though, if I can help it, he'll never stand still long enough to give them the chance!). The nails attached to the pine 2-by-4 provide additional protection. Saws trying to cut through the wood may hit the nails, causing them to jam, break, or lose teeth. The combination of nails in wood makes cheap, yet effective, armor—though, granted, it's not pretty.

FIGURE 14-6 *Top view showing actuation arm.*

Final Words

Despite his appearance, *Chew Toy* is well engineered. Making a robot from available, inexpensive parts does not mean the design is poor. In designing *Chew Toy*, attention was paid to the overall layout, to the center of gravity, and to giving the robot the ability to right itself from any orientation. The latter feature was a major design challenge. Paying heed to how the parts fit together, the location of the center of gravity, and the envelope of the robot in order for it to roll properly and right itself was an intricate problem.

We took care not to repeat the mistakes of others. No blob with wheels that had everything encased in a box for us! We wanted the components to fit together intelligently for maximum utilization. The design allows its separate parts to perform a secondary function, such as the axle being an internal support for the batteries and the motors adding additional support to the robot's overall structure. This result came from playing around with all the parts, trying different configurations, and finding the best way to fit it all together.

Conceptually, we focused on three things: good overall design for maximum offensive and defensive capabilities, ease of driving for effective movement in the arena, and the crowd-pleasing effect of *Chew Toy*. The overall design is solid. It overcomes the majority of ways robots lose in combat. Most robots don't lose as a result of bad armor; instead, they lose because they are flipped over, something internal or external breaks on impact, or they become hung up on something due to insufficient ground clearance.

Chew Toy's electronics are well cushioned against impact damage within the ammo box that has additional welded steel. *Chew Toy*'s arm can be used to right it should it be flipped, and its weapons should prove effective in combat. Although an opponent could strike the exposed wheels, they are large and provide in excess of an inch of ground clearance, which is enough to drive over grass with no difficulty. When in action, *Chew Toy* is hard to stop—it is still fully mobile and has a chance to break free even if it runs over a wedge or a lifter gets underneath. Its weapons are designed to rip chunks off other robots and drive over the debris without slowing. In the initial drive tests, grass and lawn hazards posed no problems.

Two items are very important in robotic combat: driving ability and pleasing the crowd. Battles have been lost due to poor control of a robot's movement in the arena. For this reason, you should test drive your robot as much as possible before competing and discover early how to compensate for odd quirks.

Pleasing the crowd is also important; if two robots are tied in a match, the vote of the crowd decides who wins. A robot with a good design, cool weapons that are entertaining to see in action, and the ability to show its abilities best are the ultimate objectives for pleasing a crowd. Some of the weapons that get the most cheers don't really do much real damage, but they impress the crowd, which is part of what this sport is all about.

Pete Miles—Building *Live Wires*

For a long time, I daydreamed about building the perfect combat robot. Since I'd watched robot competitions on TV religiously and had built several winning mini sumo bots, I figured I could easily build a combat machine. When I read an invitation from The Learning Channel (TLC) on the Seattle Robotics Society e-mail service asking for contestants for the premiere season of *Robotica*, I decided to build my first real combat warrior.

I gathered together some friends to help build the bot and submitted an application for the show. A week later, I got an e-mail back from TLC saying I'd been accepted to enter my robot into their show—however, I had only six weeks to construct my machine. At that point, all my friends backed out except Dave Owens. Although this meant Dave and I had a much smaller team that we'd originally expected, we decided to move forward with our project anyway.

Step 1: Making the Sketch

The first thing we did was go into the conference room at my office, break out the dry markers, and start sketching out ideas about what our robot should look like and how it would adhere to the contest rules. Before long, Dave and I realized we had two different ideas about how to build our robot. I wanted to focus on basic defensive skills and general performance characteristics, and Dave wanted to focus on weapon systems—buzz saws, pokey spikes, flipping arms, and high-kinetic-energy spinning disks to rip apart the opponents. My goal was to have a robot that had a solid body, wouldn't get stuck on anything, could run upside down, and could be fixed quickly. To my view, there was no point in having a weapon since you didn't get points for damaging opponents. *Robotica* is all about speed, agility, and strength; it's not a kill-your-opponent event.

During the first few days of the design process, Dave and I went back and forth on offensive vs. defensive capabilities. Eventually, we decided to postpone the weapons discussion until we could get the basic body designed.

Once we decided to settle down and just start building, we laid out the general goals for our bot. We wanted a robot that would be fast, strong, four-wheel drive, highly maneuverable, able to run upside down when flipped on its back, and able to be fixed quickly. The driving factor behind these requirements was *Robotica*'s figure-8 race, which would require that our machine meet all these criteria if it were to compete effectively. And, of course, we had one final agreed-upon requirement: we didn't want to spend a lot of money.

Step 2: Securing the Motors

These goals were a pretty good start, but none of the details had been worked out. For instance, when we said we wanted a *fast* robot, we really didn't know what *fast* meant in this context. Since the robot's speed is a function of the motor speed, we decided our first step should be getting the motors; we could design the robot around them.

We decided to use cordless drill motors in our bot. My friend Larry Barello, a FIRST competition mentor, recommended that we use Bosch or Dewalt drill motors. After some searching, we found a Bosch 18-volt cordless drill that had a stall torque of 430 in.-lb., and a no-load speed of 500 RPM. Some quick calculations showed that with 8-inch diameter wheels, our robot would top out at 12 MPH, which is pretty quick for a robot.

After spending $400 on the first two drills, we decided to get the rest of them from a local Bosch repair facility. We now had the replacement part numbers and all we needed was the electric motors and gearboxes. Why spend the extra money on the case, batteries, and the drill body and chuck since we were not using them?

Step 3: Adding Wheels

Next we had to figure out how to drive the wheels. I originally wanted to use timing belts to drive the wheels, but I decided to go with regular chains and sprockets because they were cheaper. From the Grainger catalog, we could see that a No. 40 chain had a maximum load rating of 1,000 pounds. With a service factor of 2 for intermittent and shock loading, this would equate to a load rating of 500 pounds. Since this was greater than the stall torque of the motors, we decided that this chain should work fine.

At this point, we ordered a whole mess of parts from Grainger: sprockets, chains, spherical pillow blocks for the four axles, and four flange mount pillow blocks for the motor mounts.

Another friend, Robert Niblock, told me about a local machine shop that builds custom racing go-karts and suggested that they might sell me some used parts. So Dave and I ran over to the machine shop to see what we could haggle over. Ken Frankel showed us his high-speed, state-of-the-art racing go-karts. They looked just like miniature Lemans or Indy racing cars. We talked for a few hours, and he sold us some of his used aluminum wheels and a dozen used racing tires, along with a set of four mounting hubs. The used racing tires were great because they were already gummed up from racing, so they provided lots of extra traction.

Step 4: Adding Motor Housings and Controllers

The next step was to build the motor housings. Cordless drill motors are not designed to be used as regular motors, so there really isn't any good mounting points on the motor and gearbox. I used a pair of calipers and reverse engineered the exterior geometry of the motors and gearboxes. Figure 14-7 shows a layout of the components used to make the mounts for the gearbox, and Figure 14-8 shows a photograph of the assembled gearboxes. The parts were machined using an abrasive waterjet. Yup, water and sand was used to cut all these metal parts. When water is pressurized to 55,000 psi and a little sand is added to it, it can cut any material known to man. One of the nice things about an abrasive waterjet is that it can cut some rather intricate features without difficulty.

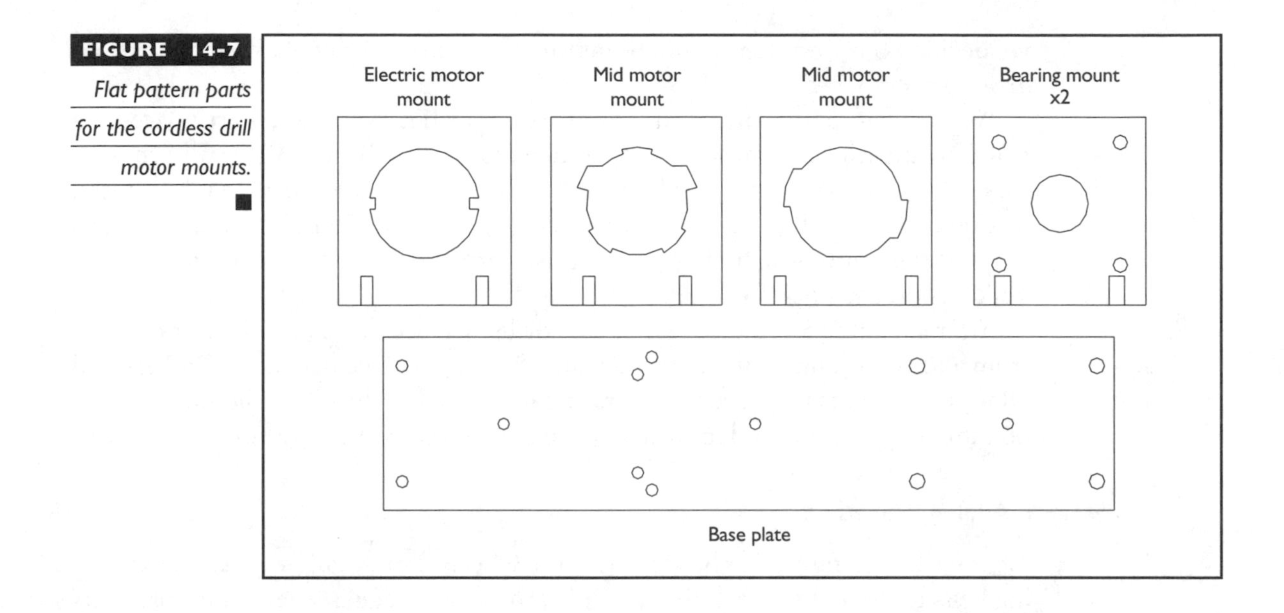

FIGURE 14-7 *Flat pattern parts for the cordless drill motor mounts.*

Originally, we wanted to use Vantec motor controllers for the robot. When I called Vantec to order one of their RDFR motor controllers, I was informed that it would take four to six weeks to arrive. Obviously, we couldn't wait that long, so I started looking around for other motor controllers. Larry Barello suggested that I look at the Victor 883 motor controllers since he has used them without any trouble with cordless drill motors in the FIRST robots he helped a lot of kids build. I checked out their spec sheets and determined that the 60 continuous amp rating

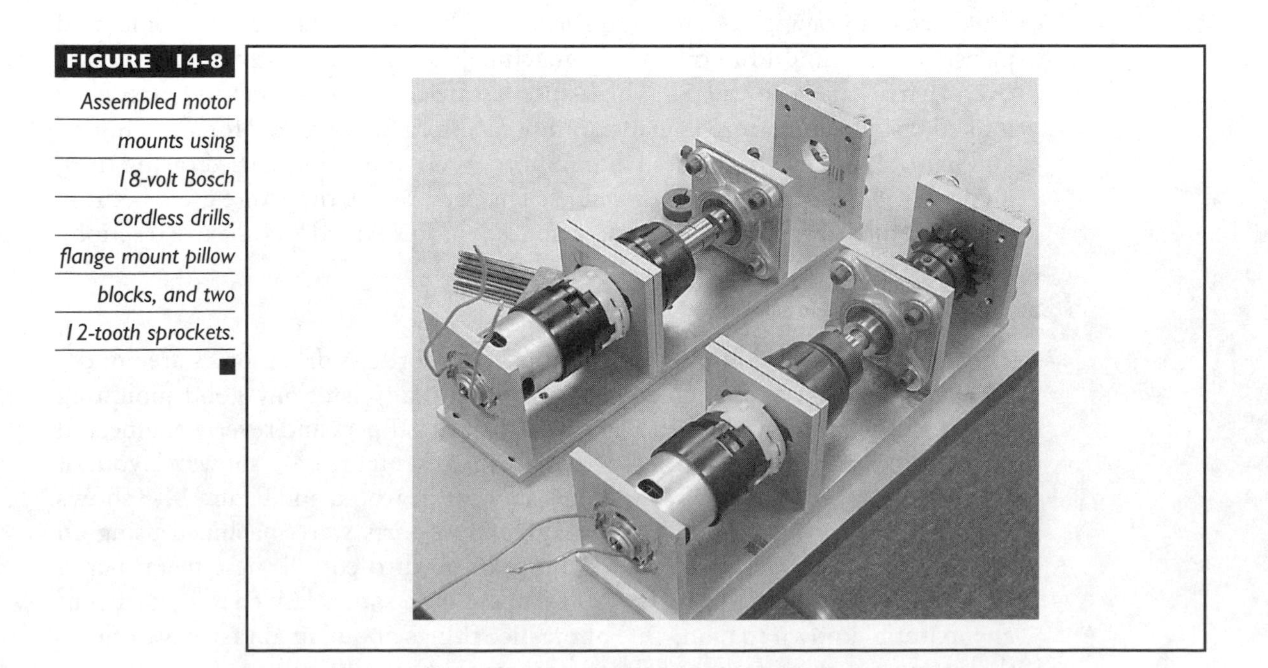

FIGURE 14-8 *Assembled motor mounts using 18-volt Bosch cordless drills, flange mount pillow blocks, and two 12-tooth sprockets.*

should be sufficient for our robot's motors. The internal resistance of the motors was measured and the calculated stall current draw would be about 110 amps. I estimated that the normal running current would be about half of the stall current (just a guess); so the Victor 883 should work, as long as I didn't push the stall current rating. I ordered three of the Victor 883's from IFI Robotics. (I needed only two of them, but I ordered a third for a spare in case I burned one out.)

Instead of having one set of batteries power both motors, we decided to have a set of batteries to power each motor. We used three 6-volt 7.2Ahr Panasonic sealed lead acid batteries to power each motor. We chose these batteries because they fit inside a 4-inch cavity requirement of our robot. They were not selected based on their capacity. Because these batteries would be used up in each match, and they were not the fast-charging type, we also purchased three battery chargers—and a total of 24 batteries for the contest. We planned on swapping out six batteries at a time between matches and recharging the batteries later. (Special note here: what ever you do, don't let your spouse find out that you spent $98 for priority shipping, and you ended up not needing the batteries the next day.)

For the radio, I went against what all the experts say. I used a regular FM radio control system. I was able to get a ground legal 75-MHz, four-channel radio from Tower Hobbies (*www.towerhobbies.com*—a great place to get R/C equipment) for $140. I didn't want to spend a lot of money for a 72-MHz PCM radio, since that was outside our budget. For servo mixing, I built a custom microcontroller-based mixing system that had a built-in failsafe feature. I didn't think I would see too much radio interference, and the mixing circuit would protect the robot with its internal failsafe feature. I also ordered two additional sets of frequency crystals in case of a radio-frequency conflict at the event.

Step 5: Layout and Modeling

The rules from the contest said that the robot must fit inside a 48-by-48–inch box. This placed a maximum geometry constraint for the robot. We decided that we wanted the robot to fit inside a 36-by-36–inch box. We laid out how the motors, gears, and wheels would look on a piece of wood (see Figure 14-9). Since the length of the motors and gearboxes was 11 inches, we couldn't directly attach them to the wheel axles. We decided to use a two-motor approach to drive all four wheels.

Because one of the goals was to make the robot a rapid maintenance design, I designed the robot to be symmetrical about the center of the robot. This way, one part could be used in four different locations in the robot. After the plywood board layout was completed, the first set of aluminum structural parts were cut out with an abrasive waterjet. A set of 1-inch-thick aluminum standoffs were cut for the pillow blocks so that the center line of the wheel axles would be at the same height of the motor mount axles. The base plate was made out of a 1/4-inch-thick piece of 1100 series aluminum. (Whatever you do, don't use 1100 series aluminum in your robots. This is one of the softest forms of aluminum you can get. I used it because I already had a big sheet of it, and I didn't want to spend any more money on the robot.) Figure 14-10 shows the next step of the fabrication process.

FIGURE 14-9 *Laying out the components on a plywood board to get a visual feel of how all the parts fit together.*

At this point, we were about four weeks into the project. With our regular jobs, we could work only for a few hours a night and on weekends. (During this time, my wife became the "Robot Widow." The only time she saw me was when I came home and went to bed and when I woke up and went to work.) We were using a pseudo-design and build-as-you-go approach with this robot. I used AutoCAD to

FIGURE 14-10 *Aluminum base frame with wheels and bearings mounted, and the motor mounts showing where they will be placed.*

design all the parts, and Dave did most of the machining work using an abrasive waterjet, drills, and mills. Once I had a new part designed, I would give the design to Dave and he would construct it. I did most of the lathe work and tapped a lot of holes. We would make a part, put it on the robot, and then update the overall layout drawings. I also used the layout drawings to gauge the size of the parts and where they should go. We didn't have the time to completely design all the parts up front and then start fabricating. Because of this approach, some parts required us to take a hacksaw to them to get them to fit together.

Step 6: Scrambling

With only two weeks before the actual contest, two members of the TLC *Robotica* team came out to shoot some film footage of the building of our robot. Up until the day they came, we scrambled to get our machine put together. Around midnight the night before the *Robotica* team arrived, we fired up the robot for the first time. It went forward about 3 feet and then reversed its path just fine—*then it died*. We were, of course, concerned about this little setback. When looking at the motors, we discovered that one of my custom-machined shaft adapters failed.

One of the primary goals of this robot was to be able to rapidly fix parts that break. So we didn't want a permanent adapter attached to the threaded output shaft of the drill motor and the sprocket shaft. What I made was an adapter that was pinned to the sprocket shaft. The other end of the adapter was threaded, and then a slot was cut down the length of the threads. The adapter was screwed onto the motor shaft, and a split collar was placed on the adapter and tightened down. I figured that this should work. Dave didn't think it would.

When we took it apart, we discovered that it did not unscrew itself off the shaft. Instead, all the threads inside the 304 stainless steel adapter were sheared off. Although my idea of making the adapter worked, ultimately the material failed.

Since TLC was coming the next morning, we put on the spare adapter and parked the robot under a table at our office. Since we left everything put together, and only disconnected the wires from the batteries, we put a sign on the robot that said "Do Not Touch—Live Wires" (the batteries were exposed and we didn't want anyone touching the robot). The next morning, everyone at work kept asking us why we named the robot *Live Wires*. After a while, I asked why everyone thought the robot was named *Live Wires*? They said the sign on the robot said not to touch *Live Wires*. I told them that wasn't the robot's name, it was a sign warning everyone to avoid getting shocked because the wires were live. Although we didn't intend that to be the name, we now had a moniker for our bot.

Figure 14-11 shows how *Live Wires* looked right before the TLC folks showed up. When they arrived with their camera running, we hand carried the robot outside, set up an empty 55-gallon drum, and put up a few traffic cones for a show. When they asked us to show off the robot, we hooked up the batteries and turned on the transmitter. At this point, I was biting my lip, expecting to see the same type of failure I saw the night before. I pushed the throttle forward, and the robot took

FIGURE 14-11

Live Wires with the motors, chains, wheels, batteries, and motor controllers hooked up prior to its first live testing for TLC.

off like a bat out of Hades. I put the brakes on, and *Live Wires* skidded to a stop. Reverse worked just fine; then I raced it around the lot, and the robot even turned on a dime. It put on a great show for the cameras for the TLC crew. It took out the 55-gallon drum, gave Dave Owens a nice ride, and nimbly ran around the traffic cones that were laid out in a slalom coarse. Our creation couldn't have worked any better.

Step 7: Building the Frame

After the great show *Live Wires* put on, we started building the frame of the robot. We used 4-inch-tall aluminum C-channels for all of the sides of the robot. Figure 14-12 shows the frame structure prior to being bolted onto the robot. You will notice that we had to cut a few notches in the bottom of the channels to account for the pillow blocks. After the frame was built, we made a set of aluminum boxes to hold the batteries in place. The last thing you want are for the batteries to rattle around inside your robot.

After the TLC guys left, we noticed that we had the same motor shaft adapter failure we had the night before. Luckily, it had held together long enough for the video taping. I still wanted a bolt-on type of solution with the threaded motor shaft, so I spent a lot of time looking at different approaches. The *proper* way would be to pin the adapter onto the shaft, but I didn't want to go that route. I decided to use the same type of mounting method the Jacobs chuck uses to attach to the motor shaft. Figure 14-13 shows this new adapter. One side is for using a removable pin

FIGURE 14-12 *The 4-inch-tall aluminum channels used for the frame of* Live Wires.

to attach to the sprocket shaft. The other side has a ½-13 right-hand thread to match the drill shaft. Down the center of the drill shaft is a 6-mm left-hand thread. (I have no idea as to why Bosch uses English and metric threads on the same part.) The left-hand thread prevents the adapter from unscrewing itself. Since the screw that came with the Bosch drill is an odd-shaped screw, I used the Bosch screws instead of trying to find my own.

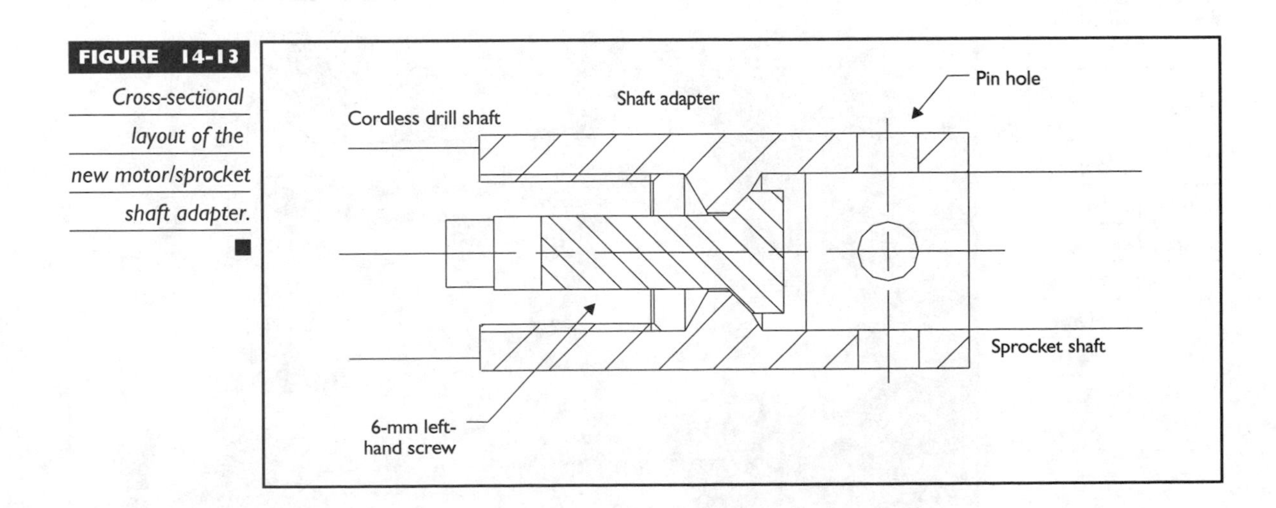

FIGURE 14-13 *Cross-sectional layout of the new motor/sprocket shaft adapter.*

Step 8: Adding a Weapon

Once the core structure was built and the new motor shaft adapters were fabricated, we had only two days left until we had to ship the robot to the TLC studios, and we still didn't have a weapon. We decided to go with a reconfigurable front-end attachment approach with the robot. We went with two types of front ends. The first is an articulated scoop/wedge and a reinforced pointed ram. The ram front end was made out of 1/8-inch steel angle irons. They were welded to form a *T*-cross section with the pointed end facing outwards and a *V* shape to pierce through opponents. The ram front end was designed for the maze event. The geometry was designed to protect the front tires and allow clearance for going over ramps and speed bumps.

The scoop/wedge was made out of 0.090-inch steel, and it was hinged to the robot body with 1/4-inch-thick steel flanges. The wedge front end articulated up and down by gravity. The geometry was designed so that there would be at least a 1/4-inch clearance from the bottom of the scoop and the ground. If *Live Wires* gets flipped on its back, the wedge will rotate to the new position, so the robot will look identical whether it is upside down or right side up.

Figure 14-14 shows a photograph of the robot with the front scoop, side walls, and the aluminum battery boxes next to the motors. In this photo, the back side is removed so that we could drill a hole to allow a finger to get inside the robot to flip a manual power disconnect switch.

Figure 14-15 shows a photograph of the inside of the final robot. Note the symmetry of all of the components inside the robot At the rear of the robot, you will notice the manual disconnect switches. The Victor 883s were mounted to the sides

FIGURE 14-14 Live Wires *showing its aluminum* C-*channel sidewalls, aluminum battery box next to the batteries, and the steel front scoop.* (photo by Kristina Lobb Miles)

FIGURE 14-15 *Internal view of* Live Wires *showing the symmetry of all of its internal components.* (photo by Clare Miles)

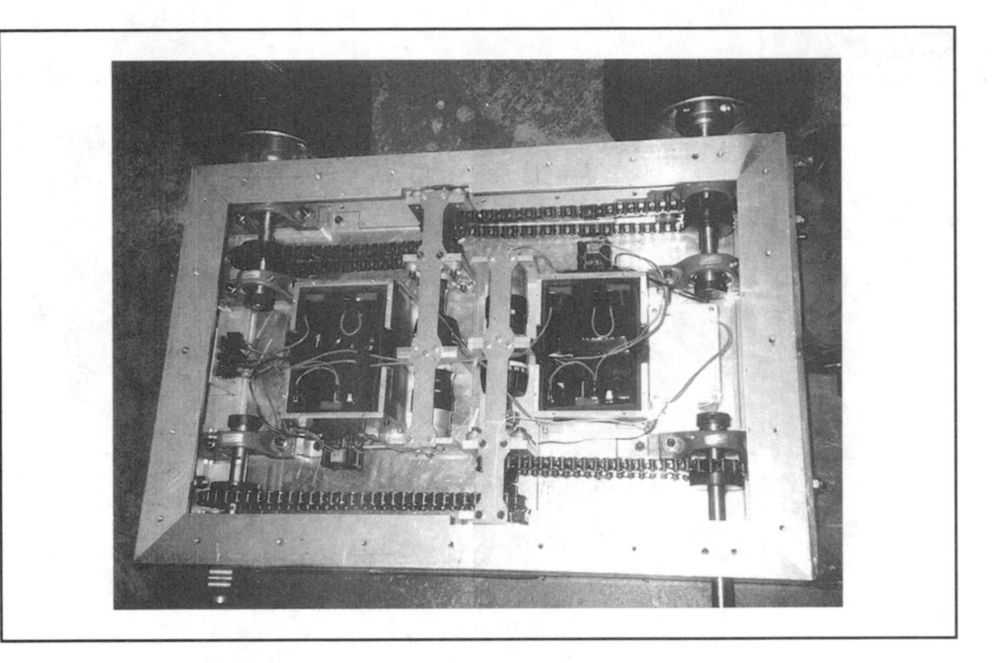

of the battery box. This entire robot was bolted together using button-head screws to allow for easy maintenance.

Finally: The Show

We stripped the robot down and shipped it to the TLC studios in seven different boxes. I decided to send it priority overnight a day before the last drop-dead day, which turned out to be a good thing because we had a major earthquake that brought everything to a standstill two days later. Luckily, the airport reopened the day before I was to leave for California, and I was able to get a flight.

Dave and I arrived at the studio about 4:00 in the afternoon. We spent that evening watching the previous contestants compete, getting interviewed for TV, and reassembling the robot. Finally, we got to bed about 3:00 in the morning. Four hours later, we went back to the studio for the weigh-in. Figure 14-16 shows a picture of the robot at the weigh-in. I was originally targeting the robot to weigh around 100 to 120 pounds, but the robot came in at a svelt 198 pounds. A bit heavier than I thought, but it was still under the 212 pounds max weight limit.

Because *Robotica* invited more robots to the show than they needed (in case of any no-shows), they had to come up with a qualification round to narrow the number of robots to 24. The first part of the qualification round was to go around the figure-8 course twice and get timed for the run. When our turn came, we put the robot in the arena and it took off. Figure 14-17 shows *Live Wires* coming around the first bend in the course. At this point, the robot started having problems. One side of the bot stopped working so it started going around in circles. This was a rather disconcerting experience. After the time expired, we gathered up

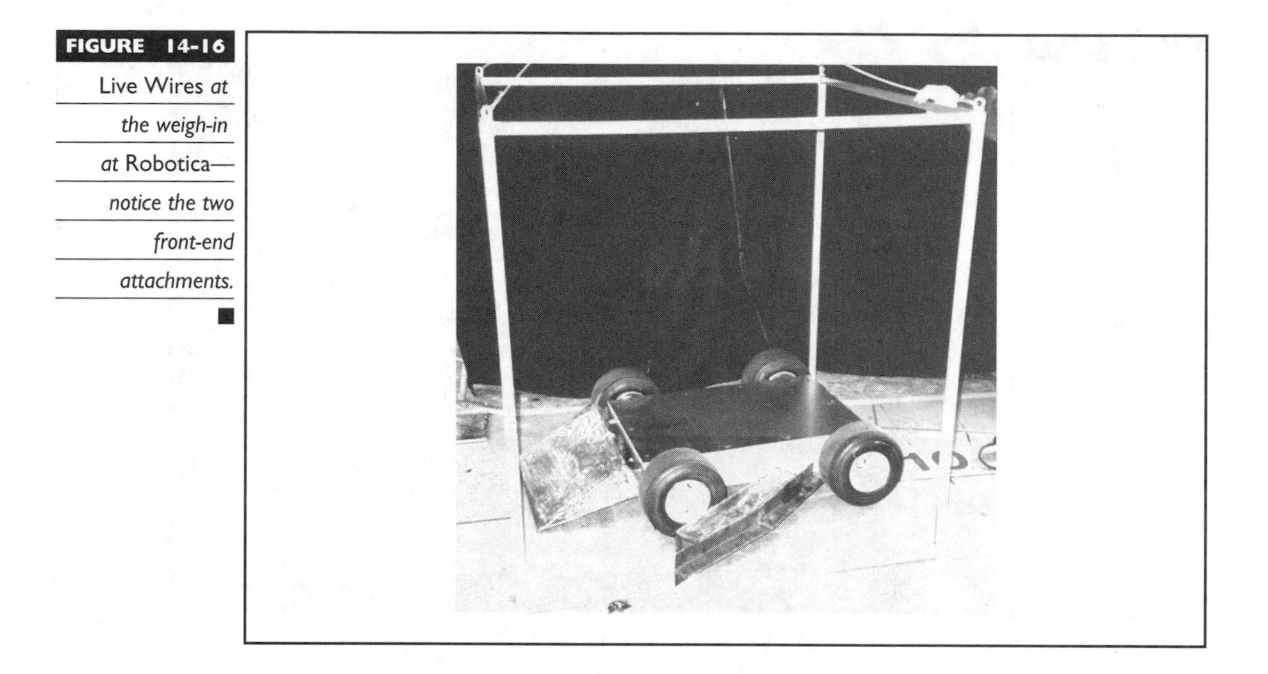

FIGURE 14-16 Live Wires *at the weigh-in at* Robotica—*notice the two front-end attachments.*

the bot and took it apart to figure out why it wasn't working. It seemed to be missing a beat when trying to drive. So we removed the 30-amp re-setable fuses that came with the Victor 883. Then we started to test drive the robot. It seemed to be running much better, and the momentary power losses seemed to go away.

FIGURE 14-17 Live Wires *running through the figure-8 qualification course at the first season of* Robotica.

During the final testing, our bot started the wire meltdown process. We had used 14-gauge wire in our robot—too small for the current going through it. We used 14-gauge wire since we had a lot of it lying around at work. We planned on getting bigger wire but never got around to it. As we drove the robot around, people kept saying, "Someone's robot is burning up," and I would grin at them and say it was mine, telling them my 14-gauge isn't quite enough for the robot. They all laughed. Then the second part of the qualification round came up. We had to drive around the figure-8 course and knock down as many cans as possible.

We knocked down the first set of cans, but the same side of the robot locked up again. This time, there was no motion and lots of the magical gray smoke was escaping out of my robot. It was a beautiful scene—gray smoke, a shuddering robot, and the smell of burnt plastic. Needless to say, we failed to qualify, and *Live Wires* failed to make it on the show.

The post-mortem on the robot showed that the 6-mm, left-hand-threaded screw sheared and caused the drill motor to seize up. After the event, I ran a calculation and discovered that this screw would shear when the torque exceeded 120 in.-lbs. I should have run the calculations before the event. *Never simply assume a part will be strong enough for the competition. Always test first!* I would have discovered this problem if I had tested the robot more before the event, but six weeks really is not a lot of time to build and properly test a robot.

Live Wires didn't do well in its first competition, but it was a lot of fun to build and it was truly heartbreaking for us to watch it fail. My experience with *Live Wires* is similar to many combat robot builders. When you take a lot of shortcuts and don't allow enough time to build the robot properly, you will run into a lot of problems. It is best to plan everything before you start, and allow plenty of time to build and test your robot.

By now, you should have enough information to get started building combat robots. It is a fun and exciting world, so what are you waiting for? Start building your robot!

chapter 15

Afterword

The Future of Robot Combat

The sport of combat robotics has boasted an almost unbelievable level of success in the past few years. Known only to a few as recently as 1996, today, a half-dozen venues host regular competition, including hit TV shows like *BattleBots*. Robots even got a lift from Jay Leno, a self-admitted gearhead who actually competed on *BattleBots* last year. "I like anything that rolls and explodes," Leno has been known to say. But even as the sport has grown into a pop culture phenomenon (you can catch references to fighting robots on TV shows from *Malcolm in the Middle* to *The Daily Show*, and combat robot toys were a popular gift last holiday season), it's not clear where the sport is headed.

Greg Munson, co-creator of *BattleBots*, sees a clear progression in robots he's had in the ring over the show's life. "From the beginning we've seen your basic home-built robots, with simple shapes like wedges that are easy for people to build in their garages. But now we're getting more exotic designs, with high-performance motors and sophisticated pneumatic systems that beef up the robot. Because, after all, the ultimate goal is destruction and killing your opponent."

Overkill's Christian Carlberg agrees, and he sees a relationship between the demands of builders and improvements in technology. "Competitors are challenging suppliers to come up with better batteries, stronger gearboxes, and lighter tires. There has been a lot of work put into reducing the weight of solid rubber tires." And while builders have traditionally relied on industrial suppliers for parts, Christian points out that groups like *Team Delta* are emerging that make electronics specifically for battling robots.

So it's clear that the robots are evolving—both in response to stiffer competition, and to the natural evolution of technology as builders master the basics and begin to innovate. One of the most important ways combat will change, then, is a proliferation of robot designs that build on and differentiate themselves from the basic spinner, wedge, and lifting arm designs that dominate the sport today. "There's always someone," Greg says, "who will come up with something entirely new. A while back, no one had ever seen a robot like *Complete Control* that scoops you, lifts you up, and flips you over backwards."

Bill Nye, the popular TV science guy and unabashed robot fan, says the field is wide open for new and innovative designs. "I have often thought that a parasite robot would be very effective. It would somehow clamp onto the enemy and then maybe drill a hole in it, then mess up his wiring. It would be out of range of the guy's weapons." Bill says such a design borrows from nature—in this case, a germ.

"A germ you can't beat up with your fists. It seems there's an opportunity for a germbot. Of course, the problem is penetrating the other guy's armor."

That's not all. Bill also predicts a rise in reconfigurable robots, which can be customized with specific weapons before each competition. "You want to go after the opponent's weakness. That requires asymmetrical weapons. In football," Bill explains, "everybody uses the same weapons. It's giant guys smashing into giant guys. You're not allowed to use glue or bombs. But imagine if you could show up with a bunch of spikes! Football would be a very different kind of game; you'd have asymmetrical combat."

And, while remotely operated combat is popular today, robots have a lot of potential for more automated competition. Competitions like *RoboCup* require the robots to use a combination of sensors and artificial intelligence (AI)–style programming to behave in competition; there's no human intervention allowed.

Will the style of competition that emphasizes robotic decision-making catch on? Most builders don't think so, citing the excitement that comes from watching a human being drive the robot. "You have to ask how entertaining will it be to have a robot go out there and try to find another robot on its own," muses *Diesector*'s Donald Hutson. "With me driving, you actually have someone to boo if I fail." *Team Blendo*'s Jamie Hyneman agrees. "The human aspect of the competition is important. You get to see this nervous guy handling the joystick, and you watch the elation on his face when it's going well and the look of dejection when he loses. It's human. Without that, it's not as interesting."

But, while builders covet their joysticks, others are intrigued by AI. Says Bill Nye, "I think it's more interesting, in a sense, when their strategy is all in their programming." Indeed, he says that this kind of technology is important as real-world industrial and scientific robots get more freedom to operate as an adjunct to humans. "It's like sending rovers to Mars. Mars is so far away that there are minutes between when we send commands and the rover reacts. It needs to be able to make decisions on its own."

What of the sport itself? Some worry that it's a fad—as so many pop-culture phenomenons turn out to be—that will fade from the public consciousness and become an obscure hobby for tinkerers. Christian Carlberg admits that the jury is still out: "This sport might be a fad that passes over the next few years, or it might grow into something as large as televised football." *Nightmare*'s Jim Smentowski explains why shows like *BattleBots* stand a good chance of becoming a staple of American life. "This is the only sport that kids and their parents in their living room, watching TV, can sit there and say 'Let's get involved in that,' and they can! Next thing you know, they'll be on the next season of *BattleBots*!"

Greg Munson is doing something about it. *BattleBots IQ* is an academic program the show has created with the help of educators and academic roboticists for high school students. "We're builders ourselves," he explains, "and so we said, 'wouldn't it be great to learn about robots in school?' *BattleBots IQ* will be an elective that students can take to apply all the math and physics and science they learn in class to build a real robot." Greg hopes that *BattleBots IQ* will be more the sum of its parts. Not only will this program encourage kids to take an interest in science and engineering, but it will teach a wide variety of skills and lead to a new generation of robot builders.

Whatever the fate of sport robotics, one thing is certain. Robots are increasingly a part of the high-tech landscape, and progress is accelerating toward the day when robots are routinely a part of everyday life. *Deadblow*'s Grant Imahara reflects: "In general, I think that robots have certainly proven themselves in industry. The next step is to bring robots into the private sector, and into people's homes. You're seeing it already with all the robotic pets on the market. In Japan, there is a movement called the Humanoid Project, which is concentrating on developing robots for the aging Japanese population as personal service robots. I think that we will begin to see this more and more as the technology becomes more advanced and more affordable."

Robots in the home? Horrors! That could be the response of a vocal minority—the same minority that abhors cell phones and turns its collective nose up at answering machines. Bill Nye isn't concerned. "There will always be Luddites who don't want to see technology evolve. The problem with that philosophy is that there's no way to draw a line. All those people—to a one—happily use lights and electricity, which makes their arguments very arbitrary. Trying to stop moving forward is just not a successful strategy; it's against human nature. It's human nature to innovate. Those that don't become food."

But perhaps no one summarizes the future better than *Biohazard*'s Carlo Bertocchini: "My guess is that in 30 years we will have robotic servants that are just as intelligent as humans. I hope they forgive us for bashing the hell out of their progenitors."

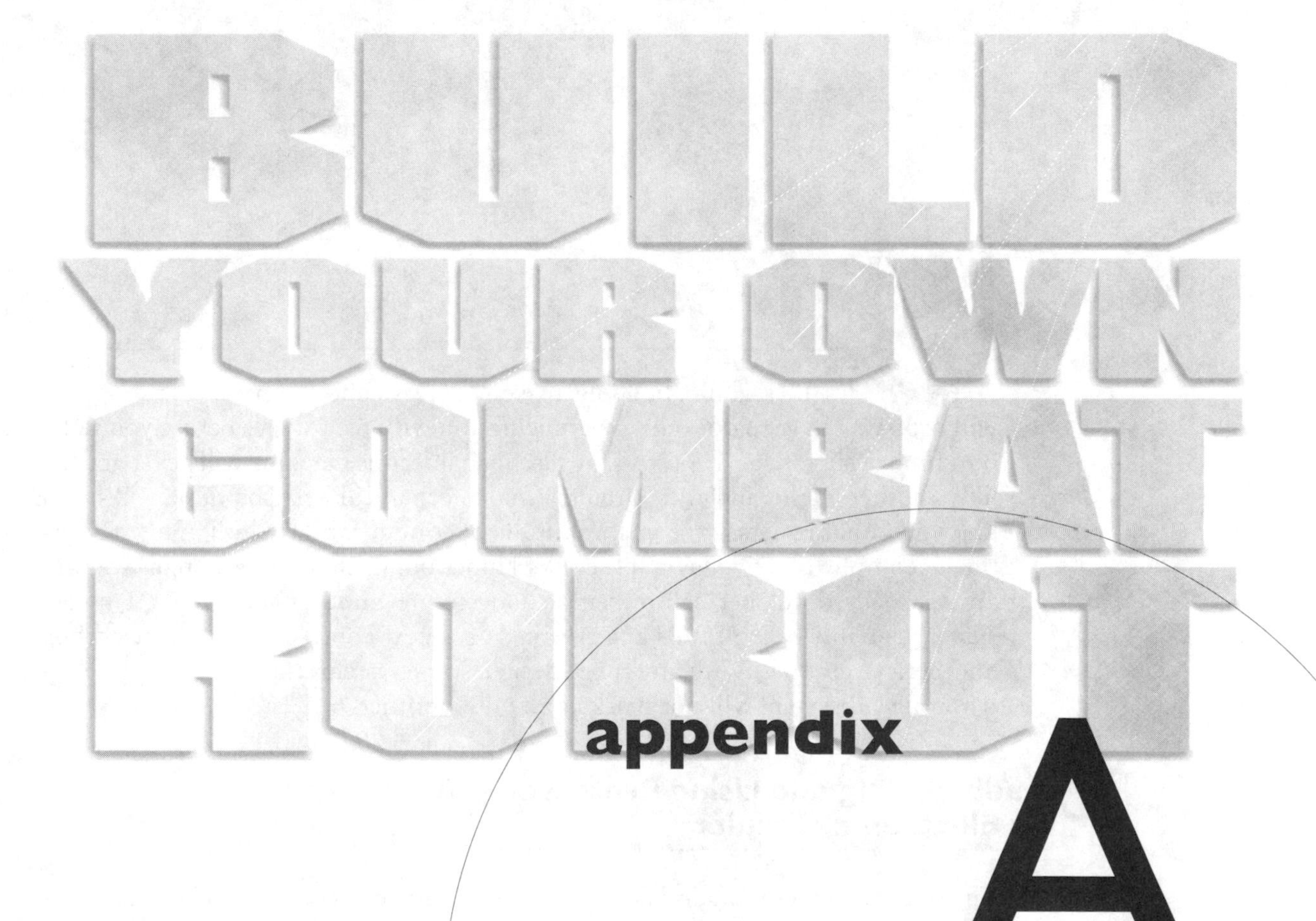

appendix A

Prototyping Electronics

UNLESS you're building an exact copy of another person's robot, you will probably have to do some experimentation with prototypes before you settle on a final design. Even if you're copying a machine, one machine will never act just like another— this includes circuitry and mechanical systems design. As you know, combat robots come in a multitude of configurations, with many circuits that accomplish every imaginable control function. Each of these systems was analyzed by its developer before being connected to another subsystem. This designer probably bench-tested and tweaked each new configuration before adding another system to it. With any robot design, some systems that perform perfectly with one subsystem will not work at all with another.

Breadboarding and Using Prototyping Boards for Electronic Circuits

The term *breadboarding* implies interconnecting a series of components in a temporary fashion to determine whether they will work together as designed. A breadboard of an electronic circuit, for example, can be built on a prototyping boards. These boards can include up to several thousand holes in which to insert standard electronic components to develop a circuit. The components are then interconnected by inserting short lengths of #22-gauge solid conductor wires in the adjacent holes.

Such prototyping boards are useful in proving out a circuit you may have seen in a magazine article, or for proving out one you designed in your head. All the major electronic supply houses carry a variety of prototyping boards, and some even contain built-in power supplies, logic indicators, and signal generators.

The use of prototyping boards can help you tweak the circuit for your particular application. Varying resistors and other components can help you narrow down a circuit to one that has the best characteristics for whatever you intend to interface it with.

At this point, you may want to start drawing out a printed circuit board (PCB) pattern by hand and *etch*, or carve with an etching solution, your own board. Computer software, such as the lower end Eagle and others, are available to lay out simple boards. PCB houses can be found on the Internet that will take Spice and Gerber (electronic circuit software) files directly as a attachment and deliver one or more etched and through-hole plated boards back to you within just a few days.

A first-time experimenter can easily make PCBs at home with simple kits obtainable from Radio Shack and many of the suppliers listed in Appendix B. "Stick-on" patterns are available for integrated circuits and transistors, as well as other components and wiring traces. These patterns are applied to a clear sheet of plastic and used as a positive mask to sensitize a treated board that is then etched with an etching solution. Similar results can use circuits from magazines and transfer them to a usable positive mask. Many computer printers and copiers can also print a mask on a sheet of plastic for conversion to a PCB.

Wire-Wrapping Prototyping

Another popular type of electronic circuit development and prototyping is wire wrapping. Just about the reverse of the prototyping boards with many small holes, wire wrapping involves the use of many headers with two rows of long, gold-plated square pins. The thin pins of the headers are inserted in a holder board and a wire-wrapping tool wraps a stripped, thin wire around a selected pin. Manual and battery-powered wire-wrapping tools are available in many electronic tool catalogs. You then cut off a desired length, strip the other end of the wire, and wrap it around another pin on another header. The pins can hold multiple wrapped wires. One bad feature of these types of boards, however, is the long pins that protrude out the back of the holder board; these can easily be bent and short to each other. This type of prototyping is best when using a series of dual in-line, pin-integrated circuits (DIP ICs).

Soldering for Robots

Soldering for robots is a bit different from the type you might use to assemble small electronic kits, especially the larger *BattleBot* types of machines. If you have experience with building kits and various experimental projects using printed circuit boards, you'll probably be pretty good at doing some of the larger and more difficult solder joints in a robot. If it's all new to you, don't despair; it's fairly easy to learn.

You can probably get by using a simple $5 soldering iron from Radio Shack for the majority of your work, but you'll soon want to buy equipment that is a bit more versatile. A soldering station made by Weller or another manufacturer allows you to vary the heat control to suit the needs of a particular job, and then hold it at that temperature. These can cost anywhere from $50 to hundreds of dollars.

Another useful soldering tool is the soldering gun. A dual-wattage gun can allow you to solder those large, high-power, cable terminal lugs, yet allow you to use lower power for circuitry. The use of a small vise also helps to hold a circuit board or ungainly wire still while you're soldering.

Three things to remember in soldering:

- **Clean** Both surfaces you intend to solder must be clean.
- **Shiny** The soldering surfaces should be shiny before soldering.

- **Not too hot/ too cold** The solder temperature must be hot enough to create a solder that will hold. You also need to protect the components you are soldering from excessive heat or you can ruin them.

If you're new at soldering, practice soldering with scrap components, wire, and metal before committing to a particular project.

Soldering Printed Circuit Boards

To gain soldering skills on printed circuit boards, you might want to find a piece of electronic equipment that's been trashed and rip it apart to solder and unsolder the parts from the circuit boards until you feel competent. You may find that unsoldering is more difficult than soldering, yet this practice will help you in learning how to apply only just enough of the hot iron's tip to the board without damaging it. Practice is really the best teacher and you don't have to worry about ruining a one-of-a-kind board.

Before embarking on any type of soldering, you should clean the soldering iron's hot tip with a wet rag or with the small, dampened sponge on a soldering station. It must be clean to do a good job. Most people like to use rosin core 60/40 solder, which is 60 percent tin and 40 percent lead, for electronic work. It is basically a tube of solder containing a tiny bit of rosin in the center. *Never* use acid core solder. Smaller 0.032-inch diameter solder is good for smaller joints; and larger, 0.050-inch and 0.062-inch diameter can be used for larger, non-circuit board joints.

Next, dab a bit of solder on the tip—that is called *tinning* the iron. Holding the soldering iron in one hand, feed a bit of solder from a reel onto the tip. The trick is to melt the solder and quickly apply it to the joint to be soldered. Use only enough to make a "tent" of the solder around the component's wire lead protruding through the circuit board's hole and neatly covering the O-shaped circuit "pad" surrounding the hole. Most soldering iron tips are of the chisel tip variety, and you want to place one of the chisel's faces flat on the surface you intend to solder to transfer the heat as rapidly as possible.

If you did it right, the tent of solder will cover the pad and taper up the wire a bit, and it will be shiny. If the solder forms a ball or is not shiny, you didn't get it hot enough. These are called *cold* joints. For printed circuits, you must be careful not to overheat the traces and cause them to lift off the board. You're working in that narrow area of getting it hot enough for a good joint but not too hot to damage the board. A 15–40 watt soldering iron, or "pencil," works best for printed circuits. Be careful to not cause "solder bridges" from one trace to another.

Another important consideration is protecting the components you are soldering from excessive heat and static electricity. *Integrated circuits (ICs)*, small transistors and diodes, capacitors, small resistors, and other smaller components can be ruined by too much heat. As with the circuit board's traces, you must keep the iron on the board and protruding lead only as long as it takes to make a clean, shiny solder joint. Tiny clip-on heat sinks can route heat away from a component. Soldering one lead of a component, and then waiting until the component cools a bit before soldering another lead, especially on ICs, helps to prevent heat damage.

Soldering Wires

Many wires are used in robot construction to run from control circuitry to motors and sensors. You should always use stranded conductor wire as opposed to solid conductor wires. The many strands allow for better flexibility and greater current carrying capacity. Before soldering the wire, strip a small amount of insulation from the wire—an amount appropriate for the particular connection.

Twist the strands slightly with your fingertips so they are held together in a slight spiral and are not splayed out. Then tin the wires with a bit of solder before soldering the wire onto a connection. To tin the wires, tin the soldering iron's tip with an excess of solder and place the heated ball of solder on the tip against the bare wire strands. As the wire heats, the solder should be sucked into the strands as you add a bit more solder. Again, too much heat can damage the wire, or at least melt or burn the insulation. Practice makes perfect.

Soldering Connectors

Connectors are used in many places on large and small robots. They connect many wires to your control modules, receivers, drivers, sensors, and many other items. Most connectors have a series of pins in rows or circular patterns. The pins usually have small cavities behind them into which wires are soldered or crimped. As always, you don't want to overheat the pins to damage the connector, but each of the pins should be tinned with a bit of solder, as should the wires to be inserted. While applying heat to the back of the pin, slowly insert the tinned wire, taking care not to have one or more strands splay out. Do this with each wire until completed. As always, practice makes perfect.

When soldering the larger wires used in combat robots and other large machines, the use of a soldering gun helps a lot. Tinning large wires that are used in terminal lugs is recommended, as more surface area of the wire is in contact with the barrel of the terminal lug, thus reducing resistance and allowing more current-carrying capacity. A large soldering gun or small torch can be used to solder copper and brass sheeting and tubing, both to each other and to wires.

Caution *Remember, good soldering takes patience for the best results. Be careful not to allow drips of solder on your clothes. For obvious reasons—it burns; but it's also impossible to get it out of some fabrics without burning (melting) it out.*

Crimp-Style Connectors

When working with any high current wiring that is subject to vibration, it is best to use crimp style connectors that screw into components such as electronic speed controllers and batteries, and terminal blocks. Soldered joints will eventually fail if the wires are allowed to vibrate. Make sure you use the right connector size with wire the gauge you are using. It is best to use the connectors that have round holes in them. The "forked"–shaped connectors should be avoided. This is because if the screw loosens a little, the connector could become loose. A loose wire can mean loses of control of your robot or could short out some of your components if it touches something it shouldn't.

Also, you should make it a practice to secure all of your wires so that they don't move around in your robot. Zip-Ties make great tie-downs for wires.

Static Sensitivity

Certain metal oxide semiconductors are used on many robots, and these can be easily damaged by static from handling. Even large, high-power Metal Oxide Semiconductor Field Effect Transistor (MOSFETs) used to make power H-bridges can be zapped by static from hands. Many of the newer ICs and discrete semiconductors have input protection diodes to carry static charges. One way to determine which products can be damaged by static is to see what material it was packaged in when you bought it. If the leads are stuck in a black foam, or the part was stored in a pink plastic package or tube, it's probably static sensitive.

One way to prevent this is to use a static band to ground you to the equipment. This is a band you strap around your wrist; it has a wire and clip you attach to the circuit to be worked on. The use of an anti-static board, such as one of the pink plastic sheets on top of your work bench, works well to eliminate static. Dry and cold days, artificial fiber clothing, and most carpeting cause a lot of static build-up. Use common sense in your robot-building area.

appendix B

Resources and References

A

As promised, following are some resources and references that might come in handy as you consider building and start building your combat robot.

Robot Competition Web Sites

- **All Japan Robot Sumo** *www.fsi.co.jp/sumo-e*
- **BattleBots** *www.battlebots.com*
- **Bot Bash** *www.botbash.com*
- **BotBall** *www.kipr.org*
- **Canada FIRST Robot Games** *www.canadafirst.org*
- **FIRST** *www.usfirst.org*
- **Northwest Robot Sumo** *www.sinerobotics.com/sumo*
- **RoboCup** *www.robocup.org*
- **RoboRama** *www.dprg.org/dprg_contests.html*
- **Robot Sumo** *www.robots.org/events.html*
- **Robot Wars (US)** *www.robotwars.co.uk*
- **Robot Wars (UK)** *www.robotmayhem.com*
- **Robothon** *www.seattlerobotics.org/robothon*
- **Robotica** *tlc.discovery.com/fansites/robotica/robotica.html*
- **Robot Society of Southern CA Competition** *www.dreamdroid.com/talentshow.htm*

Electric Motor Sources

- **Astro Flight, Inc.** *www.astroflight.com*
  (310) 821-6242
  Extremely efficient DC motors

- **RAE Motors** *www.raemotors.com*
  (815) 385 3500
- **Pittman Motors** *www.pittmannet.com/*
  (215) 256-6601
  Good selection of gearmotors
- **MicroMo Motors** *www.micromo.com/*
  (813) 822-2529
  Quality DC motors
- **National Power Chair** *www.npcinc.com*
  (800) 444-3528
  Great source for large DC motors for robots
- **Leeson motors** *www.leeson.com*
  (715) 743-7300
  Electric motors, gearmotors, and drives
- **C & H Sales** *www.candhsales.com*
  (800) 325-9465
  The must-have catalog
- **Herbach Rademan** *www.herbach.com*
  (800) 848-8001
  Marlin P. Jones *www.mpja.com*
  (800) 652 6733
  Good catalog, many items
- **Servo Systems** *www.servosystems.com*
  (800) 922-1103
  Good catalog, lots of motors(1)DC Actuator Vendors
- **Ball Screws and Actuators Co.** *www.ballscrews.com*
  (800) 882-8857
  Ball screws to make your own, plus actuators
- **Duff-Norton Co.** *www.duffnorton.com*
  (800) 477-5002
  Wide variety of linear actuators
- **Motion Industries** *www.motionindustries.com*
  800-526-9328 or 205-956-1122
  Nook Industries, Inc. *www.nookind.com*
  (800) 321-7800
  Wide variety of motion products
- **SKF Specialty Products** *www.skf.com*
  (800) 541-3624
  Good selection of DC actuators

- **Warner Electric** *www.warnerelectric.com*
  (800) 234-3369
  Long-time supplier of linear motion products

Battery Suppliers

- **Hawker** *www.hepi.com*
- **Panasonic** *www.panasonic.com/industrial/battery/industrial/*
- **Sanyo** *www.sanyo.com/industrial/batteries/index.html*
- **Power Sonic** *www.power-sonic.com*
- **Planet Battery** *http://www.planetbattery.com*

Electronic Speed Controller Vendors

- **4QD (UK)** *www.4qd.co.uk*
  Good high-power ESCs for vehicles and robots
- **Duratrax** *www.duratrax.com*
  (217) 398-6300 (Hobbico)
  Maker of R/C car speed controllers. Check with Hobbico or your local hobby store (see HiTec RCD).
- **Futaba** *www.futaba.com*
  (256) 461-7348
  All types of R/C equipment and ESCs.
- **HiTec RCD** *www.hitecrcd.com*
  (858) 748-8440
  Maker of R/C car speed controllers. Check with Hobbico or your local hobby store (see Duratrax).
- **Innovation First (used with "FIRST" robots)** *www.ifirobotics.com*
  (903) 454-1978
  "Stout Victor 883 ESCs."
- **Novak** *www.teamnovak.com*
  Maker of R/C car speed controllers. Check with your local hobby store (see also Traxxas).
- **Tekin** *www.tekin.com*
  Maker of R/C car speed controllers. Check with your local hobby store.

- **Traxxas** *www.traxxas.com*
  (888) 872-9927
  Maker of R/C car speed controllers. Check with your local hobby store (see also Novak).
- **Vantec** *www.vantec.com*
  (800) 882-6832
  Supplier of most speed controllers for combat robots.
- **Robot Power** www.robot-power.com
  (253) 843-2504
  Supplier of the OSMC motor controller.

note *Some of the preceding suppliers request that you go to your local dealer or Web site first.*

Remote Control System Vendors

- **Futaba** *www.futaba-rc.com*
  Futaba radio control systems
- **Hitec RCD** *www.hitecrcd.com*
  Hitec radio control systems
- **Airtronics** *www.airtronics.net*
  Airtronics radio control systems
- **IFI Robotcs** *www.ifirobotics.com*
  Isaac remote control systems
- **Tower Hobbies** *www.towerhobbies.com*
  Wide seelction of radio control systems
- **Best RC** *www.bestrc.com*
  Wide selection of radio control systems
- **Hobby People** *www.hobbypeople.net*
  Wide selection of radio control systems

Mechanical Systems Suppliers

- **PIC Design** *www.pic-design.com*
  (203) 758-8272
  Small gears, belts, pulleys, and clutches

- **Winfred M. Berg** *www.wmberg.com*
  (516) 599-5010
  Precision mechanical components
- **Small Parts, Inc.** *www.smallparts.com*
  (305) 557-8222
  Small supplies, metal stock, and fasteners
- **Grainger** *www.grainger.com*
  (805) 388-7076
  A one stop shopping place for most anything you will need to build a robot
- **McMaster-Carr** *www.mcmastercarr.com*
  (562) 692-5911
  Great catalog—one stop shopping place for most of anything you will need to build a robot
- **Donovan Micro-Tek** *www.dmicrotek.com*
  (805) 584-1893
  Micro stepper motors
- **Gates Rubber Company** *www.gates.com*
  (303) 744-1911
  Machinery and automotive rubber belts

Electronics Suppliers

- **Ace R/C (purchased by Thunder Tiger)**
  (816) 584-7121
  Servo test equipment
- **Advanced Design**
  (602) 544-2390
  "Robix" PC-driven, R/C servo-powered robot arms
- **Allied Electronics** *www.alliedelec.com*
  (800) 433-5700
  Good catalog
- **Digi-Key** *www.Digi-key.com*
  (800) 344-4539
  Reliable parts source

- **Effective Engineering** *www.effecteng.com*
  (619) 450-1024
  R/C animatronic gadgets
- **Scott Edwards Electronics** *www.seetron.com*
  (520) 459-4802
  R/C interfaces
- **Jameco Electronics** *www.jameco.com*
  (800) 831-4242
  Good catalog
- **JDR Microdevices** *www.jdr.com*
  (800) 538-5000
  Test equipment, parts
- **MCM Electronics** *www.mcmelectronics.com*
  (800) 543-4330
  Miscellaneous electronics
- **Mondo-Tronics/The Robot Store** *www.robotstore.com*
  (800) 374-5764
  Miscellaneous hobby robot kits
- **Mouser Electronics** *www.mouser.com*
  (800) 346-6873
  Good catalog
- **Pontech** *www.pontech.com*
  (714) 642 8458
  PC-driven, 4 R/C servo board
- **Radio Shack** *www.radioshack.com*
  (800) 442-7221
- **Ramsey Electronics** *www.ramseyelectronics.com*
  (800) 446-2295
  RF and video kits and equipment
- **Precision Micro Electronics**
  (512) 814-6843
  Accessory switch, elevon, and V-Tail mixers
- **Lynxmotion, Inc.** *www.lynxmotion.com*
  (309) 382-1816
  Robot Sumo parts supplier, also many different types of robot kits, electronics, and parts.

Microcontroller Suppliers

- **Parallax, Inc.** *www.parallaxinc.com*
  (888) 512-1024
  Basic Stamps
- **Acroname, Inc.** *www.acroname.com*
  (720) 564-0373
  BrainStem microcontrollers
- **Netmedia, Inc.** *www.basicx.com*
  (520) 544-4567
  BasicX Microcontrollers
- **Savage Innovations** *www.oopic.com*
  OOPIC microcontrollers
- **Gleason Research** *www.handyboard.com*
  (800) 265–7727
  Handy Board microcontrollers
- **BotBoard** *www.kevinro.com*
  BotBoard microcontrollers
- **Microchip, Inc.** *www.microchip.com*
  (800) 437-2767
  PIC microcontrollers
- **Atmel, Inc.** *www.atmel.com*
  (408) 441-0311
  AVR microcontrollers

Reference Books

- *Applied Robotics,* by Edwin Wise (Prompt Publications, 1999)
  A good overview of basic experimental robotics.
- *The Art of Electronics*, by Paul Horowitz and Winfield Hill (Cambridge University Press, 1989)
  This is the electronics bible; a must-have for anyone building electronic circuits.
- *Build Your Own Robot,* by Karl Lunt (A K Peters, 2000)
  A great all-around reference for advanced small robot building.
- *Mobile Robots,* by Joe Jones and Anita Flynn (A K Peters, 1999)
  A good intermediate book for mobile robot building.

- *Robots, Androids, & Animatrons,* by John Lovine (McGraw-Hill, 1997)
  A good introduction to experimental robotics.
- *The Robot Builder's Bonanza,* by Gordon McComb (McGraw-Hill, 2000)
  A great first book on experimental robotics.
- *Robot Riots,* by Alison Bing and Erin Conley (Barnes & Noble, 2001)
  An overview of battling robots and contests.
- *Electric Motor Handbook,* by Robert Boucher (Astroflight, 2001)
  Excellent book on electric motors.
- *Mechanical Engineering Design,* by Joeseph Shigley (McGraw-Hill, 1988)
  Mechanical engineers' bible for machine design.
- *Machinery's Handbook, 26th Ed.*, by Erik Oberg (Industrial Press, 2000)
  A must-have for all machinists.
- *Fundamentals of Machine Component Design,* by Robert Juvinall and Kurt Marshek (Wiley & Sons, 1999)
  Excellent book on machine design.
- *Programming and Customizing the Pic Microcontroller,* by Myke Predko (McGraw-Hill, 1998)
  Excellent book on using and programming the PIC Microcontroller.
- *Programming and Customizing the Basic Stamp,* by Scott Edwards (McGraw-Hill, 1998)
  Excellent book on using and programming the Basic Stamp.
- *Design of Weldments,* by Omer Blodgett (Lincoln Electric Company, 1993)
  Probably the best book available on weldments.

Robotics Organizations

- **Atlanta Hobby Robot Club (AHRC)** *www.botatlanta.org*
- **Chicago Area Robotics Group** *www.robotroom.com/Chibots*
- **Dallas Personal Robotics Group (DPRG)** *www.dprg.org*
- **Homebrew Robotics Club (San Francisco Bay Area, CA)**
  www.augiedoogie.com/HBRC
- **Phoenix Area Robot Experimenters** *www.parex.org/index.html*
- **Portland Area Robotics Society (PARTS)** *www.portlandrobotics.org*
- **Robotics Society of Southern California (RSSC)**
  www.dreamdroid.com/default200.htm

- **Rockies Robotics Group** *www.rockies-robotics.com*
- **San Diego Robotics Society** *www.sdrobotics.tripod.com*
- **San Francisco Robotics Society (SFRS)** *www.robots.org*
- **Seattle Robotics Society (SRS)** *www.seattlerobotics.org*
- **Triangle Amateur Robotics (Raleigh)** *www.triangleamateurrobotics.org*

Other Robotics Resources

- **Arrick Robotics** *www.robotics.com/robots.html*
- **Robot Books** *www.robotbooks.com/robot-design-tips.htm*
- **Robot Combat** *www.robotcombat.com/tips.html*
- **www.robotcombat.com/tips.htmNuts and Volts Magazine** *www.nutsvolts.com*
- **Robot Science & Technology Magazine** *www.robotmag.com*
- **Open Source Motor Controller (OSMC) project** *www.groups.yahoo.com/group/osmc/*

appendix C

Helpful Formulas

FOLLOWING are formulas that you might find helpful in calculating drive, timing belt, and V-belt centerline distances.

Chain Drive Centerline Distances

When calculating the center distance, the first step is to estimate the center distance between the two sprockets, as shown in Figure C-1. Start with a distance in which you would like the sprockets to be spaced.

The center distance is in terms of number of pitches, so divide the physical distance by the chain and sprocket pitch. For example, if you are using a #40 chain that has a 1/2-inch pitch, and the first estimate center distance is 12 inches, the first value of *C* is 24 pitches (24 pitches = 12 inches / [1/2" inch per pitch]). If the large sprocket has 20 teeth and the small sprocket has 10 teeth, chain length from Equation 1 (from Martin Sprocket and Gear Incorporated Catalog No. 60, 1987) is 63.106 pitches long. Now chains can only be in integer pitch lengths, so you either round this number up or down to the nearest integer. In this case, since the final value is closer to 63, you will use this value in Equation 2 to determine the final center distance. The final center distance is now 23.947 pitches long. To convert this back into actual inches, multiply this value by the pitch length. In this case, you are using a 1/2-inch pitch; thus, the center distance is 11.974 inches.

$$L = 2C + \frac{N+n}{2} + \frac{0.1013(N-n)^2}{4C}$$

C is equal to shaft center distances in pitches, *L* is the chain length in pitches, *N* is the number of teeth of the larger sprocket, and *n* is the number of teeth of the smaller sprocket. You can see that these formulas can be rather complex.

When building the actual robot, if you use a center distance value that is slightly larger than the theoretical center distance, it might not be possible to assemble the

FIGURE C-1 *Sprocket center distances*

chain due to the high tension. If this is the case, add another master chain link to the system.

$$C = \frac{L - \frac{N+n}{2} + \sqrt{\left(L - \frac{N+n}{2}\right)^2 - 8\left(\frac{N-n}{2\pi}\right)^2}}{4}$$

Timing Belt Centerline Distances

The pulley centerline distances are computed in a similar manner to how the centerline distances are computed with chain drive systems. The first step is to determine a belt length. Equation 3 shows this relationship from the Martin sprocket and gear catalog.

$$L = 2C + 1.57(D + d) + \frac{(D - d)^2}{4C}$$

C is the center distance, L is the belt length, D is the pitch diameter of the larger pulley, and d is the pitch diameter of the smaller pulley. Unlike equation 1, the center distance and belt length are not in terms of pitches, but they are actual distances. The pitch diameter of a timing belt pulley is always larger than the outside diameter of the teeth on the pulley. The initial center distance, C, is estimated based on preliminary robot designs. Belt lengths come only in finite lengths. Once you determine the value for the belt length, you have to compare this with the available belt lengths for the particular belt type. Then select the belt length that is closest to the one calculated here. With this new belt length, you then need to calculate the actual center distances of the pulleys. Equation 4 shows the relationship that Martin uses to calculate these distances.

$$C = \frac{b + \sqrt{b^2 - 32(D - d)^2}}{16}$$

$$b = 4L - 2\pi(D + d)$$

V-Belts

As with timing belts, a V-belt length is first estimated based on an initial sheave center distance estimate. Spotts (Spotts, M.F., *Design of Machine Elements*, Prentice-Hall, 1985, pp 292) shows the relationship for determining the belt length in Equation 6 and then determining the actual center distance in Equation 7.

$$L = \pi R_1 + 2\sqrt{C^2 + (R_2 - R_1)^2} + \pi R_2$$

$$C = \sqrt{\frac{1}{4}(L - \pi(R_1 + R_2))^2 - (R_2 - R_1)^2}$$

L is the belt length, C is the center distance, R_1 is the radius of the smaller diameter sheave, and R_2 is the radius of the larger diameter sheave. The actual belt length must match available belt lengths for the particular V-belt.

Index

C

D

F

J

K

L

S

T

U

V

W

INTERNATIONAL CONTACT INFORMATION

AUSTRALIA
McGraw-Hill Book Company Australia Pty. Ltd.
TEL +61-2-9417-9899
FAX +61-2-9417-5687
http://www.mcgraw-hill.com.au
books-it_sydney@mcgraw-hill.com

CANADA
McGraw-Hill Ryerson Ltd.
TEL +905-430-5000
FAX +905-430-5020
http://www.mcgrawhill.ca

GREECE, MIDDLE EAST, NORTHERN AFRICA
McGraw-Hill Hellas
TEL +30-1-656-0990-3-4
FAX +30-1-654-5525

MEXICO (Also serving Latin America)
McGraw-Hill Interamericana Editores S.A. de C.V.
TEL +525-117-1583
FAX +525-117-1589
http://www.mcgraw-hill.com.mx
fernando_castellanos@mcgraw-hill.com

SINGAPORE (Serving Asia)
McGraw-Hill Book Company
TEL +65-863-1580
FAX +65-862-3354
http://www.mcgraw-hill.com.sg
mghasia@mcgraw-hill.com

SOUTH AFRICA
McGraw-Hill South Africa
TEL +27-11-622-7512
FAX +27-11-622-9045
robyn_swanepoel@mcgraw-hill.com

UNITED KINGDOM & EUROPE (Excluding Southern Europe)
McGraw-Hill Education Europe
TEL +44-1-628-502500
FAX +44-1-628-770224
http://www.mcgraw-hill.co.uk
computing_neurope@mcgraw-hill.com

ALL OTHER INQUIRIES Contact:
Osborne/McGraw-Hill
TEL +1-510-549-6600
FAX +1-510-883-7600
http://www.osborne.com
omg_international@mcgraw-hill.com